AGRICULTURE AND THE ENVIRONMENT

AGRICULTURE AND THE ENVIRONMENT

Papers presented at the International Conference on Agriculture and the Environment 10–13 November 1991

Edited by

C.A. Edwards, M.K. Wali, D.J. Horn and F. Miller
The Ohio State University, Columbus, OH 43210, USA

Reprinted from *Agriculture, Ecosystems and Environment,* Vol. 46 Nos. 1–4 (1993)

ELSEVIER   **Amsterdam – London – New York – Tokyo**   **1993**

ELSEVIER SCIENCE PUBLISHERS B.V.
Molenwerf 1
P.O. Box 211, 1000 AE Amsterdam, The Netherlands

ISBN 0-444-89800-X

Printed in The Netherlands

CONTENTS

The global need for sustainability in agriculture and natural resources

Clive A. Edwards*,Mohan K. Wali

The Ohio State University, Columbus, OH 43210, USA

Abstract

It is now widely accepted that the need for seeking strategies for the sustainability of food production, and renewable and non-renewable natural resource use, is beyond question. Although the worst case predictions of human population explosion have not been realized, there seems little doubt that there is already a significant increase in numbers; at least 90 million people are added to the world population annually. Unless there is major international cooperation in addressing the problems associated with population control, it is predicted that the global human population will reach more than 14 billion by the year 2050. The provision of adequate food, fuel and space for such an increased population will be unachievable.

Until the 1980s, global increases in food production exceeded the concomitant growth of human populations however, progressively, agriculture is becoming unable to meet the world-wide per capita needs for food. A significant proportion of the world's population, particularly in developing countries, does not receive enough calories. These problems are accentuated by factors such as world-wide reductions in soil fertility, the accelerating degradation of land that is suitable for food production through soil erosion, the world-wide trend for migration of human populations from rural habitats to cities, the increasing rates of economic growth and the resultant increased demands for food in developing countries, extremely rapid rates of global deforestation, the poor quality and availability of water, and potential climatic changes due to increasing deforestation and industrial pollution.

Possible solutions to global sustainability in agriculture and natural resources must involve an integration of ecological, sociological, cultural, and economic considerations, as well as mandated international and national policies. First, particular emphasis is needed on important ecological factors such as the maintenance of all forms of plant and animal biodiversity, better understanding of biogeochemical cycles, development of systems of integrated pest, weed and disease management, design of fully integrated sustainable lower chemical input farming systems, adoption of methods of increasing the efficiency of energy use, increasing conservation of all forms of natural resources, and better methods of forestry management and preservation. Second, these multiple ecological approaches must be linked with socio-economic activities in international human population control, national management of domestic animal production, and greater education and involvement of women in farming and resource conservation. Finally, there must be major changes in public policies both internationally and nationally in the control of land use, the promotion of all aspects of proper natural resource use, the regulation of chemical environmental pollution, gases and effluents, and the promotion of sustainable agriculture, on both fragile and fertile soils particularly those in developing countries, with much greater vigor than we do now

*Corresponding author.

Introduction

The environmental lexicon, which has often developed well-intentioned metaphors and catchy phrases for environmental problems and their solutions, that are quick and painless, has been expanding rapidly in the last few decades. Concomitantly, environmental scientists have, over the years, accumulated impressive databases that describe the extent and seriousness of environmental problems. For the most part, these have targeted single issues; for example, the dimensions of the problems of human population increases and demographics, air pollution, water pollution, toxic and non-toxic wastes, depletion of natural resources, and related issues.

However, these environmental problems, each of which is serious, are closely interrelated and the overall disturbance to ecosystems world-wide is staggering in scope (Wali, 1987, 1992a). Many common themes to focus on environmental issues collectively have been proposed but no term seems to have attracted the attention and approval of scientists, lay citizens and politicians alike than that of 'sustainable development' (Wali, 1992b). Put simply, the use of this term leads inevitably to the following question: given the present utilization patterns of resources and their magnitude, can the use of renewable and non-renewable resources be sustained, temporally and spatially? The conceptual form and implications of sustainable development are now being discussed seriously by industrialists, economists and the social scientists as well.

More rare, and usually controversial, are detailed suggestions of how sustainable development can be achieved. Therefore, it was a matter of gratification that the June 1992, United Nations Conference on Environment and Development, in Rio de Janeiro, was spearheaded by a politician and two business leaders. Variously referred to as the 'Earth Summit', the 'Rio Conference', or the 'Global Forum', it was an unprecedented gathering of world leaders, government representatives and members of non-governmental organizations. By all accounts, this meeting will emerge as historic, but its specific successes can be 'only as effective as the governments choose to make them' (Haas et al., 1992). The issue of global sustainability permeates all of its 27 principles. Specifically, sustainable development requires: 'reducing or eliminating unsustainable patterns of production and consumption', and promoting 'appropriate demographic policies' (Parson et al., 1992).

The series of papers presented in this Volume concentrates on issues that are related to the sustainability of agriculture, in relation to overall environmental considerations, protection of natural resources and overall sustainable development. To attain a better appreciation of the concept, we provide here an overview of those factors that threaten global sustainability and examine possible solutions that could avoid its further endangerment.

Factors endangering global sustainability

Rates of population growth

A review of predictions of the human population explosion over the past 50 years show that the worst case scenarios were not realized and the overall global human population growth rates have slowed somewhat. Nevertheless, the increase in numbers is still staggering, with more than 90 million people added to the world annually. No dramatic solutions to this problem seem to be in sight, and it is likely that the world's population could nearly triple to as many as 14 billion before levelling off.

From 1992 to 2000, human populations in the developing countries are expected to increase by about 2% per annum, compared with about 0.5% annually in industrialized countries. By the year 2000, nine out of ten of these new citizens will live in the developing countries where food is scarcest. (World Bank, 1988). Moreover, the per capita income growth across the developing world was more than 3% between 1973 and 1980, and incomes will probably increase even further between now and the year 2000, thereby resulting in yet greater demands for food. It is likely that additional and unpredictable factors could exacerbate the expected increase in food shortages even more. There is, therefore, an urgent need to better understand the critical factors that influence such food shortages if we are to ensure a chance of maintaining global food sustainability.

Fortunately, some levelling off in human population growth has been noted in recent years. Even in the developing countries, the growth of populations has tended to decrease from an annual rate of 2.5% from 1965–1973, to about 2.0% per annum from 1973 to 1986, and this rate is predicted to fall even further to about 1.9% annually between 1990 and 2000 (World Bank, 1988). Over the same period, the average annual rate of increase of populations in industrialized countries has been much lower and has fallen from about 1% per annum 1965–1973, to 0.65% from 1973–1986. Although it was predicted to fall to as low as about 0.4% per annum from 1990 to 2000, this does not appear to have occurred very uniformly. In the U.S., for example, the population growth rates have increased, from 1.77 children per woman in 1975, to 2.01 in 1991. In spite of these slight increases, overall figures imply that while there will be only relatively small human population increases in the countries with abundant food, much larger increases will occur in the developing countries, many of which already have serious food shortages.

Moreover, the distributions of human populations are extremely uneven nationally, regionally and globally and are expected to become even more uneven and aggregated with time. Indeed, it is expected these changed populations will become even more unevenly distributed owing to the current dramatic movements of populations, from rural habitats to cities, all over the world. The United Nations has predicted that, by the year 2000, 19 of the 25 largest cities in the world will be in the developing nations (United Nations, 1987). Those 19 cities, which had a total population of 58 million in 1950, are predicted to reach a total population of 261 million people by the year 2000. Such trends in greater urbanization of populations will have major impacts and serious implications upon both the demand for food and its availability, as there will be fewer farmers to produce the needed food. Clearly, such rates of increase in human populations and their changed distributions cannot be sustainable, without altering the patterns and methods of food production significantly, and vastly improving the facilities and political will for the distribution of food internationally.

The need for sustained food production

Since the Second World War, there has been an unprecedented growth in food and agricultural production which, until recently, has kept pace with population increases. On a global scale, the rate of production of food has increased at an average annual rate of about 2.5% over this period. This corresponds with an average annual per capita increase in food production of about 0.6% since the 1950s (World Bank, 1986). Such increases in food production exceeded the concomitant growth of the world human populations until the 1980s. However, in the mid-1970s, although overall food production continued to increase at a relatively linear rate, the annual increase in per capita food production has tended to level off, barely keeping up with the more rapid increases in human populations. What is worse, any apparent global adequacy of the food supply is misleading, because of its unequal distribution across the world, so that a large proportion of the world's population remains malnourished. For example, 34% of the human populations in developing countries (700–800 million people) do not receive enough calories to maintain an 'active working life'. Such shortfalls in food production are greatest in areas such as Sub-Saharan Africa, which have a rapidly declining per capita food production. Food production per capita is also slowing down dramatically in other parts of the world such as Asia, where concern continues to mount on the sustainability of the 'Green Revolution', owing to continued and accelerating degradation of the agricultural resource base (Herdt, 1988; Conway and Banbier, 1990).

In the 1970s, some people thought that continued global development was possible if the associated price of gradual environmental degradation was acceptable. At that time, most developing countries pondered on the hard choice between development and environmental protection. However, it soon became increasingly obvious that such an assumption was far from true. Indeed, unplanned exploitation and overstressing of the environment has been one of the main causes of widespread famine and impoverishment in many parts of the world.

Although global food production has continued to increase at an average annual rate of about 2.5%, there are increasingly strong indications that this growth rate is beginning to decline world-wide due to many causes. In particular, there seem to be finite limits on how much can be achieved in food production on fertile soils by high chemical input agricultural technology, and these limits are likely to be accentuated in the future. This means that more of the poorer and more fragile soils situated in unfavorable climates will be used for food production despite their limited potential. Unless alternative food sources or production methods can be found, it could well be that limitations on global food production will inevitably be the driving force for national and/or global programs of population limitation. These limitations, viz., the potential food production, associated not only with the currently occurring declines in the natural resource base, but also with the availability of suitable land for food production, will pose very serious questions for global agricultural and environmental sustainability.

Effects of rates of economic growth

Even in many developed countries, the rates of economic growth have been relatively poor in recent years. In developing countries, although the rates have increased relative to earlier years, they have been even lower. In most developing countries, agriculture is a major source of income. Hence, when there is poor performance in agricultural production, it is almost inevitably associated with low levels of economic growth, as agriculture contributes a large proportion of financial returns to the gross national product in most developing countries. Economic growth, even within a single country, is extremely variable from year to year and fluctuates greatly, but tends nevertheless to increase overall in many developing countries, albeit relatively slowly. Between 1950 and 1975, annual economic growth ranged from 1.7% in South Asia to 5.2% in the Middle East and averaged 3.4% across all of the developing countries. In most developing countries, where a large proportion of consumer income is used for food, any future increases in income which seem likely to occur, will inevitably cause significant increases in the demand for food both in quantity and variety (Barkin et al., 1990; Hitzhusen, 1993). Clearly, such demands will exert even greater pressures on the available food supplies and increase stresses on the natural resource base.

Lack of regional self-sufficiency

Many regions of the world have had considerable difficulty in maintaining their per capita food production to keep pace with their population growth. For example, South Asia, West Asia, North Africa and Sub-Saharan Africa have been unable to maintain adequate overall per capita food production in recent years, and agricultural production per capita has actually begun to decline in all of these regions except South Asia. Even South Asia, Afghanistan, Bangladesh and Nepal are beginning to experience some declines in their per capita food production. Among the Latin American countries, the poorest per capita food production is in Central America, although Bolivia, Peru and Guyana have also begun to experience declining per capita agricultural production.

Large gaps between the food supply and human demand have been filled in the past by movement of surplus food from developed countries to developing countries but this has been spasmodic, has serious limitations and is a short-term strategy. It is essential for overall global sustainability, that food production be accelerated by all means feasible in those countries with low per capita production. This includes changed crop production, improvements in traditional cropping practices, and avoidance of serious current land misuses such as, cultivating sloping land and overstocking available land with livestock that further degrade the natural resources.

Availability of land for crop production

, More than 75% of the world land surface is not arable, because it is either too dry, stony, has an adverse climate, or all three. Thus, less than 25% of the total land area has physical conditions that are suitable for crop production. Of this land area, 13% has a low productive capacity, 6% is moderately productive and only 3% is highly productive (U.S.A.I.D., 1988). Additionally, owing to deforestation (Robinson et al., 1990), desertification (Grainger, 1990) and soil erosion (Brown, J.W., 1988), the land available for food production globally is decreasing at an alarming rate. For example, of 329 million ha of the total land area in India, as much as 175 million ha can be considered to be degraded in one form or other. Each year globally, 6 million ha of drylands are added to the 1.3 billion ha that have already been moderately or severely desertified. It has also been estimated that the productive potential of about 7 million ha of deforested land is lost annually, 3 million of these to soil erosion and 4 million to desertification (Dover and Talbot, 1987). Another 20 million ha of land are degraded annually to the point of becoming unprofitable. In Africa, on average, ten times more plant nutrients are being removed annually from soils than are being returned through inputs in the form of organic or inorganic fertilizers. It has been estimated that more than one-third of the world's available agricultural land is currently over-cultivated. Even in industrialized countries like the U.S., many soils are being eroded 10 – 20 times faster than they are being replaced. The Food and Agriculture Organization of the United Nations has predicted that, by the year 2000, the current total area of rain-fed agriculture in developing countries will

decrease by 18%, and total productivity by 29% owing to soil erosion, nutrient and organic matter depletion, pollution and deteriorating soil structure.
In total area, a net loss of land of about 540 million cultivatable hectares,
from that presently available, has been predicted by the year 2000 (Dover
and Talbot, 1987). In South America, the losses will be only about 10% but
could be as high as 30% in Central America, about 16% in Africa and up to
36% in Asia.

Soil erosion

Average estimates of soil erosion indicate that about 4 billion tons of soil
are lost annually in the United States and about 25 billion tons lost each year
on a global scale. Soil erosion tends to be much greater in fragile tropical soils
than in the more fertile temperate soils. For example, it has been estimated
that soil erosion in Africa and South America is occurring currently at annual
rates of about 7 t ha^{-1} compared with about 0.8 t ha^{-1} in Europe (Sanchez,
1976). Few parts of the world are without soil erosion. In Guatemala, 40% of
the productive capacity of the land has been lost to erosion, and some areas
of the country have been abandoned because agriculture is no longer economic. In Turkey, planners estimate that 75% of the land is affected, 54%
severely so. In much of Haiti, there is no topsoil left at all. Clearly, rates of
soil erosion depend partly on the nature of the soil and climate, but even more
on the intensity of cropping and cultivation, lack of fallow periods, and adoption of unwise practices such as cropping steeply sloping hillsides (Lal, 1991).
Most erosion occurs through wind-driven soil losses in the arid regions but is
related intimately with water erosion in the semi-arid and humid tropics. At
particular risk from water erosion are soils with more than 1200 mm precipitation per year. Soil erosion is accelerated greatly when the soil organic matter and structure are poor. Control of soil erosion is critical to long term global
sustainability because soil erosion results not only in serious losses of a key
natural resource, but also in significant reductions in agricultural productivity and off-site harm to water quality (Hall and Hall, 1993).

Problems associated with deforestation

The world's tropical forests are being lost extremely rapidly (Poore, 1989).
Estimates vary, but it has been suggested that between 25% and 40% of all
original tropical forests had already been lost by 1980 (Brown, J.W., 1988).
If present trends continue, about 12% of the tropical forests that remained
world-wide in 1980 will have been lost by the year 2000 (Gradwohl and
Greenberg, 1988; Robinson et al., 1990). In the developing world, ten trees
are cut down for each one that is replaced, and in Africa 29 trees are lost for
a new one that grows. In some regions, losses of forests have been even higher.
For example Haiti has lost 40%, Paraguay 39% and Brunei 35% of the trees

that they had in the 1980s. The most recent estimates are that more than 140 000 km^2 of tropical forests were lost in 1989, i.e. about 27 ha min^{-1}. Forests have been cleared for a variety of purposes in addition to growing crops, including use as pasture for animal production, harvesting of wood for fuel and for industrial or even political reasons (Mahar, 1989). Most of the claims that clearing of forests increases agricultural productivity are based on theoretical assumptions (Gregerson and McGaughey, 1987), but these claims have seldom been substantiated.

Estimates of the total extent of tropical forests that have been lost worldwide differ greatly, and range from 12 million ha, to 21 million ha lost annually (Myers, 1980), of which about half are believed to be due to the practice of 'shifting agriculture' or 'slash and burn' agriculture. These latter practices involve cutting and clearing areas of forest for crop production, usually by burning, but such areas can only support crops from 1 to 3 years. Globally, about 720 million ha of forests are under such 'slash and burn' systems currently. If a sufficient fallow period is left between cropping periods, plant nutrients can be replenished and sustainable agricultural production maintained. However, in many tropical soils, fallowing for even as long as 15 years may be insufficient for the levels of nutrients, particularly soil phosphorus levels, to recover (Arnason, 1982). But, with growing human population pressures, there is a strong indication that the essential fallow regenerative period will be progressively decreased, so that soil fertility will also decline substantially, leaving inadequate time for the forest to regenerate, leading ultimately, to total environmental degradation.

Thus, deforestation can create many adverse side-effects including increased rates of soil erosion, loss of soil fertility through nutrient runoff, eutrophication of water bodies and potential climatic changes through its effects on global carbon dioxide concentration. It is essential that deforestation be controlled and recognized as an urgent issue (Brown, J.W., 1993).

Poor water quality and availability

Sources of clean drinking water are in short supply in many parts of the world. Throughout the developing world, water is polluted by bacteria and chemicals which cause disease and even death. For example, in India 70% of all surface waters are polluted; the Jamuna River through Delhi picks up an enormous load of 20 million l of raw sewage and 20 million l of industrial wastes daily. It is also polluted with agrochemicals. On average, four out of five people in developing countries have no sanitary facilities and as a result 80% of childhood deaths result from water-borne disease (Brown, L.A., 1988). Wherever agrochemicals are used, they leach readily through tropical soils and contaminate groundwater. The increased use of pesticides and chemicals that have been used to increase productivity significantly has been particularly true in the 'Green Revolution' era. These problems are equally or more serious in rice paddies which use large amounts of pesticides and often combine rice production with fish farming; the fish take up the chemicals into their bodies and are then used as food. Even in developed countries, such as the United States, a significant proportion of drinking water becomes contaminated by pesticides and fertilizers (Baker, 1993) and there are strong pressures for controlling this (Hornsby et al., 1993; Shuyler, 1993; Logan, 1993).

Global and regional climatic changes

There is increasing evidence that human activities are changing the chemical composition of the atmosphere in ways which may impact global climate in the future (Mathews, 1990). Thus, there is a consensus among scientists that rising concentrations of atmospheric gases could bring about global climatic changes. For example, the atmospheric levels of carbon dioxide have increased by more than 25% since the beginning of the industrial era. Most of the carbon dioxide emissions arise from the current patterns of energy use, about 90% of which comes from burning carbon-based fuels world-wide. There is evidence that tropical deforestation is also a major source of high levels of carbon dioxide in the atmosphere. The possible atmospheric effects of increased levels of carbon dioxide on climate will likely be accentuated by other gases such as methane, nitrous oxides and chlorofluorocarbons (CFCs) which are also increasing rapidly in the upper atmosphere and may exert together as great an impact as carbon dioxide (Orr, 1993).

It still remains to be confirmed fully that these gases are influencing climate in any significant way but there is sufficient evidence to make it imperative to take some form of action as an insurance measure (Poore, 1989).

Potential solutions to sustainable development in agriculture and natural resource management

There seems little doubt that the strategies needed to achieve sustainable development will have to be both regional and people-based, as well as international and capital-intensive. These solutions must integrate ecological, sociological, cultural, economic and policy considerations if they are to be effective. A development strategy which is sustainable in an environmental sense may not be sustainable in an economic sense. Conversely, a strategy which seems to be economically and financially sustainable may not be environmentally sustainable in the long-term. Finally, both economic and environmental sustainability are dependent upon support from national and international policies designed to maintain them.

Clearly, the prime need is an integrated approach derived from a combination of the ecological, sociological, cultural, economic and political aspects to make solutions viable. and, based on an agreed definition of sustainable development (Wali 1992a). The approaches must balance urgent, current needs with long-term sustainability.

Ecological aspects of sustainability

Natural ecosystems

The maintenance of biodiversity is important. It has been estimated that although taxonomists have identified less than 2 million species of plants and animals, and as many as 30 million different species may exist. Currently, species are lost at an alarming rate particularly from the tropics, as direct and indirect results of deforestation. Tropical forests, as many scientists have noted, contain over half of the world's overall biological diversity and many of these species are very localized. As resources to conserve diversity are limited, we must use the most cost-effective means. It may be better to conserve many marginally-endangered species than to concentrate on a few critically-endangered ones. Some scientists predict that several hundred species are being lost every year and as many as 25% of the world's species could be lost in the next 25–50 years. Such losses have resulted in immense 'biotic impoverishment' of the world's ecosystems (Woodwell, 1990) and could have major impacts upon future ecological and technological advances.

The management of forests is extremely important. National governments can promulgate policies that eliminate distorted economic incentives which encourage mismanagement of forests, such as, the granting of property titles in return for forest clearing, encouraging below-cost timber sales, and promoting cattle production. International institutions must encourage such reforms which will simultaneously relieve the pressure on the remaining tropical forests and help to bring about their sustained use. Forest conservation is not sufficient. It must be accompanied by aggressive, ecologically-sensitive programs of reforestation, especially where demands for fuelwood and new land to produce crops are high. These measures will be costly and goals will only be achieved if they are accompanied by education, training and broad non-governmental participation in the planning process.

The conservation of non-renewable energy sources based on fossil fuels is very important to global sustainability. Energy strategies should emphasize an overall global reduction in the use of fossil fuels through extensive energy-efficiency programs, especially aimed at transportation and the production and use of power. These conservation efforts must be supported by greater efforts to introduce renewable energy sources such as solar, wind and water power as well as methane and ethanol production. Research on nuclear energy should be pursued in order to determine whether designs for nuclear reactors can be developed that would be safer, restore public and investor confidence, and minimize the need to dispose of radioactive wastes and contaminate the environment. Research that seeks other forms of potential energy supply such as fusion should be supported.

In oil-importing countries, i.e. in the vast majority of developing countries, energy requirements take the largest share of national foreign exchange. At the household level, low-income urban families can spend up to 30% of their budgets on cooking fuel. (This is particularly the case in poor African and Asian countries, but is also true for low-income households in many richer countries.) In rural areas of developing countries, the collection of wood and agricultural residues for cooking may be one of the most time-consuming of daily burdens, and in some regions of acute fuelwood shortages, eating patterns have been changed by the scarcity of fuel.

Agricultural ecosystems

The maintenance of biodiversity is extremely important in agroecosystems as well. The problems associated with lack of biodiversity are much greater on farms using intensive food production practices in developed countries and in the Asian countries that adopted similar practices in the 'Green Revolution'. Growing crops in continual monoculture or biculture leads inevitably to much greater pest, weed, disease, fertility and soil erosion problems than the mixed cropping systems common in developing countries. It has been suggested that biodiversity in agroecosystems is one of the two most important common factors in crop production if we wish to grow crops with the minimal use of manufactured inputs (Altieri, 1993; Edwards et al., 1993).

Before the development of inorganic fertilizers in the 19th century, the growing of crops depended completely upon the biological recycling of nutrients from plant residues and animal manures. Because of the increases in crop production that resulted from the use of inorganic fertilizers, ecological research into the important biogeochemical cycles mediated by living soil organisms has been relatively poorly studied. However, knowledge of the contribution of legumes and nitrifying bacteria to the supply of nitrogen and of vesicular-arbuscular mycorrhizae, to the availability of phosphorus for plant growth in recent years, has demonstrated the enormous potential of soil ecological research for increasing soil fertility and crop production.

The need for maintenance of adequate organic matter in soils is equally important for sustained soil fertility and structure, and the biological control of pests and diseases. Organic matter has many impacts on soils, in addition to its role as a source of nutrients for plant growth (Edwards et al., 1993). Unfortunately, organic matter degrades much more rapidly under tropical than under temperate conditions and organic matter levels tend to decrease progressively in many soils when there is continuous cropping. Conservation tillage makes an important contribution to the improvement of the organic matter status of soils and to lessening the amount of soil erosion. It is extremely important that ecological research should emphasize methods of preserving the organic matter status of soils.

The integration of arthropod pest, disease and weed management is one of the most important ways of contributing to sustainable food production (Edwards et al., 1991). This involves maximal use of cultural and biological control techniques to control pests (Zalom, 1993; Harris, 1993). These practices must be integrated with the use of minimal amounts of pesticides based on economic pest injury levels (Higley and Pedigo, 1993), if maximum crop production with minimal environmental impact is to be achieved (Pimentel et al., 1993). In many developing countries, the availability of pesticides which are expensive and are fossil fuel-based, is limited; so pest management in these countries must depend mainly upon use of resistant crop varieties and rotations combined with other cultural and biological inputs. What is needed urgently, is a true integration of indigenous, locally-developed cultural techniques with the use of new specially-developed varieties of crops grown in innovative combinations and rotations, involving agroforestry wherever possible, and biological control techniques (Yaninek and Schulthess, 1993).

One of the most important issues in food sustainability is the development of integrated nutrient management systems that maximize inputs of nutrients from cultural and biological sources, combined with minimal inputs of inorganic fertilizers (Edwards and Grove, 1993). In the 'Green Revolution', greatly increased yields were achieved through the use of improved crop varieties and considerable inputs of inorganic fertilizers. There are serious doubts now whether these practices are sustainable on many tropical soils. A much better understanding of the contribution of biological and cultural inputs, such as legumes, agroforestry and vesicular-arbuscular mycorrhizae to nutrient availability for crop production, is probably an important key to increasing productivity of the poorer tropical soils.

Finally, it is important to design integrated sustainable agricultural systems (Schaller, 1993). In traditional agriculture, crop and animal production were linked, crop and crop residues were used to feed domestic animals, and the animal manures and feces were used as nutrient sources. In most developed countries, intensive agriculture crop production has been largely separated from animal production, so these important biological links have been broken. Progressively, this is happening in developing countries, particularly in South America. Wherever possible systems re-establishing these links must be adopted.

It is extremely important to combine, not only crop and animal production but also to incorporate into crop production the issues discussed earlier in this section: i.e. maintenance of biodiversity; maintenance of soil organic matter; use of integrated arthropod pest, disease and weed management; into whole integrated farming systems (Edwards et al., 1993). These systems should avoid reliance on expensive fossil fuel-based inputs and maximize biological and cultural inputs into crop production. They should combine indigenous well established and effective practices with innovative methods, and new crops and varieties that respond well to lower chemical inputs. They must fit well into the local environment and needs, and respond to the socioeconomic requirements of the region (Buttel, 1993; Grove and Edwards, 1993). Such systems are needed in developed countries as well as developing ones.

Sociological, cultural and economic aspects of sustainability

There is no doubt that the central problem in sustainable development is human population control. In many developing countries, population pressures on the land threaten not only the environment but also national security, as people migrate in search of sustenance, often creating territorial disputes and overstressing resources. An international goal should be to eliminate as much hunger as possible by the year 2000 through increased food self-reliance and more efficient distribution. Although population pressures affect the planet as a whole, they must be addressed individually by each nation and its citizens. Countries must make their own assessments about targets for population levels and growth; ordering their development priorities and incentives accordingly. Developed nations can offer much-needed technical support and experience in birth control and family planning to help developing nations and individuals achieve their goals. The problems of achieving adequate birth control globally are great and complex. They are complicated by lack of education, ethnic and religious customs. An achievable target could be to achieve zero population growth by 2050. The main thrusts in attaining this target would be education, healthcare and access to family planning.

Because of its complexities, the problem of overpopulation clearly calls for initiatives aimed at: (1) universal access to family planning by the year 2000 (this has been estimated to require a global expenditure rising to as much as $10 billion by the year 2000); (2) greatly increased research to provide a wide array of safer, cheaper and easier birth control technologies; (3) giving priority to investment in education for women and in bringing women into full economic and political participation; (4) improved mass communication aimed at increasing support for family planning.

It is essential for sustainable development that there should be education and active involvement of women. Much of the farming and other labor in developing countries is supplied by women. They often have less access to education, political representation and recognition than men. Indeed, sustainable development is unattainable in many developing countries without the active role of women.

"The reason is brutally simple: Women perform the lion's share of work in subsistence economies, toiling longer hours and contributing more to their family income than male relatives, but are viewed as 'unproductive' in the eyes of government statisticians, economists, development experts, and even their husbands" (Jacobson, 1992).

Public policies to promote sustainability

Economic incentives are essential to achieving energy efficiency. These can take many forms such as increases in the taxes on gasoline and fossil-fuel based products and the adoption of a carbon dioxide emissions fee applicable to users of fossil fuels. All developed nations should be urged to adopt such practices.

In the 1990s, in both the industrial and developing countries, broad new energy strategies are required. All policies should (1) promote least-cost energy planning, (2) promote energy conservation and efficiency in production, transmission, and use, (3) employ the cleanest, least-polluting technologies available, (4) undertake a major program of coordinated research development and demonstration projects to ensure the rapid development of renewable energy sources, and (5) overhaul policies such as pricing, regulations, taxes, research support, in support of the above. In Third World countries, cooperation and investment by developed-countries should put the emphasis on capacity-building, that is, on the training, research and analysis, and institution-building that increase national self-reliance.

It is critical that there be national and international policies to control land use. Primarily, they must be aimed at controlling or reversing deforestation. There must be a reversal of policies that promote deforestation and creation of policies that take the pressure off forest areas, provide incentives to maintain them. It seems probable that both market incentives and government regulations will be needed. These need to be supported by international thrusts that could substitute payment of debts of developing and impoverished countries in return for greater forest protection.

National and international policies should promote sustainable agriculture by avoiding subsidies for chemical inputs or even taxing them, and providing incentives to the greater use of biological and cultural resources (Faeth, 1993).

These activities should be linked to policies that support natural resource conservation by governments and international institutions. Without such international commitments the global environment and natural resources needed for food production and economic development will continue to deteriorate. Most of the discussions at the Rio de Janeiro Conference in 1992 were directed at how this can be achieved by changes in policy, both nationally and internationally.

Industrial countries must make major investments to improve their own performance in regulation of environmental pollution and degradation. Developing countries must, in their own interest, incorporate sound environmental practices increasingly as part of their development programs. To be able to achieve this target resolving their debt deficits is crucial. Developed countries will also need to make a special effort to expand flows of funding as well as information and technology to developing countries if needed. Most important are investments in: (1) global environmental priorities; (2) slowing population growth; (3) protecting the ozone layer; (4) limiting greenhouse gas emissions; (5) preserving biodiversity, and many other non-global environmental requirements. Because of lack of funding and scarcities of other resources, developing countries are otherwise unlikely to act to minimize environmental pollution.

Conclusions from the 1991 Agriculture and Environment Conference

At the 1991 International Conference on Agriculture and the Environment, a group of participants (The group members, led by Mohan K. Wali, were Sandra Brown, Charles A.S. Hall, Frederick J. Hizhusen, John D.H. Lambert and Keith L. Wilde) summarized the premises, objectives and approaches needed to achieve global sustainability. Based on our current knowledge, they considered the following to be the premises of sustainability:

(1) Populations of people and domestic animals are high and increasingly spatially concentrated.

(2) Prime agricultural lands are already developed in virtually all countries.

(3) The development of additional lands for agricultural purposes requires substantial investments, e.g. fertility, availability of water, drainage, irrigation, and erosion control.

(4) Changes in climate, e.g. a temperature increase, could alter the extent and distribution of land suitable for agriculture.

(5) Agricultural practices have been responsible for considerable environmental degradation, locally and globally.

(6) Traditional agricultural systems, some of which were sustainable, are disappearing. They are being replaced by systems of farming that are highly dependent on finite fossil fuels and attendant technologies.

(7) Non-fossil fuel inputs of industrial agriculture are finite (e.g. water, phosphorus).

(8) The trend of industrialization is toward increasing global interdependence in food, energy, and environmental quality.

(9) Lack of hard currency in the developing countries encourages exploitation of land and increases pressure of exports thereby creating unsustainable conditions.

(10) Global indebtedness in the developed countries reduces capital investments, development assistance, and technical aid to developing countries. The debt service is large.

(11) There is simultaneous breakdown of large political structures increas-

ing trans-boundary problems. The solutions to global sustainable development will have to be multi-national.

(12) The trend towards specialization and fragmentation of knowledge is a deterrent to the development of policies for sustainability.

To meet the needs for sustainability, the following sustainable development objectives are imperative:

(1) Sustained yield, production of food, fodder, fiber, and biomass fuels, on a per capita basis.

(2) Reducing dependency on petrochemical and other finite resources.

(3) Improved environmental quality.

(4) Improved per capita income and greater equity in its distribution.

We must adopt the following approaches to achieve sustainability:

(1) A thorough integration of ecological and economic and policy paradigms.

(2) Detailed examination, analysis, documentation and promotion of representative sustainable agricultural systems (e.g. upland paddy fields, terrace cropping systems).

(3) A clear delineation of types of natural resources and environmental services and its implications in resource management (e.g. farms, fisheries).

(4) Analyses of energy flows, biogeochemical cycles, and population dynamics to seek an appropriate balance between technology and population.

(5) Incorporation of safe minimum standards in sustainable natural resource use.

(6) Monetarizing services based on environmental issues.

(7) A rigorous and systematic definition of alternative property rights and entitlement, their origins and implications for sustainability.

(8) A need for on-farm research where the farmer is a recognized 'research associate'.

(9) The use of synthetic integrative tools such as simulation modelling to examine multiple factors simultaneously.

(10) A search for an integration of appropriate disciplines in education and research.

(11) The establishment, maintenance, updating and dissemination of ecological and economic databases (where possible social, cultural).

The above criteria for a base for deriving the global principles of sustainability and sustainable development.

To achieve true sustainability, and a global sustainable development, there has to be a pervasive shift in current world perspectives from the concept of humans over nature, to a more ecological perspective of which humans are an essential part. Such an ecological perspective will lead inevitably to appropriate political and behavioral changes. Therein lies the challenge to future environmental research and education.

References

Altieri, M., 1993. Ethnoscience and biodiversity: key elements in the design of sustainable pest management systems for small farmers in developing countries. Agric. Ecosystems Environ., 46: 289–304.

Arnason, T., 1982. Decline in soil fertility due to intensification of land use by shifting agriculturalists in Belize, Central America. AgroEcosystems, 8: 27–37.

Baker, D.B., 1993. The Lake Erie Agroecosystem Program: water quality assessments. Agric. Ecosystems Environ., 46: 197–215.

Barkin, D., Batt, R.S. and DeWalt, B.R., 1990. Feed Crops vs. Feed Crops: Global Substitution of Grains in Production. Lynne Rienner, Boulder, CO, 244 pp.

Brown, J.W., 1988. Poverty and Environmental Degradation: Basic Concerns for U.S. Cooperation with Developing Countries. World Resources Institute, Washington, DC, 69 pp.

Brown, L.A., 1988. The Changing World Food Prospect. The Nineties and Beyond. World-Watch Paper No. 85, WorldWatch Institute, Washington, DC.

Brown, S., 1993. Tropical forests and the global carbon cycle: the need for sustainable land use patterns. Agric. Ecosystems Environ., 46: 31–44.

Buttel, F.H., 1993. The sociology of agricultural sustainability: some observations on the future of sustainable agriculture. Agric. Ecosystems Environ., 46: 175–186.

Conway, G.R. and Banbier, E.B., 1990. After the Green Revolution: Sustainable Agriculture for Development. Earthscan, London, 205 pp.

Dover, M. and Talbot, L.M., 1987. To Feed the Earth: Agroecology for Sustainable Development. World Resources Institute, Washington, DC, 88 pp.

Edwards, C.A. and Grove, T., 1991. Integrated Nutrient Management for Crop Production. In: Toward Sustainability. National Academy of Sciences, Washington, DC, pp. 105–108.

Edwards, C.A., Thurston, H.D. and Janke, R., 1991. Integrated Pest Management for Sustainability in Developing Countries. In: Toward Sustainability. National Academy of Sciences, Washington, DC, pp. 109–133.

Edwards, C.A., Grove, T., Harwood, R.R. and Colfer, J.P., 1993. The role of agroecology and integrated farming systems in sustainable agriculture. Agric. Ecosystems Environ., 46: 99–121.

Faeth, P., 1993. An economic framework for evaluating agricultural policy and the sustainability of production systems. Agric. Ecosystems Environ., 46: 161–174.

Gradwohl, J. and Greenberg, R., 1988. Saving the Tropical Forests. Earthscan, London, 207 pp.

Grainger, A., 1990. The Threatening Desert: Controlling Desertification. Earthscan, London, 369 pp.

Gregerson, H. and McGoughey, S.E., 1987. Social Forestry and Sustainable Development. In: D.D. Southgate and J.F. Disinger (Editors), Sustainable Resource Development in the Third World. Westview Press, Boulder, CO, pp. 9–20.

Grove, T. and Edwards, C.A., 1993. Do we need a developmental paradigm for sustainable agriculture? Agric. Ecosystems Environ., 46: 135–145.

Hall, C.A.S. and Hall, M.H.P., 1993. The efficiency of land and energy use in tropical economies and agriculture. Agric. Ecosystems Environ., 46: 1–30.

Harris, P., 1993. Effects, constraints and the future of weed biocontrol. Agric. Ecosystems Environ., 46: 273–287.

Haas, P.M., Levy, M.A. and Parson, E.A., 1992. Appraising the Earth Summit. Environment 34(8): 6–11, 26–33.

Herdt, R., 1988. Increasing Crop Yields in Developing Countries. American Economics Association, New York, 54 pp.

Higley, L.G. and Pedigo, L.P., 1993. Economic pest injury level concepts and their use in sustaining environmental quality. Agric. Ecosystems Environ., 46: 233–243.

Hitzhusen, F., 1993. Land degradation and sustainability of agricultural growth: some economic concepts and evidence from selected developing countries. Agric. Ecosystems Environ., 46: 69–79.

Hornsby, A.G., Buttler, T.M. and Brown, R.B., 1993. Managing Pesticides for Crop production and Water Quality Protection Practical Grower Guidance. Agric. Ecosystems Environ., 46: 187–196.

Jacobson, J.L., 1992. Gender Bias: Roadblock to Sustainable Development. Worldwatch Paper 110, pp. 1–60.

Lal, R., 1991. Soil Research for Agricultural Sustainability in the Tropics. In: Toward Sustainability. National Academy of Sciences, Washington, DC, pp. 66–90.

Logan, T.J., 1993. Agricultural best management practices for water pollution control: current issues. Agric. Ecosystems Environ., 46: 223–231.

Mahar, D.J., 1989. Government Policies and Deforestation in Brazil's Amazon Region. World Bank, Washington, DC, 87 pp.

Mathews, J.T., 1990. Preserving the Global Environment: The Challenge of Shared Leadership. W.W. Norton, New York.

Myers, N., 1980. Conservation of Tropical Moist Forests. National Academy Press, Washington, DC, 220 pp.

Orr, D., 1993. Agriculture and global warming. Agric. Ecosystems Environ., 46: 81–88.

Parry, M., 1990. Climate Change and World Agriculture. Earthscan, London, 157 pp.

Parson, E.A., Haas, P.M. and Levy, M.A., 1992. A summary of the major documents signed at the Earth Summit and the Global Forum. Environment, 34(8): 13–15, 34–36.

Pimentel, D., McLaughlin, L., Zepp, A., Lakitan, B., Kraus, T., Kleinman, P., Vancini, F., Roach, W.J., Graap, E., Keeton, W.S. and Selig, G., 1993. Environmental and economic effects of reducing pesticide use in agriculture. Agric. Ecosystems Environ., 46: 257–272.

Poore, D., 1989. No Timber Without Trees: Sustainability in the Tropical Forest. Earthscan, London, 252 pp.

Robinson, M., Gradwohl, J. and Greenberg, R., 1990. Saving the Tropical Forests. Earthscan, London, 207 pp.

Sanchez, P.A., 1976. Properties and Management of Soils in the Tropics. Wiley, London.

Schaller, N., 1993. The concept of agricultural sustainability. Agric. Ecosystems Environ., 46: 89–97.

Shuyler, L.R., 1993. Non-point source programs and progress in the Chesapeake Bay. Agric. Ecosystems Environ., 46: 217–222.

United Nations, 1987. Population Images (2nd Edn.). United Nations, New York.

U.S.A.I.D., 1988. Proceedings of the Bureau for Science and Technology, Office of Agriculture; Agricultural Development: Today and Tomorrow. Devres, MD.

Wali, M.K., 1987. The structure, dynamics, and rehabilitation of drastically disturbed ecosystems. In: T.N. Khoshoo (Editor), Perspectives in Environmental Management. Oxford and IBH Publishing, New Delhi, pp. 163–183.

Wali, M.K. (Editor), 1992a. Ecosystem Rehabilitation: Preamble to Sustainable Development. SPB Academic Publishing, The Hague, Netherlands.

Wali, M.K., (Editor), 1992b. Ecology of the rehabilitation process. In: Ecosystem Rehabilitation: Preamble to Sustainable Development. SPB Academic Publishing, The Hague, Netherlands, 1: 3–23.

Woodwell, G.M., (Editor), 1990. The Earth in Transition: Patterns and Processes of Biotic Impoverishment. Cambridge University Press, New York, 285 pp.

World Bank, 1986. Poverty and Hunger, Issues and Options for Food Security in Developing Countries. Washington, DC, 88 pp.

World Bank, 1988. World Development Report 1988. Washington DC, 76 pp.

World Commission on Environment and Development, 1987. Our Common Future. Oxford University Press, Oxford, 383 pp.

Yaninek, J.S. and Schulthess, F., 1993. Developing environmentally sound plant protection of cassava in Africa. Agric. Ecosystems Environ., 46: 305–324.

Zalom, F., 1993. Reorganizing to facilitate the development and use of integrated pest management. Agric. Ecosystems Environ., 46: 245–256.

Agriculture, Ecosystems and Environment, 46 (1993) 1–30
Elsevier Science Publishers B.V., Amsterdam

1

Issues in global agricultural and environmental
sustainability

The efficiency of land and energy use in tropical economies and agriculture*

Charles A.S. Hall[a,**], Myrna H.P. Hall[b]

[a]*SUNY College of Environmental Science and Forestry, Syracuse, NY 13210, USA*
[b]*Department of Forestry, SUNY College of Environmental Science and Forestry, Syracuse, NY 13210, USA*

Abstract

Technological optimism about economic processes, and especially the food production required to feed the world's growing population, must be tempered severely by constraints caused by limited land and fossil energy availability. We examine this hypothesis explicitly using several empirical assessments and several computer models, one of which was specifically designed to convey complex information to a non-technical population. We use data from various countries for our analyses but present more detail for Costa Rica. Our analyses find little support for the idea that development can take place without commensurate energy investments. We also extrapolate through the Year 2050 to project the amount of land and fossil energy that would be required to meet Costa Rican food demand, under a range of assumptions about population growth rates, land distribution, national food self-sufficiency, fertilizer availability, erosion rates and relative prices of coffee and fertilizer. Our results suggest that during the first quarter to half of the next century, it will be increasingly difficult, and eventually impossible, to feed a growing Costa Rican population no matter what (realistic) assumptions are made. The most important variables are the price of fertilizer exports and whether land is used principally for domestic crops or for export crops that are traded for food.

Introduction

Malthusian concerns about the world's carrying capacity for humans continue to be alleviated to a large degree by increases in agricultural productivity. Many investigators attribute these increases to technological innovations, which they sometimes assume to be of almost infinite potential. Others see

---

*To be reprinted in Revista Agrociencia for the Symposium Agriculture Sostenible Montecillo, Ed. de Mexico. Based in part on contributions from: Lois Levitan, Department of Natural Resources, Cornell University, Ithaca, New York, USA and Tomas Schlichter, CATIE, Turrialba, Costa Rica.
**Corresponding author.

serious and increasing constraints. The global human population continues to grow, most good agricultural land is completely developed, and the petroleum-derived basis of most agricultural technology is finite and likely to become increasingly uncertain and expensive. Hence it is important to examine carefully what the future holds for continued agricultural expansion, and to make realistic global plans for feeding people.

The issue is especially important because there are many who continue to argue that there are essentially no limits to the number of people that the earth can support, or who at least believe that 'technology' by itself will be able to deal with the problems of both increasing human populations and resource degradation. This is an important issue, for most technologies developed and implemented in the past have had a heavy energy dependence (Hall et al., 1986). However, the consensus of many writers on energy and the environment is that the world community must reduce consumption of fossil fuel resources (oil, coal, natural gas), in order to achieve 'sustainability' and to halt the deterioration of the global environment.

According to the Bruntland Report of the World Commission on Environment and Development (1987) global energy demands will increase from 40 to 500% in the next 40 years, as developing countries, with ever-increasing populations, attempt to achieve a standard of living comparable with that of people of the developed world, where one-fifth of the world's population consumes over 50% of the world ' s commercial energy (Gibbons et al., 1989). This implies a per capita increase of about ten-fold in energy and that the prospect for 'sustainability' is pessimistic (unless we are to accept 'sustainable poverty') because of the need for energy for development.

Some authors are highly optimistic that economic development can proceed without an increase in energy usage. Here are some examples of what we perceive to be a very common set of attitudes:

"the world can, in effect, get along without natural resources" (Solow, 1974).
"developing technologies to exploit renewable resources more economically... improving energy efficiency is our best hope" and "Investment in energy efficiency can help us reduce fossil-fuel use without sacrificing economic growth" (Gibbons et al., 1990).
"... in the future, cheap water transportation ... will transpose what are now deserts into arable lands" (Simon, 1981).
"The cost-effectiveness of efficiency as the most environmentally benign source of energy is well-established" (Bruntland Report of the World Commission on Environment and Development, 1987).

Their hope is based on the following evidence:

The predictions in 1973, in the face of fuel shortages, that "the high rate of growth in energy demand would continue, since a nation's energy consumption was inextricably linked to its economic development" have not materialized largely due to "the massive, unanticipated contributions of energy-efficient technologies" (Gibbons et al., 1989).

"Impressive gains in energy efficiency have been made since the first oil price shock in the 1970s. During the past 13 years the energy component of growth fell significantly in many industrial countries as a result of increases in energy efficiency averaging 1.7 percent annually between 1973 and 1983. And this solution costs less..." (Bruntland Report of the World Commission on Environment and Development, 1987).

"The feasibility of increasing energy efficiency in the industrial nations was demonstrated during the oil shortages of the 1970s. After 1973, as rising fuel prices promoted more efficient use of energy, GNP per capita continued to grow while energy consumption per capita had actually declined!" (Mazur, 1991).

This paper explores the degree to which the above statements are true, and also examines other data for evidence that there is improving energy or resource-use efficiency. If the statements are not true, or only approximately true, then their continued promulgation is extremely dangerous because they divert attention from other approaches for dealing with the problems, and thereby create complacency about the continued increase in human populations and the implications of resource depletion, mismanagement and pollution.

This paper is divided into three general parts. The first defines the terms we use, and considers some very general aspects, of what is called efficiency. The second focuses on the efficiency of general economic production and considers data for several countries both on food and economic production per unit of fossil fuel input, especially in Costa Rica. The third part considers overall efficiency in agriculture and the ability of agriculture to feed the growing population of Costa Rica. The hypothesis examined throughout is whether there is any evidence to date that technology (by itself) improves the efficiency of the economies or agriculture of developing countries. Ultimately the question we are interested in is whether technology will provide sufficient food for future generations of people in the tropics and subtropics or whether the simple ratio of numbers of people, to quantity and quality of resources, will be a more powerful factor in determining the future.

Part 1. Definition of efficiency

Efficiency is a word often used in, for example, economic considerations, but the word must be defined explicitly for each use, and one kind of efficiency may be derived only at the expense of another. The most common use of efficiency is economic efficiency, which refers to the quantity of goods or services obtained per consumer cost (Samuelson, 1976). Neoclassical economic theory is based in part on the premise that unregulated markets should produce maximum economic efficiency, because groups of producers who can generate a given product at lower prices (for example, an agricultural product in optimal climate and on optimal soils) will be able to undersell all other producers. In theory, the net effect is regional specialization for products that are best produced there, and cheaper goods and services for all consumers.

But efficiency can have other meanings as well. Labor efficiency (or productivity), means the amount of goods or services a laborer can produce in an hour. Classical economists such as Adam Smith and Karl Marx were extremely interested in this question since they perceived the value of something as being a product of the labor that went into its production. According to Smith (1789), "among a nation of hunters, if it usually costs twice the labor to kill a beaver as to kill a deer, one beaver should be worth two deer". (This can be compared with the neoclassical perspective that the value of something should be determined by what it can command in the market.)

In this century, labor productivity (usually defined as the inflation-corrected cost of value-added product generated per hour of labor) has increased a great deal in the industrial nations, for example, by a factor of about eightfold in the US from 1905 to 1973 (Hall et al., 1986). This allowed capitalists to increase wages while still making substantial profits. Most investigators have attributed this increase in labor productivity, and the greater productivity of US labor compared with European labor during that time period, to technology alone (see e.g. Denison, 1979). However, Cleveland et al. (1984) and Hall et al. (1986) showed that the relationship can be explained better by considering only the energy used per worker hour, since labor productivity and energy use per worker hour had almost identical curves for the period 1905–1984 (when the US stopped gathering data). Likewise, the economic growth of most countries is correlated closely with quality-corrected energy use (Hall et al., 1986; Paruelo et al., 1987; Kaufmann, 1992).

Thus, an equally valid index of productivity is energy efficiency, or productivity per unit of energy used, since more than 99% of the energy used in the process of economic production is not human labor but industrial use (i.e. from fossil fuels, nuclear and hydropower) (Hall et al., 1986). In US agriculture, during the period 1920–1973, for example, while labor efficiency was increasing (because of an increasing use of energy per hour of labor), industrial energy efficiency was decreasing because of a decreasing amount of labor used per unit of energy (Hall et al., 1986). It also means that about 10 kcal of fossil fuel are used for each kilocalorie of food consumed by an American (Steinhart and Steinhart, 1974). The net effect is that about a gallon of petroleum is required to feed each US citizen each day. Thus, the labor productivity of a US farmer is very high, but the efficiency of industrial energy used in agriculture is very low compared with earlier times (Hall et al., 1986). The increase in the quantity of industrial energy used per worker was especially important in the US from 1900 to 1973 because energy was much cheaper than in most other countries.

Agriculture in the developing countries is considered 'inefficient' but, in fact, where crops are raised more by people and less by fossil fuels, these agricultural systems are actually very efficient when examined as output per unit of industrial energy used. This may be an especially important consideration in the future when, eventually, fossil fuel availability will be seriously constrained and human labor very plentiful.

Capitalism and efficiency

Thus the high economic efficiency of capitalist countries is largely a result of their very high use of fossil fuel. Traditionally, this was possible in the US because fuel was cheaper, due to abundant coal and petroleum resources per capita, a well-developed oil industry and a well-developed infrastructure. Thus, in many ways, capitalism is 'efficient' because it gives large economic incentives for people to exploit resources rapidly, often with relatively little regard for the long-term consequences including energy depletion and generation of pollution.

However, the industrialized capitalistic systems have not fared well since 1973. Most people think of the impact of the 1973 oil shock principally in terms of increases in the price of oil. Instead, the major effects resulted from the internationalization of oil prices, after US oil production (formerly the world's largest) began to decline, and the US began purchasing oil on international markets at the same price as everyone else. During the 1970s, a large proportion of US corporations suddenly became internationalized, and production tended to move overseas (especially to the Pacific Rim countries) where labor was cheap and the cost of energy was, more or less, the same as in the US. In a few short years, the US lost its role as the most important industrial manufacturer in the world. Some countries with cheap labor suddenly capitalized on this overseas movement of industrial technology, and the hourly inflation-corrected take-home pay of US workers has declined by about 20% since 1973. Few people think about these things in terms of energy, but energy availability and prices were the drivers of both the early US dominance and the later loss of that position.

The present 'capitalization' of the world

At this time, the political structure of the world is changing rapidly. Most previously communist or centrally planned nations are abandoning these approaches to economics and are embracing 'free market' or 'capitalist' economics, often with little knowledge or experience of what capitalism means or its relation to resources. The principal reason given normally in support of

Table 1
Human population densities of various parts of the world in 1990

Region	Number of people (per square mile)	Population growth rate (%)	
		1980–1990	1990–2000
Northern America	33		
United States	69	1.0	0.7
Canada	7	1.0	1.0
Middle America	121	2.4	2.2
Costa Rica	155	2.7	2.6
Carribean	360	1.2	1.4
Europe	266	0.3	0.3
East Asia	293	1.2	1.2
South Asia	289	2.3	2.0
World	101	1.7	1.6

Data from Statistical Abstracts of the United States 1990.

this change in political approach is that the most economically successful nations (the US, Japan, Germany, Canada) appear to be capitalist countries. To some degree, of course, this is true, but the importance of material resources and their relation to the number of citizens should not be underestimated. The US and Canada, each with large areas of excellent agriculture, have human population densities that are only about 10 (both) to 20% (US alone) of those of Europe, Central America, the Caribbean and southeast Asia (Table 1). Consequently, the US and Canada are able to export food to more than 100 countries (Brown, 1989). Japan has a much higher population density and relatively few intrinsic natural resources, yet is also relatively successful economically. This has often been used to argue that natural resources are not important for the well-being of a nation. However, Japan imports an average of four million barrels of oil each day. In addition, Japan, like the US, has increasingly 'exported' its energy-intensive industries, such as metal manufacturing, to other countries, giving the appearance of increasing efficiency.

Part 2. The efficiency of economies

In 1987, each person in developing countries consumed an average 18 million BTU in commercial fuel and an additional 8 BTU in biofuels compared with more than 130 million BTU in western Europe and 305 million BTU per person in the US (Levine et al., 1991).

While total energy consumption in industrialized countries grew at a rate of 5% between 1980 and 1985, the growth rate for developing countries was 22% (Gibbons et al., 1989). Consequently their share of world commercial energy use grew from 14% in 1973 to 22% in 1987. Their energy demand growth averaged 5.3% per year compared with 0.7% in the OECD countries and 3.2% in the USSR and eastern Europe combined (Levine et al., 1991).

Many 'optimistic technocrats' believe that energy efficiency especially in end-use technologies, is the solution to reducing worldwide energy consumption, improving the energy/GDP ratio, reducing pollution and providing the Third World with the energy it needs to develop industry and move its citizens toward at least "the comfort level enjoyed by Western Europeans in the 1970s" (Reddy and Goldemberg, 1990). It is hard to argue with the desirability of that goal.

However, given the observed large rate of increase in energy use in developing countries this is an ambitious mandate and it seems difficult to offer the world the opportunity "to enjoy its current prosperity and mitigate the worst of global warming without even thinking about CO_2" (Cherfas, 1991). But those who advocate these perspectives (other than Goldemberg) provide us with few data or specific detailed analyses to examine where efficiency gains can be most effectively implemented in developing countries. They also tend to provide broad generalizations about alternative renewable energy resources without regard for their energy cost or environmental impact, which is considerable (Hall et al., 1986).

Some investigators are less optimistic. According to Levine et al. (1991) economic and social development will not occur in the world's lower income countries unless they get substantially more energy inputs than they now have. "If we look at the position of the industrialized countries, it seems likely that their energy consumption will increase comparatively slowly over the remainder of this century. They will be utilizing high cost energy and maximizing efficient energy use. In the developing countries the prospect is very different" (Ghosh, 1984). Given these divergent views we believe that empirical analyses are in order.

One particularly glaring problem is that optimists often base their optimism for developing countries on the (insufficiently analyzed) record of certain developed (OECD) countries. For example, between 1973 and 1985 the energy/GNP ratio (uncorrected for fuel quality) fell by one-fifth in the OECD countries, but during this time many of these countries were experiencing a pronounced and prolonged recession. This recession by itself was responsible for at least half of the reduction of energy use and also contributed to improved efficiency by contributing to the shutting down of lower-efficiency capital equipment, factories and mines (Hall et al., 1986).

Methods

It is difficult to classify all developing countries into one homogenous group. They differ widely in their economic, social and political structures and in their economic bases and prospects for future development. "However there remains an underlying homogeneity which comes into focus when they are compared with the industrialized countries. Fundamentally, they have low living standards compared with industrialized countries; expanding populations and rapidly increasing energy needs" (Baxendall in Ghosh, 1984; Smith, 1976; Seligson, 1984). We tested the hypothesis that technology has increased the efficiency with which energy (the main input) is turned into wealth (i.e. GDP or food), by examining statistical ratios of the relationship between outputs and inputs. If there has been an increase in production, not associated with an increased use of energy then we should see an improvement of output per unit input of fuel or other resources. Although this is also the technique most commonly used by those we criticize we add corrections that we find essential for the analysis (see fuel quality below).

Our analysis uses fertilizer input as a rough index of total energy input into agriculture, since fertilizers typically constitute 20–30% of all energy input to agriculture (Steinhart and Steinhart, 1974; Hall et al., 1986). We asked whether agricultural productivity (based on tons of grains and pulses) per unit of fertilizer had increased.

For purposes of illustration we focus on Costa Rica, a country whose problems of population growth, resource depletion and agricultural production we have studied in some depth.

Empirical tests of increases in efficiency

We found, for all countries that we have examined, and in contrast to the technological optimists' prediction, that the ratio of food output to fertilizer input has been declining sharply since 1950 for the more developed nations (Fig. 1). In all cases, nations were moving towards the replacement ratio, that is towards a situation where there was about 10–20 tons yield of food per ton NPK input. This reflects approximately the ratio of total weight of NPK in the food produced and the fact that worldwide we are now adding about the same quantity of industrially derived nutrients to our fields as we are removing in crops (Food and Agriculture Organization (FAO), 1981). In other words, global food production at today's levels requires the input of very high levels of chemical fertilizers. It is not known whether fertilizer efficiency will decline even further due to the saturation relation of grain production to fertilizer.

Costa Rica

In Latin America, energy consumption has grown somewhat faster than GDP, especially during the recession of the 1980s (Levine et al., 1991) and the energy/GDP ratio for Costa Rica bears this out (Fig. 2). Whether the GDP/total energy ratio (which indicates no improvement in the GDP/energy ratio between 1973 and 1985) or the GDP/quality-adjusted energy ratio

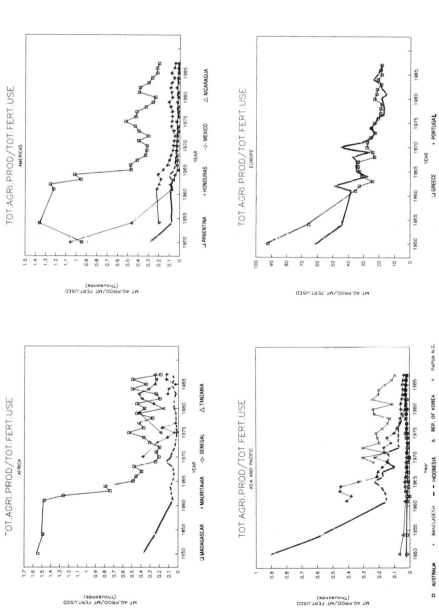

Fig. 1. Energy efficiency of agricultural production in various countries, defined here as the total production (in tons) of grains and pulses (normally the major foods produced) per unit input of commercial fertilizer (in tons). As agriculture expands and 'modernizes' the national yield per unit of fertilizer declines until it matches about the level of nutrients removed in harvests. This reflects the declining relative importance of traditional fallows, the lower response of cultivars to each unit of fertilizer, and, in some cases, degradation of yield potential because of erosion and pollution.

ENERGY USE

(a)

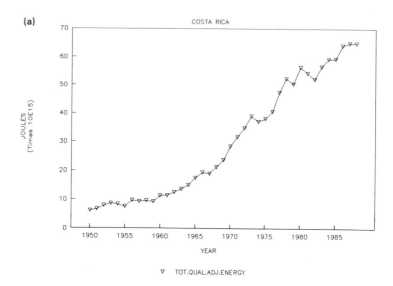

ENERGY/GDP

(b)

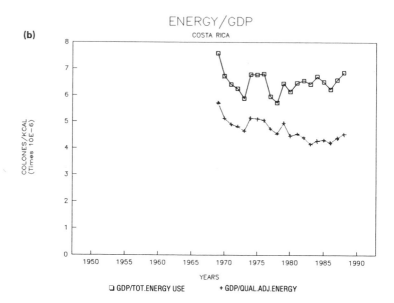

Fig. 2. (a) Energy efficiency of the Costa Rican economy, defined as annual GDP (million of colones) per annual energy use. (b) □, GDP per total kilocalories of energy without a quality correction. +, GDP per quality-corrected energy, where the hydroelectric kilocalories are multiplied by three to give a minimum estimate of its higher relative quality for economic production compared with oil. There has been no increase in energy efficiency over this time period, and a decline when quality corrected values are used.

is evaluated (which indicates a decrease in efficiency during the same time period) three conclusions are obvious: (1) GDP per unit energy input is not as high today as it was in 1970; (2) the GDP/energy ratio is very sensitive to oil price shocks (1973 and 1979); (3) there has been no increase in energy efficiency in Costa Rica during this period of tremendous incentives for such improvements.

Discussion

We have also tested more generally the hypothesis that economies are becoming more efficient. Although a number of investigators have argued that energy efficiency in the US has declined since 1920, our own analysis has shown that most of the nominal improvement in the efficiency (as measured by GNP/kcal of energy input to the US economy) is probably due to factors other than increases in what most people would perceive of as efficiency (Hall et al., 1986).

Because the energy/GDP ratio alone cannot be used to measure real energy efficiency it provides little information on the strength of the relationship between energy use and economic activity (Kaufmann, 1992). Some researchers specializing in energy use have identified parameters which strongly affect the variation in the fuel/GDP ratio. These include the types of fuel consumed, whether fuel is consumed directly by the household versus the productive sectors, the product mix and fuel prices. Uncertainties and incompleteness in record-keeping may also complicate things. An assessment of how these factors affect the economies of the developed world follows.

Fuel quality

Fuels are said to have different 'quantities' which means that they vary in the amount of useful work they can do per unit of heat content. Estimates vary on the amount of economic work that can be done per unit and type of fuel, but engineering studies, that compare the amount of different fuels required to do the same task in a factory, i.e. thermal efficiencies have shown that a heat unit of gas or oil can do from 1.5 to five times more useful work than a heat unit of coal, and that a heat unit of electricity can do from one to 20 times that of coal, depending on the task (Adams and Miovic in Hall et al., 1986). Economic studies (that is statistical studies examining economic output of various sectors of a particular economy) indicate that oil can do 1.3–2.5 times more useful economic work per heat unit than a heat unit of coal, and that a heat unit of electricity can do between 1.5 and 16.8 times more useful economic work, depending on the task, than a heat unit of coal (Cleveland et al., 1984; p. 55 in Hall et al., 1986). Obviously these values are very crude, but all show the importance of high quality fuels.

From 1929 to 1972, the fuel/real GNP ratio for the US declined by about half. During the same period the proportion of petroleum in the US fuel budget increased from 29.8 to 79.5%. Seventy-two percent of the variation in the US energy/real GNP ratio between 1929 and 1983 can be explained by changes in the type of fuel consumed, if we assume that oil is a conservative 1.3 times more effective in producing wealth per energy unit than coal. Since 1972, primary electricity, which has the highest fuel quality, has contributed an increasing proportion to the US (and Costa Rican) fuel budget. This change implies that changes in fuel mix by themselves are sufficient to explain decreases in the energy/GDP ratio over time (see Hall et al., 1986). Kaufmann (1992) also found that this was the most powerful explanation for the decreasing energy/GNP ratio of Japan, the UK, France and Germany. These countries are, in general, using more high quality fuel because oil is cheap and nuclear power subsidized, and because high quality fuels are useful in mature economies.

Household fuel consumption

The energy used to operate an industrial machine has a great impact on the output generated by a nation. The amount of fuel consumed by households, however, has little effect on national production. In other words, it should not be considered as contributing to production or used in calculating efficiency. The crude fuel/real GDP ratio, however, accounts for all fuel consumed by a society, including purely consumptive activities such as moving a very large car on a vacation trip. Thus, the way in which people spend their money has a significant effect on the energy costs of goods and services when that computation is made at the national level. Put in another way, a gallon of oil sold at the gas pump will produce about $1.20 of GNP, but if used in industry will produce $10–$20 of GNP.

Direct fuel use by the household sector can be measured by the percentage of GNP accounted for by personal consumption of fuel (e.g. gasoline and home-heating oil). The fuel/GNP ratio is very sensitive to this factor in that a 1% change in this factor will produce about a 15% change in GNP. This factor accounts for 24% of the variation in the US energy/GNP ratio between 1929 and 1983 (Hall et al., 1986). This ratio declined sharply during periods of decline in household fuel purchases (1941–1945 and 1973–1984). Conversely, it increased during 1945–1950 and 1966–1970 (Hall et al., 1986). This is true of other countries (Kaufmann, 1992). Likewise, international differences in the relative portion of household fuel consumption explains about half of the international variation in energy efficiency (Hall et al., 1986; Kaufmann, 1992). This was the second most powerful explanatory factor in the decreasing fuel/real GNP ratio found by Kaufmann.

Fuel price

The third most powerful of the four factors studied by Kaufmann was fuel price, a factor easy to discriminate due to the fluctuations in price of fuel in the 1970s and 1980s.

Product mix

Lighter industries, such as software development, real estate, insurance, stock brokering, savings and loan activities, often important in mature economies, contribute substantially to GNP while tending to use proportionally higher quality fuel, i.e. electricity. Relative increases in this factor have increased the apparent efficiency of US manufacturing as, for example, steel for US automobiles is increasingly produced in Korea, and for Japan, which exported its aluminum industries in recent decades.

Barriers to increased energy efficiency

It is a matter of faith among many economists and politicians that some manifestations of capitalism or free markets will somehow transform magically poor economies, such as those found in many areas of eastern Europe or Latin America, into wealthy ones like the US or Japan. It is astonishing to us that large numbers of people accept this approach without asking questions about whether there is the resource base to support the increased economic activity (Hall et al., 1986).

The problem is exacerbated by the recent deification of free market approaches and associated neoclassical economics. It is not our intent to review here the many difficulties and failures of the neoclassical model (e.g. Hall, 1990, 1992). Suffice it to say that we are impressed with the degree to which the central tenets of the neoclassical model have been developed almost entirely theoretically, rather than through hypothesis generation and testing, by which science normally attempts to derive truth or reality. Scientists in developing nations should make sure that they ask for sound testing and empirical validation of the economic models that are often imposed upon them by lending institutions. They should also be aware that the payment of the debt, often incurred through the process of development, or more realistically the interest on that debt, is often through the destruction of the natural resource basis and heritage of the debtor nation, as has been well documented for forest products in southeast Asia (Repetto, 1988), the soils in the undulating pampas of Argentina, and indeed most resources in debtor nations.

It is not clear that energy efficiency is still improving in the developed nations. "The recent record of the United States in improving energy efficiency is not good. Between 1974 and 1983, U.S. industry reduced unit energy consumption 20 percent. But at the same time, Japanese industry achieved a reduction in unit energy consumption of 50 percent. Due in part to an expansion of the number of U.S. cars and trucks, consumption of petroleum products here for transportation is setting new all-time records" (Abelson, 1988). Indeed, as more energy-efficient automobiles were developed as a consequence of rising oil prices, American car owners increased their leisure driving, approximately canceling out the efficiency savings (Cherfas, 1991). Meanwhile there was no overall improvement in auto fuel efficiency in technically sophisticated western Europe, because they, too, had increased their leisure driving (Cherfas, 1991). This does not augur well for those who place their hopes in energy efficiency. Cherfas (1991) warns "Lifestyle changes could eat into everything you think you've saved". From 1986 through 1988 the energy/GNP ratio, even uncorrected for the factors we raised, again leveled off, perhaps reflecting a resurgence of energy-intensive industries (Walton, 1991.

Barriers in developing countries

If 'energy efficiency' is all that's needed to launch the countries of the developing world on a course of economic growth while at the same time reducing environmental impact, then why are they not making such investments, instead of allowing energy demand to increase faster than their GNP? The following factors appear important:

(1) Governments have higher priorities, e.g. the pressing needs of a growing population for housing, health care and agricultural development. These are seen as much more important than the development of 'efficiency measures'.

(2) Government officials are keyed to growth. To them cutting back on fuel use means cutting back on growth of GNP.

(3) Subsidizing energy prices, often done in order to stimulate the economy, obviously does not encourage efficient energy use by citizens.

(4) There are no effective policy and planning organizations focusing on energy issues. For example, in 1979, the National Energy Commission was created in Costa Rica, but it contributed little to understanding or solving the country's energy problems (Hartshorn et al., 1982).

(5) Funding for energy efficiency improvements is even harder to find than funding for increased supplies. Funds for new power projects tend to come from outside the country, whereas efficiency measures would have to be undertaken by the industrial sector and private citizens, who already have heavy taxes (to provide social services and service the national debt).

(6) People everywhere, are anxious to have material goods, especially those that give them more freedom (i.e. mobility) even if the equipment is not of the most energy-efficient technology. So long as the government subsidizes gasoline or diesel fuel, personal auto use will not decline.

(7) The ultimate barrier is that of limited capability and resources. A coun-

try like Costa Rica is highly vulnerable to the price swings of international markets, is staggering under the load of foreign debt, and does not have the capital to throw out old habits and bring in new ones, whatever the long-term benefits might be.

(8) Population growth and its attendant requirements for infrastructure swamp most other government activities.

Other factors that can contribute to reduction in efficiency

What explains the rapid increase in energy consumption in the developing world? The most important factors in Costa Rica for example are: (1) pressures of population growth to build and run new infrastructure, (2) migration from rural to urban areas, (3) increased electrification in the rural sector, (4) outmoded technologies that cannot be replaced cheaply, (5) increased use of the automobile, (6) depletion of resource bases, (7) costly exploration, development and extraction of additional energy supplies, and (8) a constant push for economic growth, which implies greater energy use. We will consider each in turn.

Population growth and the building of infrastructure

The population of Costa Rica is estimated at 2.87 million people in 1988 (EIU, 1989–1990) and is growing at an annual rate of 2.6%. Although the country is not crowded there are about three people per hectare of good crop land. Infant mortality has dropped from 38 births per thousand in 1973 to 14 per thousand during 1990 (EIU, 1989–1990). One-half of the population is under the age of 15. Population growth requires the building of an energy-intensive infrastructure, especially in rural areas. Construction of housing, roads, communication systems, health facilities, water supply systems and schools and, of course, new energy plants requires huge investments of energy directly and in the fabrication of the necessary building materials — concrete, steel, and asphalt to name a few. Most of these costs are paid for by the government.

Costa Rica, the most socially responsible country in central America, has developed an extensive social security system since the 1940s, encompassing unemployment benefits, health care, pensions and housing. The degree to which the country invests heavily in social welfare is indicated by a literacy rate of 89.8% in 1982. In that same year 76% of the population qualified for free health care. Since 1982, however, cutbacks in government spending, enforced under International Monetary Fund programs, have weakened the social welfare system as has the necessity to accommodate an increasing number of refugees from neighboring countries fleeing hostilities at home (EIU, 1989–1990).

Thus, the demands of a growing population for services is one of the main sources of the high rate of growth in energy consumption in Costa Rica over the last 40 years. Relatively little investment capital was left to invest in, e.g. improved efficiency.

Rural to urban migration

In 1987, 45% of the Costa Rican population was urban, compared with 33.2% in 1960 (EIU, 1989–1990). In rural areas, energy needs (primarily for cooking) are met with biomass (fuel wood and charcoal) whereas in urban areas new arrivals must rely on commercial fuels such as propane. Biomass fuels, which are not considered in the energy/GDP ratio, made up 40% of the total energy used in Costa Rica in 1981. The reliance on wood products has declined as a factor of rural emigration but also due to the decline in availability. Since 1940 Costa Rican forests have gone from 67% of total land area to only 17% (Sader and Joyce, 1988).

Rural electrification

In 1975, Costa Rica drew up the National Rural Electrification Plan designed to bring electricity to 90% of the rural population by the Year 2000. As part of this initiative the Costa Rican government completed the Arenal Dam around 1980. This dam, about 40 m high and built of earth and rocks, generates 150 000 kW of primarily base-load power. Additionally it provides irrigation water to the dry province of Guanacaste. Unlike many huge dam projects such as those in Brazil's Amazonia, its purpose was to meet demand, not create it (Reisner and MacDonald, 1986). It is part of a well thought-out effort to avoid the urban migration problems being experienced in many other countries of Latin America.

In spite of the enlightened social goals of this project, its transmission grid took years to complete due to lack of financing. Jorge Figuls, chief of planning for the Instituto Costarricense de Electricidad says, "Rural electrification is actually bad business for ICE. It may end up costing a lot more to distribute the power [over Costa Rica's mountainous terrain] from Arenal than it cost to build the dam and powerhouse" (Reisner and MacDonald, 1986).

Outmoded technologies

Examples of highly inefficient use of energy in developing countries are not hard to find, and the continual pressure of an expanding population makes it very hard to retire inefficient capital equipment, since all of it is needed. Older power plants consume 18–44% more fuel per kilowatt hour of electricity produced, and suffer transmission and distribution losses 200–400% higher than that of modern equipment in industrialized countries (Levine et al., 1991). Many industrial processes in developing countries consume far more energy per unit of output than those in advanced economies (Levine et al., 1991). This is especially true in such industries as steel, concrete, fertilizer, and pulp and paper production. The largest single industry in Costa Rica is food processing, and chemical products, textiles and metal processing follow in importance. These industries are all heavily dependent on imported materials. Without further investigation it is not possible to comment on the efficiencies of particular industrial processes in Costa Rica. Since demand for energy has increased by 10% per year since 1986, the country is now resorting to four diesel-powered electricity generators (EIU, 1989–1990), which will increase the country's already significant reliance on foreign oil and generate electricity much less efficiently than hydroelectric power plants.

Throughout Latin America one sees many older cars, buses and airplanes, the kind of 'gas guzzlers' that are becoming outdated in the industrialized countries of the North. Often this is exactly where they originated because people from developing countries are anxious to buy the 'rejects' of the industrialized countries and the industrialized countries are anxious to sell. The managing director of an East German factory making Wartburg cars insisted that investment by Opel/GM in a new plant should not prevent him from selling his immensely inefficient automobiles to Lithuania, regardless of the lingering environmental cost (Levine et al., 1991). Costa Ricans keep their old vehicles running, since they often cannot afford to replace this form of 'outmoded technology'. Simply taking a ride in Costa Rica shows that many Costa Ricans cannot invest in even new piston rings for their older cars and trucks.

Transportation

Two important differences exist in end-use consumption between countries of the developing and the developed world. In the former, the energy use for transportation, industry and agriculture take a much greater share than in developed countries (Ghosh, 1984; Levine et al., 1991). In Costa Rica during the 1970s, the increase in the transportation sector caused total energy consumption to grow much more rapidly than either the residential–commercial–public or industrial–agricultural sectors. By 1976, petroleum consumption began a dramatic increase that has continued irregularly to the present. The phenomenon was due, in part, to government subsidy of diesel fuel and a fiscal policy that heavily favored the import of diesel vehicles despite the price increases in oil.

Tourism, one of the economy's growth areas, contributes substantially to growth in the transport sector. One-third of export earnings, after coffee and bananas, come from tourism. Earnings were put at $110 million in 1987, and numbers of tourists increased by 20.2% in the first 9 months of 1988 compared with a year earlier. In the same period, the number of cruise ship arrivals rose from 58 to 96. Although figures are hard to come by tourists tend to use far more transportation than locals.

Depletion of resource base

Increased use of energy-intensive technologies to compensate for declining resource quality may contribute to increased energy consumption in Costa Rica. Costa Rica is an agrarian society. Seventy-four percent of exports are agricultural commodities (Hartshorn et al., 1982; Reisner and MacDonald, 1986).

Large tracts of land that were once replete with magnificent tropical forests, and subsequently, fertile agricultural land, are now barren, eroded and worn out. Erosion adversely affects crop productivity by reducing the availability of water, nutrients, and organic matter, and, as the topsoil thins, by restricting rooting depth (Larson et al., 1983; Pimentel et al., 1987). Many of the soils found in the southern hemisphere were once capable of maintaining fertility indefinitely under a system of slash and burn agriculture with 5- to 18-year fallow periods, during which time the forest regenerated (Hall et al., 1985). In Costa Rica, human population pressure, land scarcity affected by land tenure patterns, and government policy which encouraged the raising of cattle for export has left Costa Rica largely deforested and its soils depleted (Annis, 1990). As the best land was cultivated, land-hungry farmers were pushed onto steep, easily eroded hillsides. On 5% and 15% slopes soil erosion rates can reach from 87 t ha^{-1} year^{-1} to 221 t ha^{-1} year^{-1} (Pimentel et al., 1976). As the soil became too depleted for crop production it was turned over to grazing cattle, under the easy loan policies generated by the 1968 Alliance for Progress. It has been estimated that for every kilogram of beef produced in western Costa Rica, 100 kg of topsoil is lost down the rivers. Farmers, in the canton of Puriscal, Costa Rica complained, as early as 1972, "The land is tired now. We used to get good crops, but now we get little, even with fertilizers." Others added that "we are trapped because the price of fertilizer keeps going up and up" (Barlett, 1982). Considered the bread basket of the country in 1935, by 1979 the region had been declared an 'emergency area' by the Minister of Agriculture and Cattle Raising, due to its eroded land base which no longer afforded many inhabitants means to a sustainable livelihood (Hueveldop and Espinosa, 1983).

Increasing amounts of fertilizers have had to be imported to bolster this degrading resource. Total fertilizers applied has increased from approximately 27 000 t in 1960 to around 90 000 t today (FAO, various years). Considering that land in cultivation of permanent or annual crops has increased only slightly since 1960, it appears that productivity is declining in inverse proportion to the amount of energy required to compensate for the decline of Costa Rica's most important resource base. Indeed coffee production per hectare has decreased from nearly 1.5 tons in 1981 to less than 1.2 in 1986, although disease has also played some part in the decline (EIU, 1989–1990). Disease and pest control require substantial inputs of oil-based pesticides. Since either all of the fertilizer and other agricultural inputs, or all of the fossil fuel from which they are made, must be imported, the country is paying a heavy energy price. In this sector of the economy it is easy to see why the energy/GDP ratio is increasing. There is a point at which the addition of fertilizers no longer increases yields. More generally, since the classic response of crops to a fertilizer is steep at first and then less pronounced, all developing countries require progressively more and more energy-intensive fertilizer per unit of crop yield.

Another example of a degrading resource comes from the shrimp industry. At one time, shrimp were the common man's food in Costa Rica as elsewhere in Latin America. Even 10 years ago large shrimp were available on the menu of even the smallest cafes, but today they are not. Earlier, in the waters off the west coast, fishermen in small boats were able to obtain abundant catches. Pesticide and fertilizer contamination of the country's waterways as well as over-fishing has caused the shrimp catch to decline in recent years (Solorzamo et al., 1991).

Exploration, development and extraction of additional energy supplies

In a country where 85% of the energy supply already is imported (79% oil, 5% electricity), where demand has been increasing at 10% per year since 1986, where the inability of even an ambitious program of dam building is unable to avoid the current shortfall in energy supply and where Costa Ricans must once again resort to diesel-powered electricity generators, there is urgent need to find domestic sources of energy. In the early 1980s, Chief of Planning for ICE, Jorge Figuls, was asked why his country had not developed its hydroelectric resources more aggressively. He responded, "We don't need it" (Reisner and MacDonald, 1986). Whilst looking with disdain on Brazil's frenetic efforts to harness megawatts of power from the Amazon, some Costa Ricans were congratulating themselves in the early 1980s on the completion of the new Arenal reservoir and power plant that could now meet the energy demands of even the dry season (Hartshorn et al., 1982).

In 1981, OLADE (the Latin American Energy Organization) estimated future energy demand as a function of GNP growth, based on the historic correlations of energy consumption and GNP in Costa Rica. They predicted a growth rate of anywhere from 3.5% to 7.0% per year and an energy requirement in 1990 of 1.4 to 2.3 million TEP (tons petroleum equivalent). In fact the annual growth in GNP, based on constant 1987 local currency, from 1981 to 1989 was only 7.1%, but their electrical demand grew at 10% per year.

How will energy supplies meet future demands in Costa Rica? Even with a 55 MW Japanese-funded geothermal plant due to come on line at Las Miravalles in 1992, and two more hydroelectric generators on the Toro River scheduled for operation by 1994, there is still a very large shortfall to be overcome.

New supply projects are very expensive to develop (Kaufmann, 1991; Cleveland and Kaufmann, 1991). Costa Rica has an active petroleum exploration project on the Caribbean slopes of the Talamancas, but the deposits are small and very difficult to access, thus requiring heavy energy investment per unit of oil retrieved. The small petroleum deposits are expected to have a potential of 10–20 million TEP (4–8 years worth gross). Preliminary studies indicate coal reserves of the order of 30 million tons (i.e. 21 million TEP) (Hartshorn et al., 1982) but the difficulty of their extraction may make the energy and hence dollar return on investment too small to attract developers.

Conclusions

Applying the empirical record of the developed world to that of the developing world is futile. The energy problems of the Third World will require a much more sophisticated analysis and solutions than the easy solutions proferred, such as increasing the use of biomass fuels. Normally the people are already very proficient in using them and their increased use will further deplete the resource base and take land away from needed food production. It will not help to: (1) suggest the use of fluorescent light bulbs at $20.00 each (1991 US dollars) in the home of every peasant for whom the cost might equal one-tenth of his yearly income; (2) build nuclear plants for which few developing countries have the scientific and technological infrastructure to support; (3) encourage the purchase of new energy-efficient automobiles so that more and more people can drive and further pollute the already black skies of such cities as Sao Paolo, Mexico City and Delhi.

In conclusion, we believe that the technological solutions offered are diverting our attentions from the pressing task at hand, and are exacerbating an already terrible situation.

Part 3. Efficiency of agriculture

Food production is a work process like any other production. Modern agriculture requires both solar and fossil energy. Neither fossil energy nor solar energy are 'free inputs', in that both fossil energy and the land from which solar energy is collected are scarce resources that cost farmers money. Land becomes scarce as populations grow, and fossil fuel becomes scarcer as a function of market demand, availability (or lack thereof) of cash or foreign exchange and, ultimately, depletion.

Table 2 gives some basic agricultural statistics and a brief history of world agriculture. From the inception of agriculture, thought to be some 10 000 years ago, until about 1960, the world's growing population was maintained essentially from low-yielding field crops alternating with fallowing periods to maintain fertility (Hall et al., 1985; Detwiler and Hall, 1988, and references therein). Yields were about 0.5–1 t ha^{-1} year^{-1}. Yields were occasionally much higher in intensive irrigated rice farming than grain. Almost all increases in production were derived from expanding the areas under cultivation. From about 1960, most of the good (flat, fertile, moist) land in the world was already being farmed — for example, most of southeast Asia, Central America, the Middle East, Europe and large parts of Africa. Although large areas were not yet cultivated these were generally areas of relatively low intrinsic productivity.

However, global food production has more or less kept up with the increases in population growth since 1960, although the land area base is no longer expanding. The reason, first clearly identified by Odum (1967), is the industrialization of agriculture (Hall et al., 1986). In all parts of the world, we are now using the energy equivalent of roughly one barrel of oil per hectare to increase yields from their earlier levels of about 1 t ha^{-1} to more recent yields of about 2–3 t ha^{-1} year^{-1}. The energy is used in many ways, such as for building and running irrigation systems and tractors, but most importantly to manufacture and distribute energy-intensive fertilizers (especially nitrogen).

The greatest hope of being able to continue to produce as much food as the world's growing population needs is normally considered to be through technology. However it is important to examine what this technology means. To date, this has meant mostly an increasing use of fossil fuel. Will future technology be any different? It is important to ask that question if we are to have a sustainable agriculture.

We believe the issue of a sustainable agriculture to be far more difficult to solve than many people think. When someone asked David Pimentel of Cornell University what he thought sustainable agriculture meant, he said: "A world of 1 billion people and shifting cultivation". What he meant was that shifting cultivation, and perhaps some riparian or paddy rice systems, are the only truly sustainable agriculture that we have had; any other agriculture is using up fossil fuels and soils. Grain crops have an asymptotic response to fertilizer, and hence respond better initially to available fertilizer additions than to later levels (Fig. 3). We hope that this pessimistic view is wrong, but it will require some very hard and different thinking to make it happen, especially in a world of ever increasing mouths.

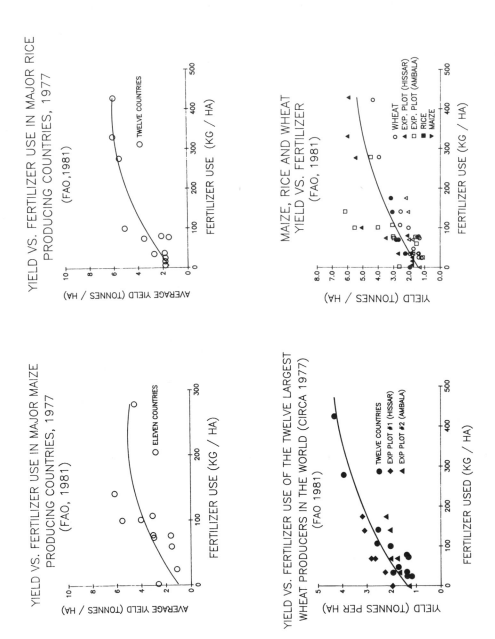

Fig. 3. Response of grain crops to fertilizer additions for maize, rice, wheat, and all combined (FAO, 1981).

Table 2a
Some basic agricultural facts

- A person needs about 2600 kcal day^{-1} to survive; this is about a million kilocalories per year or 286 kg grain;

- Grain fields yield from 0.5 to 5.0 (occasionally more) tons of grain per hectare per year. Higher yields almost always require fertilizers, pulses (i.e. beans) are lower yielding, as are most vegetables and of course animal yields are much less

- One can feed from two to 20 people per hectare per year, assuming the people eat only grain.

- Yields have continued to increase with time (Table 2b)

- Yield is a function of area cultivated, fertilizer use and fallow period (see Fig. 4b)

Table 2b
A short history of agricultural yield

Time	Yield (t ha^{-1})	Regeneration (type)	Fossil fuel use (Barrels oil eq. ha^{-1})
5000 B.C.–1960	0.5–1.0	Fallow	0
1960–1990	1.0–2.0	Fertilizer	1
1990–?	?	?	1+?

The patterns of food production over time

One way to think about modern agriculture is summarized in Fig. 4a, which plots isopleths of yield as a function of both intrinsic site quality and fossil fuel derived inputs. In general, the trajectory of a particular site, or nation, over time is from the lower right toward the upper left. Yields may increase, but if the fossil fuel derived inputs are removed, yields tend to fall to levels below their original value because site quality has declined. Although the intrinsic quality of most of our major agricultural soils has declined substantially (Pimentel et al., 1976), this is not reflected in yields because of increasing inputs. The interaction of fallow, the use of lower quality land with expansion and fertilizer inputs on yield is summarized in Fig. 4b.

Costa Rica and agricultural production

We chose Costa Rica to examine the future of agricultural production because it is considered to be relatively prosperous among developing nations, with good agricultural land, which has enabled it to earn much of its foreign exchange from the sale of agricultural products abroad. We also chose it because it has been severely affected by the price crises following the 'oil shocks' of 1973 and 1979, and because it has a database that is reasonably complete.

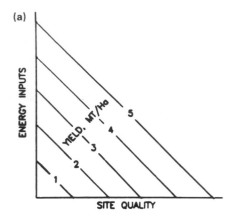

ENERGY & RESOURCE QUALITY FOR AGRICULTURE

ENERGY & RESOURCE QUALITY FOR AGRICULTURE

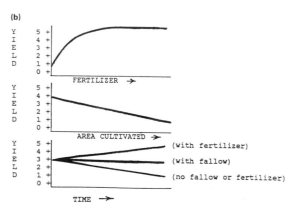

Fig. 4. (a) Agricultural production vs. site quality and energy inputs (from Hall, 1992). (b) Effects of fertilizers, area cultivated and fallowing on yield.

Finally, Costa Rica is a country with valuable and unique natural forests that are being rapidly converted to crops and pastures (Sader and Joyce, 1988; Flores Rodes, 1991).

Costa Rica is a nation without fossil energy resources, so it must import all petroleum products and derivatives as well as most manufactured goods and these goods use fossil energy. Costa Rica, like most less developed countries, has been increasing its use of imported oil dramatically, so that in many respects its economy is becoming as industrialized as are the traditional 'industrialized' countries. However, Costa Rica was affected severely by the energy price shocks of the 1970s because its dependence upon fossil fuels had increased in the years after 1945. For example, in 1970, only 4% of each working person's wage was spent on petroleum imports, whereas by 1982 the equivalent of nearly half of each person's wage went to purchase imported petroleum. During this time, the proportion of all export earnings needed to pay for petroleum imports increased from 5% in 1970 to 21% in 1982. Meanwhile, the mean cost of a kilogram of nitrogen fertilizer increased ten-fold, even after correcting for inflation. Since about 1985, petroleum prices relaxed somewhat, except for a brief surge ln the last half of 1990 during the war with Iraq. Finally, Costa Rica, in the 1980s, imported about 20% of its food calories. All of these imports have been 'paid' for with increasing debts. As oil prices have relaxed in the 1990s trends have reversed to some degree.

Methods

We have derived a series of computer models to examine the possible future of Costa Rican agriculture. The procedures and results of these simulations are reported in Levitan (1988). They involve a computer simulation (projected onto a screen) showing a three-dimensional view of Costa Rica and its mountainous terrain, with forested areas shown in green and agriculture and pasture in yellow. This central picture is surrounded by ten small 'graphlets' giving numerical values over time for the population of people and cows, the area in annual and perennial crops, the amounts of fertilizers used, the yields per hectare and per ton fertilizer applied, the total food calories produced, the total calories required to feed the growing population, some basic economic parameters (e.g. money gained from coffee sales, money cost of imported fertilizer) and land use patterns. The simulation starts in 1940 and, over time, the green area in the central image changes to yellow as agriculture and pastures expand, and the small graphs chronicle the changing populations, the changing nature and yields of agriculture, and the changing patterns of land use.

Results

We found that to produce all of the food required from domestic crops, given the current rate of population growth (2.6% per annum), and using our 'present day' estimates of yield per hectare all of the 1 million ha suitable for production of annual crops in Costa Rica would be needed to produce the calories needed to support the domestic population by about 2030 (Levitan, 1988) (Fig. 5). Since only 325 000 ha are currently in annual cultivation, this would mean that nearly 700 000 ha of the land now used as relatively high quality pasture and woodlands would need to be brought under cultivation, as well as the equivalent of all of the more than 80 million ha now used for export crops. In addition to these severe ecological ramifications, this scenario would reduce the amount of land available for perennial crops and pas-

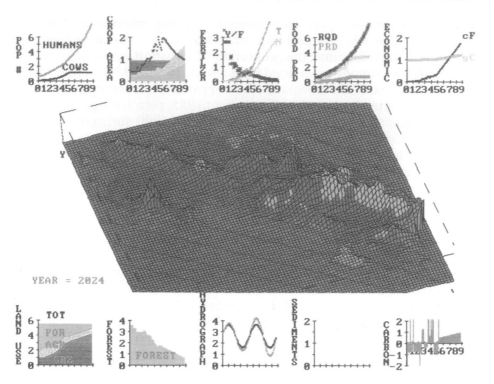

Fig. 5. Final frame of computer simulation of agricultural development in Costa Rica as presented at this Conference. The central picture represents undisturbed forest (dark) versus agriculture and pasture (lighter). The initial frame, for 1940, showed 67% forest. The final frame (held constant since 1983) showed 17% forest. The small graphs around the margin show important variables from 1940 to 2024. Empirical data are used to 1990 ('5' on graphs) and then a simulation is run that attempts to maximize food production. Populations in millions; Crop areas represented as annual; perennial, and total area of (best) land class I, yield per hectare in black points; Total and nitrogen fertilizer in million tons, yield per fertilizer in hundred tons food per ton fertilizer; food required for entire human population (black) and produced in nation (light blue); national cost of fertilizer and other agricultural inputs and gains from international coffee sales (in millions of 1985 dollars); land areas in millions of hectares; simulated expansion of floods and droughts by deforestation; estimated carbon gain (+)/release (−) to atmosphere, in millions of tons, assuming CO_2 fertilization of forests. Note that the initial gain of carbon is lost when higher inventory forests are later cut.

ture by about 25%. This would limit the production of exports severely and make it extremely difficult to pay for the large quantities of fertilizer required. The productive potential is much less if the potential effects of erosion are included. At some point it will be impossible to feed a growing population no matter what resources are available.

Thus, Costa Rica will not be able to maintain its current food production strategy unless crop yields are able to surpass levels beyond those now attainable, regardless of the affordability of fertilizer. It should be noted that the empirical results for 1940–1990 are such that, although crop yields per hectare have increased from less than 1–2 to about 3 t ha^{-1}, yields per ton of fertilizer will fall from more than 100 to about 20 t food per ton of fertilizer. Thus, by this measure agriculture is becoming much less energy efficient. If new technology is to be more energy efficient it will have to reverse this trend.

Discussion

A very important question for the future of Costa Rica is to decide 'what policies might be viable in a future of uncertain energy prices, or more accurately, uncertain ratios between the cost of fossil fuels (and their derivatives) and the monetary value of exports'. Our energy-sensitive simulation model for Costa Rican agriculture can be used to explore possible options for mitigating the effects of increasingly costly production inputs. The basic assumptions underlying this analysis are that agricultural technology is not independent of energy systems, that fossil energy is likely to become increasingly scarce, and that the ramifications of this future scarcity are reflected poorly in the market (Hall et al., 1986). Under what conditions is it beneficial to maintain a high level of fossil energy flow-through in Costa Rican agriculture and, in particular, into which type of crops should the energy investment be directed?

Economic vs. energy analysis

Many economists believe that prices are the best indicator of energy scarcity or lack, and many believe that governments should intervene as little as possible. Other scientists believe that information and insights about economic systems, beyond what is found through conventional economics, can be gained by applying ecological systems analysis to the interrelationships between natural and socio–economic systems. We have developed a series of hypotheses about the relationship of energy to fundamental economic processes and the ways that changing energy regimes impact economic processes. One important tool in this approach is to compare the physical or monetary return of an economic process such as agricultural production, with the amount invested (Cleveland et al., 1984; Paruelo et al., 1987).

We conclude that if, in fact, new agricultural technology continues to be energy intensive and if the supply of fossil fuel is in some major way finite, then the only way out of this dilemma for most developing countries is to invest in incentives for population control. All other strategies appear only to delay the inevitable population/land pressure, in the case of Costa Rica, by a decade or so. Appropriate incentives might be negative taxes for more than two children or improved social security for people with few children or greater opportunities for women. Perhaps a global system of social security would be the best investment that we could make for the future of feeding the world's growing populations from uncertain resources. We perceive that population control in the US, where each new person adds far more environmental impact due to the higher standard of living, is just as critical an issue (Hall et al., 1993).

In addition, population stability, if it should be attained, would give Costa Rica far more options for having some chance of continuing to supply a decent material standard of living for future Costa Ricans and their excellent system of social benefits. A failure to achieve population stability will almost certainly mean that Central America will be far more beholden to the US than at any time in the past, and may mean that the US will choose not to suffer the resource, financial and environmental consequences of attempting to continue to make up the difference between the food that Costa Rica can produce and the food that it needs.

To any serious student of the problem it is clear that we have only a decade or two to respond. While we hope that technology can help, we see no evidence of its impact so far, and believe that the technological optimists are seriously contributing to the problem by promising false solutions.

Acknowledgements

We would like to thank Will Ravenscroft for many of the analyses and David Pimentel and an anonymous reviewer for critical comments.

References

Abelson, P.H., 1988. Need for long-range energy policies. Editorial. Science, 240: 856.

Annis, S., 1990. Debt and wrong-way resource flows in Costa Rica. Ethics Int. Affairs, 4: 107–122.

Barlett, P.F., 1982. Agricultural Choice and Change, Decision Making in a Costa Rican Community. Rutgers University Press, New Brunswick, NJ, 196 pp.

Brown, L., 1989. Reexamining the world food prospect. In: L. Brown (Editor), State of the World. Norton, New York, pp. 41–59.

Bruntland Report of the World Commission on Environment and Development, 1987. Our Common Future. Oxford University Press, New York, 400 pp.

Cherfas, J., 1991. Skeptics and visionaries examine energy savings. Science, 251: 154–156.

Cleveland, C.J., 1991. Natural resource scarcity and economic growth revisited: economic and biophysical perspectives. In: R. Costanza (Editor), Ecological Economics: The Science and Management of Sustainability. Columbia University Press, New York, pp. 289–317.

Cleveland, C.J. and Kaufmann, R.K., 1991. Forecasting ultimate oil recovery and its rate of production: incorporating economic forces into the models of M. King Hubbert. Energy J., 12: 17–46.

Cleveland, C.J., Costanza, R., Hall, C.A.S. and Kaufmann, R., 1984. Energy and the United States economy: a biophysical perspective. Science, 225: 890–897.

Denison, E.F., 1979. Explanations of current declining productivity growth. Surv. Curr. Bus., 79: 1–24.

Detwiler, R.P. and Hall, C.A.S., 1988. Tropical forests and the global carbon budget. Science, 239: 42–47.

EIU Country Profile, 1989–1990. Costa Rica. 1990. EIU, London.

Food and Agriculture Organization, Fertilizer Yearbook, FAO.

Food and Agriculture Organization, Production Yearbook. FAO.

Food and Agriculture Organization, Commodities Yearbook, FAO.

Food and Agriculture Organization, Forest Products Yearbook, FAO.

Food and Agriculture Organization, 1981. Crop production levels and fertilizer use. FAO fertilizer and plant nutrition bulletin. Villa delle terme di Caracalla, Rome.

Flores Rodes, J.G., 1991. Economics, policy and natural resource issues in Central America. International workshop on Ecology and Economics, Turrialba, Costa Rica, CATIE, Turrialba, 60 pp.

Ghosh, P.K., 1984. Energy Policy and Third World Development. Greenwood Press, Westport, CT, 392 pp.

Gibbons, J.H., Blair, P.D. and Gwin, H.L., 1989. Strategies for Energy Use. Sci. Am., 261: 136–143.

Hall, C., 1990. Sanctioning resource depletion: economic development and neo-classical economics. Ecologist, 20: 61–66.

Hall, C., 1992. Economic development or developing economics: what are our priorities? In: M.K. Wali and J.S. Singh (Editors), Environmental Rehabilitation. Vol. 1. Policy Issues. Elsevier, Amsterdam, pp. 101–126.

Hall, C.A.S., Detwiler, R.P., Bodgonoff, P. and Underhill, S., 1985. Land use change and carbon exchange in the tropics: I. Detailed estimates for Costa Rica, Panama, Peru and Bolivia. Environ. Manage., 9: 313–334.

Hall, C.A.S., Cleveland, C.J. and Kaufmann, R., 1986. Energy and Resource Quality: The Ecology of the Economic Process. Wiley Interscience, New York, 577 pp. (reprinted 1991 by University Press of Colorado).

Hall, C., Pondius, G., Ko, J.Y. and Coleman, L., 1993. The environmental impact of having a baby in the United States. Populations and Environment, (in press).

Hartshorn, G. et al., 1982. Costa Rica, Country Environmental Profile, A Field Study. Tropical Science Center, San Jose, CA, 123 pp.

Heuveldop, J. and Espinosa, L. (Editors) 1983. El Componente Arbóreo en Acosta y Puriscal, Costa Rica. Centro Agronomico Tropical de Investigacion y Ensenanza, Turrialba, 122 pp.

Kaufmann, R., 1991. Oil production in the lower 48 states. Reconciling curve fitting and econometric models. Resources and Energy, 13: 111–127.

Kaufmann, R.K. and Cleveland, C.J., 1991. Policies to increase US oil production: likely to fail, damage the economy, and damage the environment. Ann. Rev. Energy, 16: 379–400.

Kaufmann, R., 1992. A biophysical analysis of the energy/real GDP ratio: implications for substitution and technical change. Ecol. Econ., 6: 35–56.

Larson, W.E., Pierce, F.J. and Dowdy, R.H., 1983. The long term threat of soil erosion to crop production. Science, 219: 458–465.

Levine, M.D., Meyers, S.P. and Wilbanks, T., 1991. Energy efficiency and developing countries. Environ. Sci. Technol., 25: 4.

Levitan, L., 1988. Land and energy constraints and the development of Costa Rican agriculture. Unpublished Master's Thesis. Cornell University, Ithaca, NY, 194 pp.

Mazur, A., 1991. Global Social Problems. Prentice-Hall, Englewood Cliffs, NJ, 207 pp.

Mellor, J.W. and Gavian, S., 1987. Famine: causes, prevention and relief. Science, 235: 539–545.

Odum, H.T., 1967. Energetics of world food production. In: D.F. Hornig (Editor), The World Food Problem. US Government Printing Office, Washington, DC, pp. 55–94.

Paruelo, J.M., Apahlo, P.J., Hall, C.A.S. and Gibson, D., 1987. Energy use and economic output for Argentina. In: G. Pillet (Editor), Environmental Economics. The Analysis of a Major Interface. Leimgruber, Geneva, pp. 169–184.

Pimentel, D., Dazhong, W. and Gianpietro, M., 1986. Technological changes in energy use in US agricultural production. In: C.R. Carroll, J.H. Vandermear and P.M. Rossett (Editors), Agroecology. McGraw Hill, pp. 147–164.

Pimentel, D., Terhune, E.C., Dyson-Hudson, R., Rochereau, S., Samis, R., Smith, E.A., Denman, D., Reifschneider, D. and Shepard, M., 1976. Land degradation: effects on food and energy resources. Science, 194: 149–155.

Reddy, A.K.N. and Goldemberg, J., 1990. Energy for the developing world. Sci. Am., 263: 110–118.

Reisner, M. and McDonald, R.H., 1986. The High Cost of Dams. In: A. Maguire and J.W. Brown (Editors), Bordering on Trouble: Resources and Politics in Latin America. Adler and Adler, Bethesda, MD, pp. 270–307.

Repetto, R., 1988. The forest for the trees: government policies and the misuse of forest resources. World Resources Institute, Washington, DC, 105 pp.

Sader, S.A. and Joyce, A.T., 1988. Deforestation rates and trends in Costa Rica. 1940 to 1983. Biotropica, 20: 11–19.

Samuelson, P., 1976. Economics. Tenth Edition. McGraw Hill, New York, 917 pp.

Seligson, M.A., 1984. The Gap Between Rich and Poor. Westview, Boulder, CO, 418 pp.

Simon, J., 1981. The Ultimate Resource. Princeton University Press, Princeton, NJ, 415 pp.

Smith, A., 1789. An inquiry into the nature and causes of the wealth of nations. Modern Library, New York, 1229 pp.

Smith, T.L., 1976. The Race between Population and Food Supply in Latin America. University of New Mexico Press, Albuquerque, NM, 194 pp.

Solow, R., 1974. The economics of resources or the resources of economics. Am. Econ. Rev., 64: 1–14.

Solórzano, R., de Camino, R., Woodward, R., Tosi, J., Watson, U., Vasquez, A., Jimenez, J., Repetto, R. and Cruz, W., 1991. Accounts Overdue: Natural Resource Depreciation in Costa Rica. World Resources Institute, Washington, DC, 110 pp.

Steinhart, J. and Steinhart, C.E., 1974. Energy use in the US food system. Science, 184: 307–316.

Walton, H., 1991. Trends in renewable energy use and conservation in the U.S. Int. J. Energy Environ. Econ., 1: 1.

Agriculture, Ecosystems and Environment, 46 (1993) 31–44
Elsevier Science Publishers B.V., Amsterdam

Tropical forests and the global carbon cycle: the need for sustainable land-use patterns

Sandra Brown[1]

Department of Forestry, University of Illinois, 110 Mumford Hall, 1301 W. Gregory, Urbana, IL 61801, USA

Abstract

Deforestation and degradation of forest and agricultural land in the tropics contribute to the increase in atmospheric carbon dioxide, although the magnitude of the flux is under much debate. There is a great potential to control deforestation and further land degradation through increasing productivity and sustainability of the tropical landscape. Increased sustainable development will not only improve the economic well-being of the people living in the tropics but it has a great potential to reduce atmospheric carbon dioxide concentrations through reduction of emissions and increase in area of land that serves as carbon sinks.

Introduction

One of the contemporary major global issues is the increase in concentration of greenhouse gases in the atmosphere and its potential to influence global climate patterns. Among the many greenhouse gases, the one of most concern is carbon dioxide because of its relatively long residence time in the atmosphere. The main source of atmospheric carbon dioxide is the burning of fossil fuels, presently contributing about 5.6 Pg C year^{-1} (1 Pg$=10^{15}$ g) (Ramanathan et al., 1987). Land-use changes in the tropics, particularly deforestation, is also another source of changes in atmospheric carbon dioxide whose magnitude is hotly debated (Table 1). The increase in carbon dioxide emissions for 1989 is mostly due to an increase in the tropical deforestation rate. For recent times, net carbon dioxide emissions to the atmosphere are equivalent to about 20–65% of those produced by burning fossil fuels. The range of carbon dioxide emissions from the tropics (Table 1) is due to uncertainties in estimates of forest biomass and deforestation rates.

What are the main causes of tropical deforestation? As population and economic growth increases there is an increased demand for food. To meet this increased demand, with little increase in yields per unit area, requires an increase in agricultural land. In the tropics, this land comes from forest lands. For example, the increase in agricultural land in a large part of south and

---

[1]Present address: Department of Forestry, University of Illinois, W-503 Turner Hall, 1102 S. Goodwin, Urbana, IL 61801, USA.

Table 1
Estimates of net rates of release of CO_2 to the atmosphere from tropical land-use change (Pg C (10^{15} g) year^{-1})

Year	Rate	Source
1980	0.9–2.5	Houghton et al., 1985
1980	0.4–1.6	Detwiler and Hall, 1988
1980	0.5–0.7	Grainger, 1990
1980	0.7–1.7	Crutzen and Andreae, 1990
1980	0.5–1.0	Hall and Uhlig, 1991
1989	1.1–3.6	Houghton, 1991

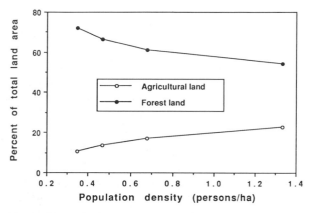

Fig. 1. Change in forest area and agricultural lands with increasing population density for south and southeast Asia (data from E. Flint, personal communication, 1991).

southeast Asia, due to increases in population density from about the 1860s to the present, is a mirror image of the decrease in forest area (Fig. 1). This increase in agriculture was historically mostly due to conversion to annual crops, although in more recent decades, perennial tree crops have become increasingly important (E. Flint, personal communication, 1991).

Most of the best lands in the tropics for relatively high yielding agriculture is already used for such, so that new lands being cleared are on increasingly marginal soils which demand larger areas to maintain yields. These marginal lands are also more susceptible to damage because they are often hilly or are of low soil fertility. This situation highlights another cause of deforestation: agricultural lands become so damaged that they are useless for food production and are abandoned, thereby forcing further deforestation. Much of the land currently being cleared of forests is replacing land that was degraded, so that the increase in agricultural land area does not equal the decrease in forest land (Fig. 2). The extent and seriousness of this problem is probably worse

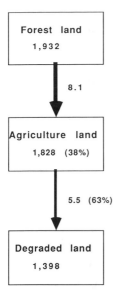

Fig. 2. Average annual net changes in the areas (million ha) of forests, agriculture and degraded lands in the tropics. The arrows represent the net loss in forests (million ha year^{-1}) and net increases in degraded lands. The numbers in parentheses are the percentages of the net loss of forests that appear as a net gain in agricultural and degraded land (modified from R.A. Houghton, personal communication, 1992).

than Fig. 2 demonstrates, and may be a more serious global problem than tropical deforestation.

Clearly there is a great potential to control deforestation and prevent further land degradation through increasing productivity and sustainability of agriculture in the tropics (Dale et al., 1993). Because most forest management agencies in the tropical world are understaffed, protecting forest lands from illegal encroachment is virtually impossible. However, if sustainable agricultural practices were available and acceptable to these people, the need to clear new forest lands could be reduced. Further research and technology development for rehabilitating degraded lands is also needed to prevent further forest destruction.

In sum, the key to further reducing tropical deforestation rates rests mainly in the hands of the agricultural community through the development of sustainable food production systems. These sustainable production systems must: (1) improve the economic well-being of the people without jeopardizing future needs, (2) use resources appropriately without degrading them, (3) use resources to contribute to equity and social justice and avoid social disruptions, and (4) use resources to optimize maintenance of cultural and biological diversity and environmental quality for the future (Ramakrishnan, 1992).

The main focus of this paper is to review the impact of human activities on the tropical forest landscape in relation to the global carbon cycle, and the main causes for these changes. Although the emphasis will be on continental tropical Asia, the comments are generally applicable to most of the tropical region. The article concludes with a discussion of the need for sustainable landscape patterns and how this may be brought about.

Changes in the tropical forest landscape

Forest clearing and its underlying causes

The last published, complete analysis of the status and rates of change of tropical closed and open forests was carried out for the period 1976 to 1980 by the Food and Agriculture Organization (FAO) (Lanly, 1982). At this time, the total rate of forest conversion to agriculture was 11.3 million ha year^{-1}, with about 66% occurring in the closed forest formations (Fig. 3). About half of the land was going into permanent agriculture, mostly due to planned development schemes, and the other half into the shifting cultivation cycle. In

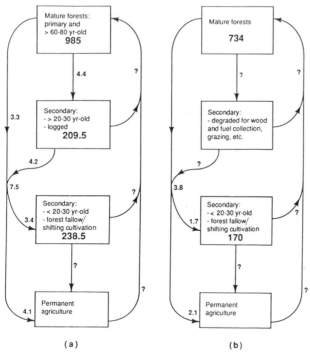

Fig. 3. Estimates of forest areas (million ha) as of 1980 and rates of conversion (million ha year^{-1}) for 1981–1985 for the tropics: (a) all closed forest formations, and (b) all open forest formations (data from Lanly, 1982).

fact, clearing by shifting cultivators (subsistence agriculture) was the single most important cause of tropical deforestation (Lanly, 1982).

Contrary to popular belief, less than half of the deforestation of closed forests originated from mature forest (referred to here as mature, not primary, based on the evidence given in Lanly, 1982); most originated from already logged-over forests. The situation for open forests was unknown.

Analysis of changes in forest areas for Peninsular Malaysia further emphasize that deforestation occurs more from previously logged forests than from mature forests. Peninsular Malaysia is one of the few tropical countries that have completed a national forest inventory for two periods of time: 1972 and 1982. Analysis of the inventories showed that only 8.6% of the mature forests were converted to non-forest, and that about 34% of previously logged hill and swamp forests were converted to agriculture or to rubber and oil palm plantations (Brown et al., 1993). However, shifting cultivation in this country is a minor problem which may account for the small loss of mature forests.

During the last decade, rates of tropical deforestation have increased by a factor of 1.4 (Table 2), but the types of forests being cleared and the types of agricultural lands being produced are not known yet. The FAO project is still in progress and should be published in 1993 (K.D. Singh, personal communication, 1991).

The underlying causes of conversion of tropical forests to other land uses are complex. It has been argued that conversion is done to satisfy local human needs rather than the needs of multinational corporations (Lugo and Brown, 1982). Others argue that much of the conversion is caused by political, economic, and social factors such as land tenure systems, transmigration programs, government subsidies (e.g. to the ranchers in the Amazon), availability of credit and technical assistance, that have little to do with the need to satisfy human needs (Schmink, 1987). Whatever the causes, there is no doubt that some strong correlations exist between human needs and actions and deforestation in the tropical landscape. These will be discussed next because

Table 2
Deforestation rates of tropical forests (10^6 ha year^{-1})

Time period	1980[1]	1981–1990[2]
Tropical America	5.6	7.4
Tropical Africa	3.7	4.1
Tropical Asia	2.0	3.9
Total	11.3	15.4

[1]Lanly, 1982.
[2]Food and Agriculture Organization, 1993.

knowledge of the pattern of these relationships will enable the present and future role of tropical forests in the global carbon cycle to be better understood.

As mentioned above (Fig. 1), agricultural expansion in the tropics in response to increased human needs for food production has resulted in a decline in the area of forests. The patterns of decline in forest cover in individual countries is likely to be similar to each other, but the magnitude of the reduction will be different reflecting differences in the long history of human use of the landscape, past and present population densities, and political, social, and economic systems. For example, countries in south and southeast Asia show that the decline was initially steep as agriculture expanded rapidly onto the better lands, followed by a levelling-off in the reduction of forest cover as only the most inaccessible and highly unsuited lands (very steep and/or high rainfall conditions) for agriculture remained (Fig. 4) (Sader and Joyce, 1988).

Accessibility to forest areas is a key factor in bringing about deforestation. Before large-scale development, rivers were a main transportation corridor into the interior of forest regions and most clearing was restricted to the zones along these rivers. Construction of transportation networks to improve economic development generally makes forest areas more accessible to clearing. Most people are aware of the increase in deforestation of the Brazilian Amazon following the construction of roads in that region (Woodwell et al., 1987). Similar responses have been found for Costa Rica (Sader and Joyce, 1988) and the Philippines (Liu et al., 1993). In Costa Rica, for example, where a

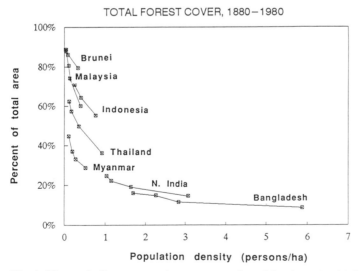

Fig. 4. Change in forest cover (as a percent of total land area) with increasing population density for several countries in south and southeast Asia (data from E. Flint, personal communication, 1991).

strong relationship was found between transportation networks and forest clearing, few tracts of forest remained in proximity to any transportation corridor by 1983 (Sader and Joyce, 1988).

Logging of tropical forests is often cited as a major cause of deforestation. As discussed above, logging in itself is not a major cause because trees are harvested selectively. However, logging leads indirectly to deforestation as the construction of logging roads once again opens up the region and increases its accessibility. This increased accessibility to subsistence agriculture leads to further deforestation as the land is slashed and burned to grow crops.

Formation of secondary forests

Secondary forests result from the impact of humans on forest lands (Brown and Lugo, 1990). They include forests resulting from abandonment of agricultural lands (permanent or shifting cultivation) and from logging, as well as those under continuous use, particularly in the dry tropical zones, such as grazing, fuel wood collection, and burning. Under this definition, most tropical countries, with few exceptions, have larger areas of secondary forests than primary ones (Brown and Lugo, 1990).

The area of secondary forests as of 1980 was significant (Fig. 3); however, this estimate is likely to be very conservative (Brown and Lugo, 1990). It is clear that such large areas deserve attention by ecologists and foresters because efforts to conserve biodiversity in the tropics may rest on how these secondary forests are managed. If secondary forests can be managed sustainably, considerable pressure will be removed from the primary forests. Because of the many values provided by secondary forests for human use, particularly for wood products (Brown and Lugo, 1990), there is a great incentive to manage them properly.

Secondary forests are also important from a carbon cycle perspective. Requirements for new permanent agricultural lands should be met from these forest types when feasible as they have smaller carbon pools than mature forests, thus contributing less carbon dioxide to the atmosphere when they are cleared. Furthermore, if they are allowed to recover they can be significant sinks of atmospheric carbon for many decades (Lugo and Brown, 1992).

Degradation of forests

Land use practices that produce the secondary forest formations described above are quite well known. What is not so well known is the insidious degradation of tropical forests often due to the widespread removal of biomass as illegal tree harvesting ('log poaching') for either timber or for fuel (Brown et al., 1991); this is often brought about by rural people supplementing their subsistence. Log poaching is particularly common in the tropical moist or seasonal forests that contain valuable timber trees. Recent analyses for south and southeast Asia (Flint and Richards, 1991; Brown et al., 1991, 1993; Iverson et al., 1993) and for the Brazilian Amazon (Brown and Lugo, 1992) suggest that very few forests in these regions have escaped the impact of this practice.

The following results from two case studies (Peninsular Malaysia and continental south and southeast Asia) illustrate the magnitude of this type of forest degradation. The two national forest inventories carried out for Peninsular Malaysia in 1972 and 1982 produced forest maps and stand tables (number of trees per hectare by diameter class) for 11 forest classes. The forests were divided into six undisturbed and five disturbed classes. The undisturbed classes included three classes of primary hill forests (superior, good and moderate based on volumes per hectare), two classes of restricted forests (poor hill forests and montane forest), and freshwater swamp forests. The disturbed classes included logged hill and swamp forests before and after 1966 and shifting cultivation/hill forest mosaics. Analysis of these inventories, using a geographic information system (GIS), showed that forest area decreased by 17%, but the overall forest aboveground biomass decreased by 28% during the decade (Brown et al., 1993). Not all forest classes lost biomass during this interval. Decreases in biomass per unit area occurred in the primary hill forests (16% decrease), logged hill forests (31% decrease), and shifting cultivation/forest mosaics (36% decrease). All other forest classes gained biomass.

These changes in biomass per unit area were highly correlated with the change in the perimeter to area ratio (P/A) of the forest polygons (Fig. 5). Forests that showed a decrease in edge per unit area (i.e. a negative change in

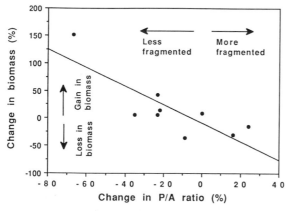

Fig. 5. Change in biomass density as a function of the change in the perimeter to area ratio (P/A) for forests of Peninsular Malaysia between 1972 and 1982. The relationship was significant ($P < 0.05$) and explained 70% of the variation between the two variables (from Brown et al., 1993).

P/A) became less accessible and gains in biomass resulted (Brown et al., 1993). Several of these forest classes that gained biomass were clearly recovering from previous logging. However, other classes were described as undisturbed, lending support to the idea that large areas of tropical forest lands are sequestering carbon and acting as carbon sinks (Lugo and Brown, 1986, 1992).

Forests that experienced a loss in biomass density during the 10 year interval had an increase in P/A. In other words, the edge of these forests increased proportionally more than the decrease in area, resulting in increased accessibility. The primary hill forests experienced the largest increase in P/A during the interval, followed by logged hill forests. For the logged forests, it is easy to suggest that the reduction in biomass density was due to increased intensity of tree harvesting. However, illegal log harvesting in these forests as well as in the primary hill forests is the most likely cause (Thang Hooi Chiew, Malaysian Forestry Department, personal communication, 1992). Large diameter trees are often removed in this illegal operation, and it is the large diameter trees that can account for significant quantities of forest biomass (more than 40% in undisturbed forests) (Brown and Lugo, 1992; Brown et al., 1993).

Geographic information systems technology has also been applied to modeling forest biomass for continental south and southeast Asia as a whole (Iverson et al., 1993). Here three spatial databases of forest biomass were produced: a potential forest biomass map without the influence of humans on the landscape, a potential forest biomass map for the forest area as of about 1980, and an actual forest biomass map as of about 1980.

The potential forest biomass map was modeled using spatial databases for those bio–physical factors believed to most influence biomass, i.e. climate (including temperature, humidity, rainfall, growing season length, moisture availability), rainfall alone, soil texture and depth, elevation, and slope. This map was then overlaid with a forest map of the region for 1980 to extract the potential forest biomass for these existing forests (potential biomass-80). Finally, the actual forest biomass map was produced by adjusting the potential biomass-80 map for the influence of human actions using a variety of data sources including actual forest inventories, human population densities, presence/absence of shifting cultivation, and advice of experts in the region (details of the adjustments are given in Iverson et al., 1993).

Without the influence of humans in the region, the landscape is capable of supporting about 176 Pg of aboveground biomass, or forests with an average aboveground biomass of 322 Mg ha^{-1} (Fig. 6). Cumulative deforestation alone, throughout human history in the region, reduced this to 63 Pg of biomass (Potential-80 in Fig. 6). Biomass density of these remaining forests increased reflecting preferential human habitation and forest clearing in drier climates (e.g. high population density in much of India and central Thailand) which support lower biomass.

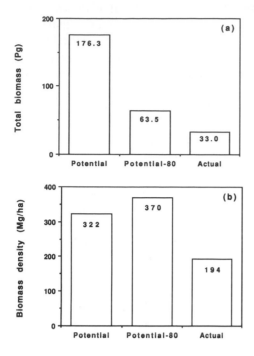

Fig. 6. Differences in (a) total aboveground biomass, and (b) biomass density for forests of continental south and southeast Asia. Potential biomass, without the influence of humans; Potential-80, reduction due to deforestation alone; Actual, reduction of Potential-80 due to further degradation from human impacts (from Iverson et al., 1993).

The degradation of forests due to processes described above have reduced the total forest biomass by an additional 48%. The net result is that in 1980 the actual amount of biomass in forests of this region was about 19% of that which the region could potentially support. Clearly, forest degradation, due to biomass removal (both commercial and illegal harvesting), is a serious problem in this part of the world. Similar human activities are likely to be having the same effect on tropical forests in other parts of the tropics (Brown et al., 1991; Brown and Lugo, 1992).

Degradation of forests has several implications for the global carbon cycle. As stated before, a major uncertainty in changes in estimates of the carbon dioxide flux to the atmosphere from tropical deforestation (Table 1) is the biomass of forests being cleared. The choice of values to use in models has been the subject of much debate for many years. Some have argued for lower estimates (Brown and Lugo, 1984; Brown et al., 1989) whereas others argued for higher estimates (Houghton et al., 1985). Results from analyses such as those described above strongly suggest that forest biomass estimates for contemporary forests should, on average, be low, whereas estimates used for models that start in the last century can be higher (see Iverson et al., 1993 for range).

Reduction in biomass of remaining forests means that these forests, when cleared for agriculture, have less biomass to burn with consequently less carbon dioxide entering the atmosphere. Although deforestation is increasing (Table 2), on average, forests are likely to contain less biomass and thus less carbon per unit area following degradation. However, the fate of the biomass removed from the forest is largely unknown but critical for determining the impact of degradation on the carbon cycle. If it is going into long-lived wood products, the net result of biomass degradation is to make forests less of a source of atmospheric carbon when they are cleared or a potential carbon sink if they remain as forests. If, on the other hand, the wood removed during degradation is being burned for fuel, the net result is the same as burning high biomass forests when they are cleared. But, if the removed wood is being used for fuel and the land remains as forests then wood removal tends to have a neutral role in the carbon cycle because the amount of carbon dioxide produced during burning for fuel is likely to be fixed via photosynthesis in recovering forests.

Finally, the indications of widespread tropical forest degradation, even in moist forests, clearly supports the notion that most tropical forests are not primary but are secondary. Large fractions of these have the potential to become significant carbon sinks through natural recovery processes (Lugo and Brown, 1992) if pressure on them could be removed by increasing sustainable economic development.

Need for sustainable tropical landscapes

Clearly, sustainable land use patterns that meet the criteria given by Ramakrishnan (1992) must be created to prevent further wanton destruction and degradation of tropical forests. To satisfy human needs for food, fiber, and fuel, an optimal mix of land uses and conversions, such as those that exist today in the tropics, will be required (Fig. 7). Neither absolute preservation of mature forests nor complete conversions to intensively managed systems can be advocated as the solution to land management problems in the tropics.

People must derive benefits from the land conversion process itself as well as from the various stages of land use, in both forests and non-forests, to maximize the full value of the resources (Fig. 7). However, values, yields, services, benefits, trade-offs, and costs to people will differ according to the type of land use produced. People will benefit most when the conversion rate from one land use to another (the boxes in Fig. 7) is not too fast, when land uses are not skipped, and when the landscape is most diverse (Brown and Lugo,

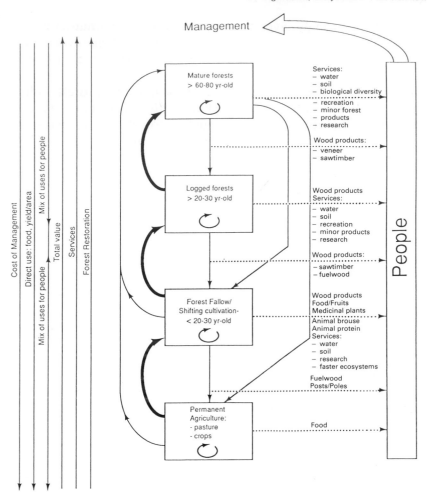

Fig. 7. A model of sustainable tropical landscapes, showing the conversion of forests to other land uses, the product and services (dotted lines) that people derive from the land and the conversion process itself, recovery of forests (arrows pointing up on the left side of boxes), and trends in the costs and benefits to people from the different land uses (trends increase in direction of arrows on the left-hand side of the diagram). Each land use (boxes) must be managed sustainably (circled arrows in boxes), can be converted to more intensive use where conditions assure success (downward arrows), or be returned to more complex systems to restore productive capacity (upward arrows). Heavy shaded arrows show those pathways most favored by natural processes. Conversions that jump over a land use stage may occur if enough energy and resources are available to overcome the costs (from Brown and Lugo, 1990).

1990). This calls for the management of tropical lands with a landscape perspective.

A move towards sustainable landscapes would not only provide direct benefits to people living in the tropics, but it would also help to reduce global atmospheric carbon dioxide concentrations. For example, carbon accumulation in biomass of logged forests and forest fallows could continue for decades if less of this land had to be cleared and burned to meet the needs for food production or to replace badly damaged lands. The need to continue clearing and degrading more forest lands may be reduced, and as many of these forests are secondary, they would be allowed to sequester carbon for long time periods. Finally, implementation of other techniques to improve sustainability of tropical agriculture such as agroforestry, soil organic matter management, and erosion control would likely lead to an enhancement or conservation of carbon pools in the tropics.

Acknowledgments

This paper is based on research supported by a contract DEFG02-90ER61081 between the US Department of Energy and the University of Illinois (S. Brown, PI). I thank A.E. Lugo and L. Iverson for their helpful comments.

References

Brown, S. and Lugo, A.E., 1984. Biomass of tropical forests: a new estimate based on forest volumes. Science, 223: 1290–1293.

Brown, S. and Lugo, A.E., 1990. Tropical secondary forests. J. Trop. Ecol., 6: 1–32.

Brown, S. and Lugo, A.E., 1992. Biomass estimates for tropical moist forests of the Brazilian Amazon. Interciencia, 17: 8–18.

Brown, S., Gillespie, A.J.R. and Lugo, A.E., 1989. Biomass estimation methods for tropical forests with applications to forest inventory data. For. Sci., 35: 881–902.

Brown, S., Gillespie, A.J.R. and Lugo, A.E., 1991. Biomass of tropical forests of south and southeast Asia. Can. J. For. Res., 21: 111–117.

Brown, S., Iverson, L. and Lugo, A.E., 1993. Land use and biomass changes of forests in Peninsular Malaysia during 1972–1982: use of GIS analysis. In: V. Dale (Editor), Effects of Land-Use Change on Atmospheric Carbon Dioxide Concentrations: Southeast Asia as a Case Study. Springer, New York, in press.

Crutzen, P.J. and Andreae, M.O., 1990. Biomass burning in the tropics: impact on atmospheric chemistry and biogeochemical cycles. Science, 250: 1669–1678.

Dale, V.H., Houghton, R., Grainger, A., Lugo, A.E. and Brown, S., 1993. Emissions of greenhouse gases from tropical deforestation and subsequent uses of the land. In: Sustainable Agriculture and the Environment in the Humid Tropics. National Academy Press, Washington, DC, pp. 215–260.

Detwiler, R.P. and Hall, C.A.S., 1988. Tropical forests and the global carbon budget. Science, 239: 42–47.

Flint, E.P. and Richards, J.F., 1991. Historical analysis of changes in land use and carbon stock of vegetation in south and southeast Asia. Can. J. For. Res., 21: 91–110.

Food and Agriculture Organization (FAO), 1993. Summary of Conclusions, Forest Resources Assessment 1990 Project, FAO, Rome, 2 pp.

Grainger, A., 1990. Modelling future carbon emissions from deforestation in the humid tropics.

In: Proceedings of the Intergovernmental Panel on Climate Change, Conference on Tropical Forestry Response Options to Global Climate Change, Sao Paulo, Brazil, 9–12 January 1990. Report No. 20P-2003, Office of Policy Analysis, US Environmental Protection Agency, Washington, DC, pp. 105–119.

Hall, C.A.S. and Uhlig, J., 1991. Refining estimates of carbon released from tropical land-use change. Can. J. For. Res., 21: 118–131.

Houghton, R.A., 1991. Tropical deforestation and atmospheric carbon dioxide. Climate Change 19: 99–118.

Houghton, R.A., Boone, R.D., Melillo, J.M., Palm, C.A., Woodwell, G.M., Myers, N., Moore, B. and Skole, D.L., 1985. Net flux of CO_2 from tropical forests in 1980. Nature, 316: 617–620.

Iverson, L.R., Brown, S., Prasad, A., Mitasova, H., Gillespie, A.J.R. and Lugo, A.E., 1993. Use of GIS for estimating potential and actual forest biomass for continental south and southeast Asia. In: V. Dale (Editor), Effects of Land-use Change on Atmospheric Carbon Dioxide Concentrations: Southeast Asia as a Case Study. Springer, New York, in press.

Lanly, J.P., 1982. Tropical Forest Resources. FAO Forestry Paper 30, Rome, 106 pp.

Liu, S.D., Iverson, L.R. and Brown, S., 1993. Rates and patterns of analysis of deforestation in the Philippines: application of geographic information system analysis. For. Ecol. Manage., 57: 1–16.

Lugo, A.E., and Brown, S., 1982. Conversion of tropical moist forests: a critique. Interciencia, 7: 89–93.

Lugo, A.E. and Brown, S., 1986. Steady state terrestrial ecosystems and the global carbon cycle. Vegetatio, 68: 83–90.

Lugo, A.E. and Brown, S., 1992. Tropical forests as sinks of atmospheric carbon. For. Ecol. Manage., 54: 239–256.

Ramakrishnan, P.S., 1992. Shifting Cultivation and Sustainable Development: Interdisciplinary study from Northeastern India. Man and the Biosphere Book Series, Vol. 10. UNESCO, Paris, and Parthenon Publishing, Carnforth, 424 pp.

Ramanathan, V.L., Callis, L., Cess, R., Hansen, J., Isaksen, I., Kuhn, W., Lacis, A., Luther, F., Mahlman, J., Reck, R. and Schlesinger, M., 1987. Climate-chemical interactions and effects of changing atmospheric trace gases. Rev. Geophys., 25: 1441–1482.

Sader, S.A. and Joyce, A.T., 1988. Deforestation rates and trends in Costa Rica, 1940–1983. Biotropica, 20: 11–19.

Schmink, M., 1987. The rationality of forest destruction. In: J. Figueroa, F.H. Wadsworth and S. Branham (Editors), Management of the Forests of Tropical America: Prospects and Technologies. Institute of Tropical Forestry, US Dep. Agric. For. Serv., Rio Piedras, Puerto Rico, pp. 11–30.

Woodwell, G.M., Houghton, R.A., Stone, T.A., Nelson, R.F. and Kovalick, W., 1987. Deforestation in the tropics: new measurements in the Amazon basin using Landsat and NOAA Advanced Very High Resolution Radiometer Imagery. J. Geophys. Res., 92: 2157–2163.

Agriculture, Ecosystems and Environment, 46 (1993) 45–54
Elsevier Science Publishers B.V., Amsterdam

Kikuyu agroforestry: an historical analysis

Alfonso Peter Castro

Department of Anthropology, Syracuse University, 308 Bowne Hall, Syracuse, NY 13244, USA

Abstract

This paper provides a historical analysis of indigenous agroforestry practices among Ndia and Gichugu Kikuyu cultivators in Kirinyaga District, Kenya. The Kikuyu have long-devised agroforestry practices such as selective cutting, protection of woodland, and intercropping, in response to competing pressures for retaining and removing tree cover. Traditional religious beliefs, common property resource tenure, and farm forestry practices contributed to the conservation of trees. It is argued that indigenous agroforestry techniques and strategies mitigated the impact of deforestation by incorporating valued multipurpose trees in local production systems. The paper briefly examines how indigenous agroforestry was modified by state forestry, agricultural, and land use policies and programs. Although communal tree management strategies have collapsed, indigenous farm forestry practices remain a resilient and viable part of contemporary land use strategies.

Introduction

In a recent article Chambers (1991) states: "A common belief is that while professionals take a long-term view of sustainability, poor rural people live 'hand to mouth' and take a short-term view. Often, the opposite is true". He contrasts the near-sightedness of most politicians, bureaucrats, businessmen, social scientists, and other experts with the long-range vision of rural dwellers who (when they have the ability to do so) devise complex and diverse sustainable livelihoods. According to Chambers, as well as a growing number of others (Brokensha et al., 1980; Warren et al., 1989; Castro, 1992), 'outsiders' can learn much about sustainable natural resource use by recognizing "the knowledge, creativity, and competence of rural people".

This paper seeks to increase understanding of sustainable land use systems in rural Africa; it presents a historical analysis of indigenous agroforestry practices among the Ndia and Gichugu Kikuyu of Kirinyaga District, Kenya. The paper describes ways in which people tried to reconcile competing pressures for retaining and removing trees. It shows how tenure relations based on a communal property rights regime, religious beliefs, and local farm forestry practices contributed to the conservation of trees before the onset of colonial rule in 1904. It is argued that agroforestry mitigated the impact of deforestation by incorporating valued multipurpose trees into local production systems. The paper briefly examines how indigenous agroforestry strategies were modified by state forestry, agricultural, and land use policies and programs. Agroforestry remains a dynamic yet continuous part of farming systems in contemporary Kirinyaga.

The importance of agroforestry in rural Africa cannot be understated (Cook and Grut, 1989; Kerkhof, 1990; Okigbo, 1990). Trees are inseparable from any consideration of satisfying basic human needs and attaining environmental stability. Understanding the historical dynamism of local agroforestry offers significant insights into the nature of sustainable production systems. As will be seen, many of the farm forestry practices in Kirinyaga have stood the test of time.

The setting

Kirinyaga in Kenya's smallest rural district, covering 1437 km^2 in Central Province. It is located on the southern slopes of Mount Kenya, the second tallest mountain in Africa (5199 m). Most of the district possesses high agro–economic potential, with fertile volcanic soils and abundant rainfall (Jaetzold and Schmidt, 1983). Kirinyaga is the homeland of the Ndia and Gichugu Kikuyu. The population has more than doubled in the twentieth century, totaling over 300 000 (Kirinyaga District, 1984).

In pre-colonial times the Ndia and Gichugu Kikuyu were similar to other 'stateless' societies of central Kenya (see Kenyatta, 1938; Middleton and Kershaw, 1965; Muriuki, 1974; Ambler, 1988; Davison, 1989). They engaged in farming and herding. Their social organization centered around kinship (clan, lineage and household) and territorial groups (age- and generation-sets). People resided in dispersed homesteads.

Three types of sources have been used to reconstruct the history of indigenous agroforestry practices: (1) European accounts of Kirinyaga at the time of contact, including colonial records; (2) descriptions from other parts of central Kenya; (3) data collected by the author in Kirinyaga during 1982–1983 and 1988. All of these sources required careful sifting and analysis. None of the elders interviewed in Kirinyaga had been adult participants in pre-colonial life, but they were familiar with many aspects of local tradition. In particular, they had first-hand knowledge of tenure and resource use patterns before land privatization and the expansion of cash-cropping in the 1950s and 1960s. This study focuses on social and historical dimensions of indigenous agroforestry, rather than detailing every aspect of local tree use. For indepth botanical identities and uses, see Leakey (1977a,b,c) and Riley and Brokensha (1988a,b).

Conflicting pressures about trees

People in Kirinyaga have long relied on trees to supply vital materials such as timber, fuel, fencing, food, fodder, medicine, and utensils. Trees serve other purposes. They offer environmental amenities — shade, windbreak, privacy — and are used as boundary markers. In the past, selected groves and trees were the sites of communal worship and local ceremonies.

The multiple uses of trees reflects both the area's diverse flora and the impressive botanical knowledge accumulated by the Kikuyu (Leakey, 1977a). In the 1930s, Leakey (1977c) recorded from Kiambu Kikuyu elders the names of over 400 different trees and plants used in pre-colonial times. Leakey pointed out that his list was "not in any way exhaustive". Knowledge about, and the use of, trees in pre-colonial Kirinyaga was not as thoroughly documented, but early colonial accounts suggested that the situation was similar (Crawshay, 1902; Embu District, no date). Indigenous knowledge about trees derived from close interaction with, and dependence on, a range of local ecozones. In Kirinyaga people utilized, to varying extents, the gradient of ecozones running down the southern slopes of Mount Kenya: afro–alpine moorland, bamboo and moist montane forest, permanently settled farmland, and sub-humid woodland and grassland (Stigand, 1913; Castro, 1983, 1988, 1991a,b; Davison, 1989).

The usefulness of trees has always conflicted with the need for agricultural land. The long process of settlement in central Kenya by Bantu-speaking peoples was accompanied by extensive clearing of primeval forest (Muriuki, 1974). Over the course of many centuries iron axes, cultivating knives, and fire gradually converted woodland into homesteads, gardens, and pastures. Europeans who ventured into Kikuyu country during the late 1800s and early 1900s commented on the lack of tree cover (Crawshay, 1902; Dickson, 1903; Von Hohnel, 1968). Gedge (1892), for example, who traveled through the southern Mount Kenya region in 1891, noted that 'extensive clearances' caused the forest to be 'some distance' from settlements and fields. Stigand (1913) described the heart of Kikuyuland as a "rolling, almost treeless, cultivated country".

The word 'almost' in the foregoing account is significant, for the Kikuyu attempted to reconcile these competing pressures to retain and to remove trees. Restraints on cutting trees were included in customary tenure rights and land use practices. These were reinforced by cultural beliefs about the nature of trees. Indigenous agroforestry practices tried to maintain some tree cover near homesteads and farmland. People did not want to halt deforestation, but to mitigate its impact. The application and effectiveness of such practices varied over time.

Communal tree management practices

The traditional landholding system in Kirinyaga can be classified as a communal property rights regime (McCay and Acheson, 1987; Berkes, 1989). Rights to land were held by "an identifiable community of interdependent users" (Feeney et al., 1990). Families derived their land rights through membership in descent groups — the patrilineal clan and the more closely defined lineage. The Kikuyu also recognized several forms of temporary land occupancy, with descent groups granting farming or residency rights to non-kinsmen. Access and use rights were also defined within the context of 'neighborhood' or local territorial groups. Thus, rights to trees were enmeshed in a web of social obligations involving kinsmen and neighbors.

By the late 1800s, communal controls were placed on the clearing of woodland by descent and neighborhood groups. Permission was supposed to be obtained from local elders and rightholders before harvesting trees or farming along the forest frontier on the slopes of Mount Kenya and on the boundaries of smaller patches of woodland such as Njukiin Forest (Castro, 1988, 1991b). People could not simply choose any convenient site to clear land. Such areas were protected as timber reserves, a practice commonplace elsewhere in central Kenya (Kenyatta, 1938; Leakey, 1977a). However, they were regarded as land banks, to be opened to cultivation as needed. In contrast, groves and trees set aside as the sites of communal worship and ceremonies were considered inviolable (Castro, 1990).

Descent groups exercised special controlling rights over large multipurpose trees on their holdings. People were required to seek the approval of lineage elders before felling such trees. A payment of a goat to the elders was expected. The trees of two species, *Cordia abyssinica* and *Pygeum africanum*, were regarded as 'clan property'. These species were highly regarded for their building timber and other uses (see Table 1). Permission was needed before cutting such trees.

Farm-level strategies to conserve trees

Several strategies were used at the household level to retain tree cover on their particular holdings. Small plots of heavily wooded land were sometimes set aside to supply building material, fuel, and other items (Routledge and Routledge, 1910; Kenyatta, 1938; Leakey, 1977a). In certain cases people placed a curse on trees in order to protect them (Nyeri District, 1914; Orde-Browne, 1925; Leakey, 1977a). Present-day practices suggest that protected stands of trees were situated in a variety of settings, including along steep ridges and at the head of narrow valleys and ravines. Such trees were cut as needed.

The protection of individual trees or small copses during land clearing was a common and widespread practice. When preparing a plot for cultivation, certain trees would be spared the axe. Before setting fire to the plot, brush and other combustible materials would be removed from around such trees. This land use strategy was motivated and reinforced by spiritual and utilitarian concerns.

Traditional religious beliefs influenced the practice of selective clearing. The Kikuyu believed that trees possessed spirits capable of intervening in human affairs (Leakey, 1977b). A tree spirit would be angered if its abode was cut and it had no nearby tree to go to. When clearing land, people were supposed to leave 'a large and conspicuous tree' at intervals to absorb the spirits from the ones harvested (Hobley, 1967; Leakey, 1977b). Such a tree was called a murema kiriti ('one which resists the cutting of the forest'), and it was not supposed to be cut or allowed to fall without a ceremony transferring the spirits to another site. If a person failed to perform the ceremony within a short time-span, he or she would be killed by the tree spirits (Leakey, 1977c). Young people who used such a tree for fuel would become sick or die, but very old men and women could burn it without danger (Hobley, 1967).

There was a practical side to selective clearing as well: it permitted a convenient supply of forest products to be available for family consumption. One of the most visible reasons why the Kikuyu spared large trees was to hang bcchivcs on thcm (Routlcdgc and Routledge, 1910). Crawshay (1902, p. 33), an early European visitor to Kirinyaga, wrote, "many a fine tree owes its existence to the bees". *Cordia abyssinica*, with sweet white flowers, was a favorite 'bee tree', but any large tree with sprawling limbs was used. Besides its value as a foodstuff, honey was highly prized for brewing and various ceremonial uses.

Certain types of trees were preferred because they supplied multiple or special products. Such species often had botanical properties — seeding profusely, easily sown or transplanted, reproduction from cuttings, a capacity for being heavily pruned or coppiced, the ability to be intercropped — which facilitated their retention. Cuttings from *Markhamia hildebrandtii, Commiphora zimmermanni*, and *Cordia abyssinica*, for example, were used to mark boundaries. *Erythrina abyssinica* and *Euphorbia tirucalli* were propagated from cuttings for living fences. Volunteer seedlings from several species, including *Kigelia africana*, were often protected.

Multipurpose trees that could be intercropped with staple foods were especially valued. *Commiphora zimmermanni* regularly served as a vine prop for yams and sweet potatoes. *Markhamia hildebrandtii* was often pruned heavily to obtain poles and firewood, then maize, beans, and other crops were planted around it. Maize was sometimes interplanted with *Cordia abyssinica* and *Erythrina abyssinica*, both deciduous species. Grain and pulses were sown under *K. africana. Croton megalocarpus* and *Albizzia gummifera* commonly furnished shade for cattle.

Tree propagation was carried out on a limited scale in the pre-colonial era. The purpose of such practices was not reforestation per se, but meeting specific needs such as marking boundaries, making fences, setting hedges, and supporting vines. Kikuyu leaders from Kirinyaga and neighboring areas told the 1929 Maxwell Committee, "Trees have only been planted since Europeans came" (Maxwell et al., 1929), i.e. planting trees in large numbers and for the purpose of reforesting farmland were colonial innovations.

Table 1 lists common indigenous agroforestry trees in Kirinyaga, including their major uses and key botanical properties. The list is not exhaustive because other species were used to varying extents. There was no single pattern of indigenous agroforestry. Rather, there were numerous variations on a similar theme of trying to incorporate trees into household production systems. This diversity reflected both the physical and social environments. The district is characterized by considerable local differences in topography, rainfall, soils, and natural vegetation. Families faced different situations regarding the availability of land and labor, the extent of their ethnobotanical knowledge, and the crop mixes possible within their particular ecozone. What is important to emphasize is that households were utilitarian in their management and use of trees.

The contemporary incidence of the species listed in Table 1 should not be

Table 1

Bridelia micanthra (mukoigo). Poles, timber, and firewood (propagation by cuttings, pruning, drought resistant).

Commiphora zimmermanni (mukungugu). Vine props, poles, utensils, quick-set hedges (cuttings, intercropping, pruning).

Cordia abyssinica (muringa). Beehives, stools, mortars, well covers, and building timber (cuttings, volunteers, pruning, intercropping).

Croton macrostachyus (mutundu). Poles, medicinal uses, and boundary markers (coppicing).

Croton megalocarpus (mukinduri). Poles, boundary marker, cattle shade (profusely seeding, fast growing).

Erythrina abyssinica (mubuti or muhuti). Living fence, medicinal uses, beads, weather indicator (cuttings, pruning, and some intercropping).

Ficus natalensis (mugumo). Ceremonial and medicinal uses (cuttings).

Kigelia africana (muratina). Fruit used as a fermenting agent (intercropping, pruning).

Markhamia hildebrandtii and *Markhamia platycalyx* (muu). Poles, building timber, and firewood (cloning, volunteers, pruning, intercropping).

Pygeum africanum (mweria). Mortars, pestles, poles, building timber, tool handles, and cattle enclosures (coppicing, pruning).

Ricinus communis (mubariki). Castor seed and oil (profuse seeds, fast growing).

This table is meant to be illustrative rather than comprehensive. It lists in alphabetical order indigenous agroforestry trees traditionally utilized in Ndia and Gichugu Divisions, Kirinyaga District. Botanical and vernacular terms (in parentheses) the species are given, followed by major traditional uses and, in parentheses also, properties of the tree which fostered its incorporation into local production systems.

Source: Leakey, 1977c; Castro, 1983; Riley and Brokensha, 1988b.

used as guides to their past distribution. Forestry programs, agrarian intensification, and other changes in land use since 1904 have influenced present-day patterns. *Commiphora zimmermanni*, for example, was 'the commonest' farmland tree in Kikuyuland at the turn of the century (Stigand, 1913). Easily propagated from cuttings, it was valued highly as a vine prop and quick-set hedge. Although still widely grown, this tree has been surpassed in local popularity and spatial distribution by exotics such as the multipurpose *Grevillea robusta*.

Indigenous agroforestry practices constituted a form of practical conservation that was highly selective in terms of species. While *Markhamia hildebrandtii, Commiphora zimmermanni*, and similar multipurpose trees might be protected, other species decreased in number. Some historical evidence and traditions suggest that species such as *Heywood lucens, Ekebergia capensis,* and *Milletia oblata* became increasingly scarce as settlements and fields expanded in the late nineteenth and twentieth centuries (Hutchins, 1907; Maher, 1938).

Several early colonial writers claimed that the Kikuyu practiced a particularly destructive form of shifting cultivation (Eliot, 1905; Hutchins, 1907; Baker, 1931). This misconception greatly underestimated the stability, sustainability, and resilience of their indigenous agriculture. The Kikuyu commonly rotated fields, with plots allowed to remain fallow for several seasons. Dense weed growth, rather than soil exhaustion per se, was frequently the main reason for taking land out of production. Such fields were brought back into cultivation long before they could revert to secondary forest. In some places plots were under almost continuous production (Kenya Land Commission, 1934). Fallowing allowed trees to regenerate to some extent, but there was no long-term rotation involving farmland and trees. Fallowing fields served as a source of forest products such as firewood, medicine, and fodder.

Indigenous agroforestry in perspective

There has been much change and continuity in local agroforestry practices during the past 100 years. The willingness and ability to maintain trees on farmland has fluctuated. Even before the onset of colonial rule, the coastal caravan trade accelerated land clearing in Kirinyaga and other parts of Kikuyu country (Castro, 1991a). Highly localized wood shortages emerged in some areas by the turn of the century (Gedge, 1892; Dickson, 1903; Routledge and Routledge, 1910). The decrease in tree cover was not necessarily synonymous with increased land degradation. It often involved only a shift in cropping patterns, with food crops replacing trees. Farming techniques such as intercropping and no-till farming made it possible to expand farm production without undermining soil management (Leakey, 1977a).

Colonial rule led to the decline of communal tree management strategies.

In the case of Mount Kenya, the government alienated the forested slopes, evicted local rightholders, and imposed its own bureaucratic administration (Castro, 1988, 1991c). Colonial institutions also eventually assumed management of the district's remaining woodland and sacred groves as local controls broke down in the face of rapid social change. In particular, increased socio–economic differentiation and the immigration of Kikuyus from other areas (themselves the victims of colonial land alienation) undermined local communal bonds (Castro, 1990, 1991b). The privatization of land tenure in the 1950s greatly curtailed descent group and neighborhood control over individual land use.

A major innovation of the colonial era was the introduction of large-scale farm forestry campaigns and the establishment of local tree nurseries to support such activities. As early as 1912, officials used demonstration plots and communal block plantings to foster reforestation. The latter was very unpopular, as the Kikuyu had no tradition of communal tree planting, and they regarded it as compulsory labor (Castro, 1983). There were suspicions about the government motives as well, fearing that state-sponsored tree growing would give the colonialists claim to the land. The introduction of fast-growing *Acacia mearnsii* (wattle) as a cash crop with home-use values met with great success among Kikuyu households in the 1930s and 1940s (Kitching, 1980). Its tanning-rich bark was sold to private traders, and families had the option of selling or keeping its wood for poles and fuel. Local tree nurseries supported the planting of other exotic and indigenous species during this period (Castro, 1983).

Both the system of local tree nurseries and the wattle industry collapsed during the Mau Mau revolt in the 1950s. Kenya received its independence in 1963. Families in the late 1950s and the 1960s were absorbed in increasing cash cropping on their newly privatized holdings. In particular, coffee and tea emerged as major commodities. State farm forestry efforts, however, received a significant boost in the late 1970s with the start of the Rural Afforestation Extension Scheme. By the 1980s, the district had its own forester engaged in promoting agroforestry and a number of local tree nurseries (Castro, 1983).

There has been much continuity with earlier farm forestry practices, as selective clearing, coppicing, intercropping of trees and food crops, woodlots, boundary trees, and other techniques and strategies remain commonplace. Fallowing has decreased, though chemical fertilizer is often used. The most successful exotic species have been ones which fit into pre-existing tree husbandry practices, such as the multipurpose and easily reproduced *Grevillea robusta* and *Acacia mearnsii*. Indigenous agroforestry practices have provided the foundation for past and recent successes with farm forestry in Kirinyaga. Households continue to rely on time-tested and resilient indigenous techniques and strategies to meet the challenges of rising population pressures and the growing demand for economic development.

Acknowledgements

Research in Kirinyaga, Kenya, was supported by the National Science Foundation, the Intercultural Studies Foundation, and the University of California, Santa Barbara. The author also wishes to thank the Kenya National Archives, Nairobi, Kenya, the Embu District Archives in Embu, Kenya, and Special Collections, Bird Research Library, Syracuse University, Syracuse, New York. The views expressed in this paper are those of the author and do not reflect the views of the supporting agencies.

References

Ambler, C.H., 1988. Kenyan Communities in the Age of Imperialism. Yale University Press, New Haven.

Baker, R.St.B., 1931. Men of the Trees. Dial Press, New York.

Berkes, F. (Editor), 1989. Common Property Resources. Belhaven, London.

Brokensha, D.W., Warren, D.M. and Werner, O. (Editors), 1980. Indigenous Knowledge Systems and Development. University Press of America, Lanham.

Castro, A.P., 1983. Household energy use and tree planting in Kirinyaga. University of Nairobi, Institute of Development Studies Working Paper, Nairobi.

Castro, A.P., 1988. Southern Mount Kenya and colonial forest conflicts. In: J.F. Richards and R.P. Tucker (Editors), World Deforestation in the Twentieth Century. Duke University Press, Durham, pp. 33–55, 266–271.

Castro, A.P., 1990. Sacred Groves and social change in Kirinyaga, Kenya. In: M.S. Chaiken and A.K. Fleuret (Editors), Social Change and Applied Anthropology. Westview Press, Boulder, CO, pp. 277–289.

Castro, A.P., 1991a. Indigenous Kikuyu agroforestry: a case study of Kirinyaga, Kenya. Hum. Ecol., 19: 1–18.

Castro, A.P., 1991b. Njukiine Forest: transformation of a common-property resource. For. Conserv. Hist., 35: 160–168.

Castro, A.P., 1991c. The Southern Mount Kenya Forest since independence: a social analysis of resource competition. World Dev., 19: 1695–1704.

Castro, A.P., 1992. Social forestry: a cross-cultural analysis. In: M. Wali (Editor), Ecosystem Rehabilitation. Vol. 1. SPB Academic Publishing, The Hague, pp. 63–78.

Chambers, R., 1991. In search of professionalism, bureaucracy and sustainable livelihoods for the 21st century. IDS Bull., 22: 5–11.

Cook, C.C. and Grut, M., 1989. Agroforestry in Sub-Saharan Africa. World Bank, Technical Paper No. 112, Washington, DC.

Crawshay, R., 1902. Kikuyu: Notes on the country, people, fauna and flora. Geogr. J., 20: 24–49.

Davison, J., 1989. Voices from Mutira. Rienner, Boulder.

Dickson, B., 1903. The eastern borderlands of Kikuyu. Geogr. J., 22: 36–39.

Eliot, C.E., 1905. The East Africa Protectorate. Arnold, London.

Embu District, no date. Kikuyu and Embu trees. In: Embu District Political Record Book (PC/CP/1/5/1). Kenya National Archives, Nairobi.

Feeney, D., Berkes, F., McCay, B.J. and Acheson, J.M., 1990. The tragedy of the commons: Twenty-two years later. Hum. Ecol., 18: 1–19.

Gedge, E., 1892. A recent exploration of the River Tana to Mount Kenya. Geogr. Soc. Proc., 14: 513–533.

Hobley, C., 1967. Bantu Beliefs and Magic, Second Edition. Cass, London (originally published 1922, with second edition in 1937).

Hutchins, D., 1907. Report on the Forest of Kenya. HMSO, London.

Jaetzold, R. and Schmidt, H., 1983. Farm Management Handbook of Kenya. Vol. II/B, Central Kenya. Republic of Kenya, Ministry of Agriculture, Nairobi.

Kenya Land Commission, 1934. Kenya Land Commission Evidence, Vol. 1. HMSO, London.

Kenyatta, J., 1938. Facing Mount Kenya. Secker and Warburg, London.

Kerkhof, P., 1990. Agroforestry in Africa. Panos Institute, London.

Kirinyaga District, 1984. Development Plan 1984–88. Kenya Ministry of Finance and Planning, Nairobi.

Kitching, G., 1980. Class and Economic Change in Kenya. Yale University Press, New Haven.

Leakey, L.S.B., 1977a. The Southern Kikuyu before 1903. Vol. 1. Academic Press, New York.

Leakey, L.S.B., 1977b. The Southern Kikuyu before 1903. Vol. 2. Academic Press, New York.

Leakey, L.S.B., 1977c. The Southern Kikuyu before 1903. Vol. 3. Academic Press, New York.

Maher, C., 1938. Soil Erosion and Land Utilisation in the Embu Reserve, Part I. Department of Agriculture, Kenya Colony, Nairobi.

Maxwell, G.V., Fazan, S.H. and Leakey, L.S.B., 1929. Native Land Tenure in Kikuyu Province. Colony and Protectorate of Kenya, Nairobi.

McCay, B.J. and Acheson, J.M. (Editors), 1987. The Question of the Commons. University of Arizona Press, Tucson.

Middleton, J. and Kershaw, G., 1965. The Kikuyu and Kamba. International African Institute, London.

Muriuki, G., 1974. A History of the Kikuyu, 1500–1900. Oxford University Press, Nairobi.

Nyeri District, 1914. Handing Over Report. Kenya National Archives, Nairobi.

Okigbo, B.N., 1990. Sustainable agricultural systems in tropical Africa. In: C.A. Edwards, L. Rattan, P. Madden, R.H. Miller and G. House (Editors), Sustainable Agricultural Systems. Soil and Water Conservation Society, Ankeny, pp. 323–352.

Orde-Browne, G., 1925. The Vanishing Tribes of Kenya. Seeley and Service, London.

Riley, B.W. and Brokensha, D.W., 1988a. The Mbeere in Kenya. Vol. 1. University Press of America, Lanham.

Riley, B.W. and Brokensha, D.W., 1988b. The Mbeere in Kenya. Vol. 2. University Press of America, Lanham.

Routledge, W. and Routledge, K., 1910. With a Prehistoric People. Arnold, London.

Stigand, C.H., 1913. The Land of Zinj. Constable, London.

Von Hohnel, L., 1968. Discovery of Lakes Rudolf and Stefanie. Vol. 1. Frank Cass, London (originally published 1894).

Warren, D.M., Slikkerveer, L.J. and Titilola, S.O. (Editors), 1989. Indigenous Knowledge Systems. Iowa State University, Technology and Social Change Program, Ames.

Agriculture, Ecosystems and Environment, 46 (1993) 55–68
Elsevier Science Publishers B.V., Amsterdam

Geologic research in support of sustainable agriculture

L.P. Gough*, J.R. Herring

US Geological Survey, Denver Federal Center, MS 973 Denver, CO 80225, USA

Abstract

The importance and role of the geosciences in studies of sustainable agriculture include such traditional research areas as, agromineral resource assessments, the mapping and classification of soils and soil amendments, and the evaluation of landscapes for their vulnerability to physical and chemical degradation. Less traditional areas of study, that are increasing in societal importance because of environmental concerns and research into sustainable systems in general, include regional geochemical studies of plant and animal trace element deficiencies and toxicities, broad-scale water quality investigations, agricultural chemicals and the hydrogeologic interface, and minimally processed and ion-exchange agrominerals. We discuss the importance and future of phosphate in the US and world based on human population growth, projected agromineral demands in general, and the unavailability of new, high-quality agricultural lands. We also present examples of studies that relate geochemistry and the hydrogeologic characteristics of a region to the bioavailability and cycling of trace elements important to sustainable agricultural systems.

Introduction

Purpose

We examine some of the research areas in which the geological sciences are contributing to the agricultural needs of both developed and developing countries. Some of these areas, such as mapping, classifying, and evaluating soils, soil amendments, and fertilizer raw minerals are traditional roles of agrogeology. Certainly the evaluation of agromineral resources, such as phosphate rock, potassium minerals, and zeolite deposits, is important and ongoing. Some not-so-traditional agrogeology roles involve studies of desertification, animal and plant inorganic element deficiencies and toxicities (e.g. selenium and molybdenum mobilization and bioaccumulation), broad-scale water quality evaluations (e.g. the National Water Quality Assessment Program—Hirsch et al., 1988), and even astrogeology (e.g. agricultural potential of moon soil—Ming et al., 1989). For the purposes of this paper we will limit

---

*Corresponding author.

our discussion to the following specific topics: (1) agromineral resource characterization and assessment (focus on phosphate rock); (2) biogeochemical cycling and agricultural practices; (3) agricultural chemicals and the hydrogeologic interface; (4) agriculture, land degradation, and the geosciences; (5) minimally processed and ion-exchange agrominerals.

The human population motivator

The exponential increase in the human population is a fact of our current existence. We can debate its sociological significance and the mechanism by which we have managed to temper the eighteenth century predictions of T.R. Malthus, but the implications that this increase has for agriculture are sobering. For example, with an estimated population of one billion by the year 2000, India must somehow double its current annual average agricultural output of less than 1 t ha^{-1} (a realistic yield using sustainable agricultural techniques with little or no irrigation or chemical fertilization). For economic reasons, a heightened concern for soil conservation and an acknowledgment that only marginal to highly degraded lands remain for agricultural expansion, India wants to increase the productivity of non-irrigated drylands from 40% of total agricultural output to approximately 60% by the year 2000 (US Dryland Farming Team Report, 1987). Success will require implementing both sustainable agricultural practices and increased agrogeologic input.

Agromineral resource assessments

In order to maintain yields, any agricultural system must apply more plant nutrient elements as fertilizers annually than are removed from the soil in foodstuffs, in order to allow for losses due to volatilization, chemical fixation, leaching, and erosion. Sustainable practices attempt to reduce these losses by returning to the soil as much organic farm residues as possible; the difference between what is returned and what is required to produce desired yields must be made up by the application of chemical fertilizers. Shacklette (1977, p. 18) gives the following example of the net loss of phosphorus (P) to the US through the export of wheat: one railroad carload of wheat weighs about 2.7 t, and this amount of wheat contains about 11 kg of P. An 'average' hectare of US soil contains about 146 kg total P in the plow zone, but only a small amount (about 45 kg ha^{-1}) exists in a form available to plants. At a yield rate of 3.4 t ha^{-1}, the crop of wheat removes about 13.4 kg of P from a hectare of soil. Erosion losses can be estimated at 11.2 kg ha^{-1}. These two losses (a total of 24.6 kg) would exhaust the available P in a hectare of soil in 2 years, in the same time that a carload of wheat was being produced on the hectare.

The US imports much of its domestic potash (K_2O) fertilizer needs but is self-sufficient in ammonium (nitrogen as NH_4) and phosphate (phosphorus as P_2O_5). For example, in 1989, the US produced only 27% of the potash needed for domestic consumption; the difference was obtained through imports from Canada, Israel, the USSR, and Germany (Herring et al., 1991). Very little of the world's nitrogen fertilizers come from traditional guano deposits but instead are produced from nitrogen in natural gas. Nitrogen fertilizers are, therefore, no longer considered to have a significant minable mineral source.

Phosphate rock (PR) is the most valuable industrial agromineral in the US. Annual sales of PR account for $1.1 billion, about one-quarter of the total value of non-construction industrial minerals. However, the value of the PR sales is only the beginning of a manufacturing process, that produces many different refined products. Figure 1 shows that of the non-energy mineral commodities produced in the US in 1990, agricultural minerals account for only about 5%, whereas construction 'minerals' (sand and gravel) accounted for 48%. From 1985 to 1990 these proportions remained nearly the same (Fig. 2) even when corrected for inflation. US PR production in 1990 was 46.3 million t of which 26.8 million t were converted into higher value products and exported, providing about two billion dollars in revenue (Herring and Stowasser, 1991).

Stowasser (1991) and Herring and Stowasser (1991) point out that as our phosphate reserves dwindle, it will be increasingly difficult to maintain our world dominance of the phosphate market, unless new beneficiation technology proves profitable and environmental concerns are appropriately and economically addressed. For example, current phosphate rock beneficiation

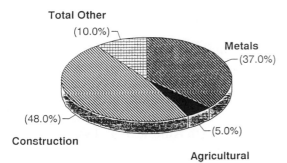

Fig. 1. Proportion of production value of non-energy mineral commodities in the US, 1990. Construction minerals refer to sand, gravel, paving stone, etc.

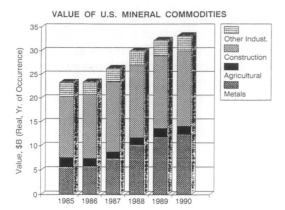

Fig. 2. The value of mineral commodities in the US between 1985 and 1990. (Data modified from USBM minerals yearbook.)

technology employs the following general scheme:

$$Ca_5(PO_4)3F + 5H_2SO_4 + 10H_2O \rightarrow 5CaSO_4 \cdot 2H_2O + 3H_3PO_4 + HF\uparrow$$

fluorapatite	sulfuric acid	gypsum	phosphoric acid	hydrogen fluoride gas
(2.1)	(1.7)	(3.2)	(1.0)	(0.08)

This reaction produces phosphoric acid, which can be easily separated from gypsum, the solid reaction product. The figures in parentheses below the reaction components are the mass ratio of that component needed or produced relative to one mass unit of phosphoric acid.

There are a number of environmental consequences of this reaction including the release of hydrogen fluoride gas (which is extremely phytotoxic) and the production of acidic gypsum (phosphogypsum). The ratio of phosphogypsum to phosphoric acid produced is about 3.2:1 and the waste heaps that result require specialized reclamation techniques because of their general infertility and low pH.

Phosphate deposits can also be compromised if the phosphate contains undesirable quantities of environmentally sensitive elements. For example, Togo, which ranked tenth in world PR (apatite) production in 1990, has reserves estimated at 130 million t and is dependent on phosphate as its major foreign exchange commodity. However, high cadmium content of Togolese phosphates has led to a decline in demand in North American and western European markets.

Herring and Stowasser (1991) predict that phosphate production could peak in this country by the Year 2000. Figure 3 shows projected US production of phosphate rock (million t year^{-1}) over the next 45 years showing both planned and potential production (the latter being the 'best case' produc-

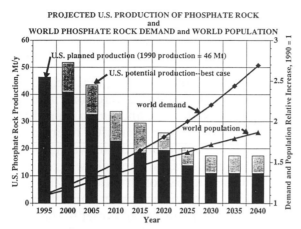

Fig. 3. Projected US phosphate rock production, megatons per year (million t year^{-1}), in the bar graph and line data of relative increases of World population growth (triangles) and phosphate demand (squares). Planned production of phosphate rock, solid pattern, is that planned and announced by companies to date while potential production, stippled pattern, assumes the best case scenario of additional production that might occur announced by the companies. World population growth and World phosphate demand are shown relative to 1990 values: 5.3 billion people (= 1) and 162 million t of phosphate rock. Projected World annual population growth is 1.7% overall between now and 2040 and includes tapering to an optimistic annual 1.5% in the final decade. Phosphate demand is projected at 2% and exceeds population growth because most arable land has already been developed, and future food growth will depend on greater crop yields from this existing crop land — something that can be achieved only with the use of fertilizers. Commodity data are from Howasser (1991) and population data from the Bureau of the Census.

tion). Also given are graphs of the relative increases of world population growth and phosphate demand. The population and demand data (from the Bureau of the Census and Bureau of Mines, respectively) are shown relative to 1990 values: 5.3 billion people (= 1) and 162 million t of phosphate rock. Projected world annual population growth is presented as 1.7% overall between now and 2040 and includes a decline to an optimistic annual 1.5% in the final decade. Phosphate demand is projected at 2% and exceeds population growth because most arable land has already been developed, and future food growth will depend on greater crop yields from existing crop land.

Whether these higher yields come solely from increased fertilizer use or from more effective nutrient management practices remains to be seen. The decreasing phosphate production trends for the US shown in Fig. 3 indicate, however, that the US will become increasingly vulnerable to phosphate import pressures. The regions of the world with the greatest potential for exploitable phosphate reserves outside the US are Africa and the middle East, specifically, Morocco, Tunisia, Libya, and Jordan (Constant and Sheldrick, 1991).

Biogeochemical cycling and agricultural practices

San Luis Drain, California

In 1982 the US Fish and Wildlife Service began suspecting that irrigation drainage practices in the San Joaquin Valley of California were contributing to the occurrence of waterfowl mortality, birth defects, and avian reproductive failures at the Kesterson National Wildlife Refuge (hereafter referred to as Kesterson). It had been only 1 year since subsurface drain water, from irrigated fields, was first pumped into the San Luis Drain and allowed to flow north into Kesterson. From the time of the drain's construction by the US Bureau of Reclamation in 1950 until 1981 only surface water run-off from fields was directed into the closed basin refuge; plans for an outlet to the San Francisco Bay were abandoned in the 1970s because of a lack of funds. By 1983, only 2 years after the start of receiving subsurface drain water, the concentration of metals and agricultural chemicals in Kesterson surface waters, sediments, and biota had risen dramatically. For example, levels of arsenic, boron, cadmium, chromium, and copper were found to be elevated in surface water of the refuge. Of particular concern, however, was the sudden dramatic increase in the apparent bioaccumulation of selenium through the food-chain, showing, for example, an increase in fish tissue of 50 to 100 times (Marshall, 1985) over surface water concentrations of 200–400 ppb, which were themselves elevated (EPA Maximum Contaminant Level for drinking water is currently 10 ppb).

A distinction must be made, however, between contaminants that are introduced through the direct result of agricultural application (such as pesticides) and contaminants that occur naturally but are mobilized and transported by agricultural practices (such as selenium and trace metals). The San Joaquin Valley agricultural drain water, that is disposed of by placement in evaporative basins (like Kesterson), deep-well injection, or river discharge, carries contaminants whose origin may be upslope and far removed from the site of actual discharge. This situation makes contaminant discharge a regional problem and the contaminants themselves a challenge to trace (Crooks, 1986; Deverel and Fujii, 1988; Tidball et al., 1989; Presser et al., 1990). Selenium is a natural 'contaminant' whose biogeochemical cycle is greatly affected by agricultural practices.

It would be naive to assume that Kesterson was an isolated incidence of selenium mobilization by agricultural practices (Deason, 1989; Port, 1989). In 1985, the Department of the Interior created the Task Group on Irrigation Drainage that was composed of representatives from the five agencies (Bureau of Reclamation, Bureau of Indian Affairs, Bureau of Land Management, Fish and Wildlife Service, and the Geological Survey (USGS)) that were either legally or remedially affected by the quality of irrigation drain water or who could offer research assistance in the definition and scope of likely problems (Deason, 1986, 1989). The legislative motivation for this concern came predominantly from the Migratory Bird Treaty Act of 1918, the Clean Water Act of 1982, and the Endangered Species Act of 1973 as amended in 1978 and 1982.

One of the first goals of the Task Group was to identify areas that had the potential to cause harmful effects to human health, fish, and/or wildlife due to irrigation drainage practices. In addition to Kesterson, the Task Group settled on 19 areas. Reconnaissance-type studies on nine of the areas were completed in 1988; similar studies on the remaining ten are currently being completed. Results of the research on the first nine areas shows that several general principles are controlling the concentration of constituents, associated with irrigation drainage, found in water, bottom sediment, and biota (Sylvester et al., 1991). These include (1) the characteristics of geologic sources of naturally occurring trace elements, (2) yearly variations in precipitation and stream flow within the areas, (3) evapo–transpiration dynamics of arid and semi-arid areas, and (4) presence of internal drainage basins and sinks that results in the accumulation of constituents. Based on these criteria, four of the nine reconnaissance areas were selected for intensive study. The detailed studies focused on the following goals: (1) to determine the magnitude and extent of irrigation-induced water quality problems, and (2) to provide the scientific understanding needed to mitigate or resolve identified problems. The working objective for each of the four detailed studies was "to determine the extent, magnitude, and effects of contaminants associated with agricultural drainage, and, where effects are documented, to determine the sources and exposure pathways that cause contamination" (Erdman et al., 1991).

Kendrick Project, Wyoming

Of the 20 areas mentioned above, Kesterson has received by far the most extensive study, not only because it was the first area to be identified as having a problem, but also because of the tremendous impact that management decisions, resulting from Kesterson research, had on agri-business in the San Joaquin valley. An example of another area that has received considerable study is the Kendrick, Wyoming Reclamation Project. Based on reconnaissance investigations that showed anomalous levels of selenium in bottom sediments (Severson et al., 1987), water, and biota (Peterson et al., 1988), the area was targeted as one of four to receive further, detailed study. Results of the detailed studies are reported in Erdman et al. (1991) and See (1991). The following highlights summarize their findings: (1) soils, in general, were not found to have abnormally high total selenium concentrations; soils developed over Cretaceous Cody Shales, however, tended to have higher concentrations than other soils; (2) selenium levels in big sagebrush (*Artemisia tridentata* Nutt.) growing throughout the area tended to be considerably higher than the norm; (3) 15% of the alfalfa samples collected from irrigated fields throughout the area had selenium concentrations greater than 4 ppm (dry weight base), levels potentially hazardous to livestock if fed over prolonged periods; incidences of selenosis within the study area, however, have not been commonly reported; (4) although selenium-containing alfalfa and surface and drain waters were high in one particular area, climatic and crop management practices were found to contribute significantly to large, as yet unpredictable temporal variation.

The results of these studies are currently being evaluated by a recently formed Kendrick Task Group, and management decisions that will affect municipalities, land-resource planning, water supply strategies, and agricultural practices will be made (R.C. Severson, personal communication, 1991).

Agricultural chemicals and the hydrogeologic interface

In 1989 the President's Water Quality Initiative prompted the US Department of Agriculture (USDA) and the US Department of the Interior to define their roles in protecting the nation's water resources, especially from agricultural chemicals. The management of agricultural chemicals and wastes to meet environmental and public safety objectives was an unfamiliar activity of these two departments and required that they work closely with each other and with the US Environmental Protection Agency. One research activity that evolved as a direct result of this initiative was the Management System Evaluation Areas (MSEA) program. Five MSEA sites have been established that represent four basic hydrogeologic settings: (1) till/alluvium, Iowa; (2) claypan over loess, Missouri; (3) valley alluvium, Ohio; (4) sand plains, Nebraska and Minnesota. All five sites are within the mid-continent's major area for corn and soybean production and have focused on studies at scales of 1–5 ha to systems analysis of aquifers covering areas of many tens of square kilometers.

There are two basic objectives of the MSEA Program[a]: "(1) determine the precise nature of the relationship between agricultural activities and ground water quality, and (2) develop and induce the adoption of technically and economically effective agrichemical management and agricultural production strategies that protect the beneficial uses of ground and surface water quality". Sustainable agricultural practices form a major component of the research emphases.

---

[a]USDA and Cooperating State Agencies, 1989, Water Quality Program Plan to Support the President's Water Quality Initiative. USDA Office of the Deputy Assistant Secretary for Science and Education, Washington, DC.

The major contribution of the USGS to the MSEA program has been geo-hydrologic site characterization and research on the fate and transport of atrazine, carbofuran, alachlor and nitrate through the unsaturated zone in the groundwater system. For example, sand-plain aquifers are hydrogeologically vulnerable to groundwater contamination from agricultural practices because their aquifers and the vadose zone above have high hydraulic conductivities. Specific studies include: (1) the transport and fate of atrazine, and atrazine degradation products, in peat-rich wetlands adjacent to corn and soybean fields; (2) the definition of a chemical mass balance for anthropogenic nitrogen by monitoring natural reactions affecting nitrogen speciation and distribution in an agriculturally impacted watershed; (3) the monitoring of depth-to-ground water using geophysical techniques such as ground-penetrating radar in areas low in clay/high in sand; (4) involvement in the development and modification of deterministic models that help define the transport of solutes through various hydrogeologic terrains; (5) making available to the USDA all USGS water quality databases in formats that are compatible with their own. The summer of 1991 was the first field season for most of the studies; published results for the preliminary phase can be expected by spring of 1992.

Agriculture, land degradation, and the geosciences

Within the last decade, there have been major advances in methodologies for remote data acquisition and analysis which have dramatically assisted studies assessing land degradation. Real interest in space-based remote sensing began in 1972 with the launch of Landsat 1 (formerly ERTS-1), but the launch in 1982 of Landsat 4, with its thematic mappers (TM) that had an instantaneous field of view of 30 m × 30 m, was of special interest to land resource scientists. This system proved superior to the multispectral scanners (MSS), in the earlier satellites, for the acquisition of data used to inventory and monitor hydrology-related trends, agricultural lands and their uses, native vegetation, pollution plumes, and mineral resources (Lo, 1986). However, Landsat MSS data are still used to evaluate plant succession, ecosystem productivity, reflectance from the soil–litter–vegetation complexes and other rangeland characteristics (Ringrose and Matheson, 1991). In India, Landsat TM data have recently been used to successfully distinguish moderately from strongly sodic soils and to map these characteristics (Rao et al., 1991). In Africa, India, and many developing countries in general, Landsat imagery is increasingly being used to determine areas of overgrazing in relation to vegetation communities, industrialization, deforestation, mining, and urbanization in order to keep abreast of burgeoning human and grazing animal populations and their impact on land use planning (Connant et al., 1983; Shiva

Prasad et al., 1990). The US has launched five Landsat satellites with a sixth one scheduled for deployment sometime in 1993.

In 1986, the French launched the first of a series of earth-observational satellites, the 'Systeme Probatoire d'Observation de la Terre' or SPOT, which are an improvement in resolution over Landsat TM by about a factor of three. Data from SPOT satellites have been particularly important in distinguishing various transition zones between vegetation communities and between agricultural crops (Jewell, 1989; Baker et al., 1991). In addition, general use of the National Oceanographic and Atmospheric Administration (NOAA) advanced very high resolution radiometer (AVHRR) satellites for monitoring arid lands and their advancement or decline has greatly assisted studies that focus on agriculture as it relates to global change issues (Prince and Justice, 1991).

Minimally processed and ion-exchange agrominerals

Unprocessed phosphatic and potassic carbonatites and igneous or sedimentary phosphatic rock can be applied directly to agricultural lands to improve soil fertility and productivity (Hammond et al., 1986). There is renewed interest in this field, particularly in developing countries. In the tropics, where weathering rates are high, the addition of minimally processed agrominerals can enhance the availability of micronutrient (trace) elements that are commonly depleted (Chesworth et al., 1983). For example, a conference in Zomba, Malawi, focused on agroecosystems which utilize raw phosphate rock instead of processed phosphate fertilizers. Reports at the conference were divided into sections on geology and resource appraisals, beneficiation, and processing and application (Commonwealth Science Council, 1987). Mechanisms to enhance the dissolution of phosphate from phosphate rock to increase availability for plant uptake included partial acidulation, micro-organism interactions, and mixing phosphate rock with ion-exchangers, peat, or pyrite. Hammond et al. (1986) report on research in sub-Saharan Africa, Latin America, and Asia using partially acidulated phosphate rock as an inexpensive, locally available source of agricultural phosphorus.

Geological soil amendments such as clay (particularly phyllosilicate minerals like montmorillonite and bentonite), volcanic materials (such as pumice, perlite, scoria), sand, rock fragments, peat, and humates, applied singly or in combination, serve to selectively improve a soil for a specific agronomic purpose (Severson and Shacklette, 1988). Zeolites (hydrated aluminosilicates) have also been used as soil amendments in agriculture (Sheppard, 1984); however, their use as ion-exchangers for the slow release of nutrients may have greater significance to sustainable agriculture (Lai and Eberl, 1986). Ion-exchange fertilizers, such as the clinoptilolite (a zeolite) that is saturated with ammonia, attempts to manage the 'capacity' factor of the soil rather than the 'intensity' factor. The saturated mineral is mixed with PR (apatite) and is applied to the soil. In the following general equation (Lai and Eberl, 1986; Barbarick et al., 1990), the ammonium serves to lower the calcium activity in the soil solution by sequestering Ca^{2+} in the zeolite after exchange for the ammonium:

$$PR(apatite) + NH_4 \text{ exchanger} \rightleftharpoons Ca \text{ exchanger} + NH_4^+ + H_2PO_4^-$$

Barbarick et al. (1990) explain: "By use of this technique, P concentrations can be buffered in the soil solution, the rate of P and N (and K using a K-exchanger) release can be controlled by varying the exchanger/phosphate-rock ratio, and the system can be renewed by adding additional exchanger, or by recharging the exchanger with NH_4 or K". These authors have demonstrated that zeolites have great potential for reducing the need for chemical fertilizers by acting to slowly release plant nutrients throughout the growing season. This ability is particularly appealing in tropical soils where P is usually fixed very rapidly. The technique would not work, however, in many arid and semi-arid soils that are high in Ca minerals (such as calcite, anhydrite, or gypsum) that would complete with apatite for exchangeable Ca.

Summary

The traditional aspects of agrogeology, which include the mapping and classification of soils, and the appraisal of soil amendments and fertilizer minerals, are augmented by other areas of the geosciences in providing information for the management of sustainable agroecosystems. This can be especially helpful when systems are being developed on lands considered to be marginal for agriculture or for the reclamation of degraded lands (Gough et al., 1991).

If world food availability is to be maintained at reasonable cost, then both vigorous agromineral resource characterization and new technologies for their prudent and environmentally sensible use are needed. Areas of technology that will become increasingly important to addressing these needs are: (1) application and timing of optimum fertilizer amounts for soil and crop conditions; (2) direct application of low-grade phosphate rock to crop land; (3) progressive, moderated release of nutrient compounds; (4) improved processing to utilize lower-grade resources; (5) improved agromineral mined-land reclamation strategies.

An alternative to reducing the requirement for phosphate is the development of agricultural methods less dependent on applied fertilizers. For example, in soils of tropical and subtropical regions, particularly those soils enriched in oxides of iron or aluminum, phosphate is strongly bound and largely unavailable to plants. In tropical and sub-tropical Alfisols most of the phosphate is bound to iron-rich, highly-weathered phases, whereas in Vertisols there is a large fraction bound to clays. Cultivation in these areas with crops such as pigeon pea, which more readily access the bound phosphate, significantly improve crop yield.

The following research and management strategies are emphasized for integrating the geosciences with sustainable agroecosystem strategies.

(1) Because of expected world population increases, agromineral resource appraisals into new production technologies are important.

(2) Investigations are needed to define best management practices for the use of soil amendments, ion-exchange fertilizers, and minimally processed agrominerals in sustainable agricultural systems (especially in rain-fed/dryland systems).

(3) International cooperation and linkages are needed for the assembly and application of GIS information to investigations of the applicability of GIS to research on problems of trace element deficiencies and excesses, erosion, salinization, flooding, and deforestation.

(4) Both basic and applied research is needed on rock-water and soil–water interface processes in the vadose zone below the plow zone, to better define the transformation, transport, and sinks for pesticides and chemical fertilizers.

(5) High-intensity agriculture is adversely affecting both surface and groundwater quality and the public is becoming more sensitive to these environmental concerns. It is of interest that in the arid and semi-arid west, the disposal of irrigation drainage water has become an important concern not only because it transports anthropogenic chemicals but also because it can mobilize naturally occurring selenium and other biologically important trace elements.

Acknowledgments

We thank Richard B. Wanty, Ronald C. Severson, James M. McNeal, and Daniel R. Muhs, USGS, for supplying background information given in some sections of this paper.

References

Baker, J.R., Briggs, S.A., Gordon, V., Jones, A.R., Settle, J.J., Townshend, J.R.G. and Wyatt, B.K., 1991. Advances in classification for land cover mapping using SPOT HRV imagery. Int. J. Remote Sensing, 12: 1071–1085.

Barbarick, K.A., Lai, T.M. and Eberl, D.D., 1990. Exchange fertilizer (phosphate rock plus ammonium-zeolite) effects on sorghum-sudangrass. Soil Sci. Soc. Am., J., 54: 911–916.

Chesworth, W., Macias-Vasquez, F., Acquage, D. and Thompson, E., 1983. Agricultural alchemy: stones into bread. Episodes, 1983: 3–7.

Commonwealth Science Council, 1987. Agrogeology in Africa. Proceedings of the Seminar. CSC Technical Publication Series No. 226, Commonwealth Science Council, London, 263 pp.

Connant, F., Rogers, P., Baumgardner, M., McKell, C., Dasmann, R. and Reining, P. (Editors), 1983. Resource Inventory and Baseline Study Methods for Developing Countries. American Association for the Advancement of Science, Publication 83-3, Washington, DC, 539 pp.

Constant, K.M. and Sheldrick, W.F., 1991. An Outlook for Fertilizer Demand, Supply, and Trade, 1988/89–1993/94. The World Bank Technical Paper No. 137, The World Bank, Washington, DC, 110 pp.

Crooks, W.H., 1986. Toxics in agricultural drainage in California's Central Valley. In: Toxic Substances in Agricultural Water Supply and Drainage, Defining the Problems. US Committee on Irrigation and Drainage, Denver, CO, pp. 1–12.

Deason, J.P., 1986. U.S. Department of the Interior investigations of irrigation-induced contamination problems. In: Toxic Substances in Agricultural Water Supply and Drainage, Defining the Problems. US Committee on Irrigation and Drainage, Denver, CO, pp. 201–210.

Deason, J.P., 1989. Irrigation-induced contamination—how real a problem? J. Irrig. Drain. Eng., 115: 9–20.

Deverel, S.J. and Fujii, R., 1988. Processes affecting the distribution of selenium in shallow ground water of agricultural areas, western san Joaquin Valley, California. Water Resources Res., 24: 516–524.

Erdman, J.A., Severson, R.C., Crock, J.G., Harms, T.F. and Mayland, H.F., 1991. Selenium in soils and plants from native and irrigated lands at the Kendrick Reclamation Project area, Wyoming. In: R.C. Severson, S.E. Fisher, Jr. and L.P. Gough (Editors), Proceedings of the 1990 Billings Land Reclamation Symposium on Selenium in Arid and Semiarid Environments, Western United States, 25–30 March 1990, Billings, MT, US Geological Survey Circular 1064, Washington, DC, pp. 89–105.

Gough, L.P., Hornick, S.B. and Parr, J.F., 1991. Geosciences, agroecosystems and the reclamation of degraded lands. In: M.K. Wali (Editor), Ecosystem Rehabilitation, Preamble to Sustainable Development. Volume 1: Policy Issues. Academic Press, The Hague, pp. 47–61.

Hammond, L.L., Chien, S.H. and Mokwuyne, A.U., 1986. Agronomic value of unacidulated and partially acidulated phosphate rocks indigenous to the tropics. Adv. Agron., 40: 89–141.

Herring, J.R., Piper, D.Z., Williams-Stroud, S., Spirakis, C.S. and Sheppard, R.A., 1991. U.S. Geological Survey research on agricultural industrial minerals in the United States — implications for exploration. US Geological Survey Circular 1062, Washington, DC, pp. 38–39.

Herring, J.R. and Stowasser, W.F., 1991. Phosphate — our Nation's most important agricultural mineral commodity and its uncertain future. Proc. Ann. Meet. Geol. Soc. Am., 21–24 October 1991, San Diego, CA, Geol. Soc. Am., Boulder, CO, pp. 299–300.

Hirsch, R.M., Alley, W.M. and Wilber, W.G., 1988. Concepts for a National Water-Quality Assessment Program. US Geological Survey Circular 1021, Washington, DC, 42 pp.

Jewell, N., 1989. An evaluation of multi-date SPOT data for agriculture and land use mapping in the United Kingdom. Int. J. Remote Sensing, 10: 939–951.

Lai, T.-M. and Eberl, D.D., 1986. Controlled and renewable release of phosphorous in soils from mixtures of phosphate rock and NH_4-exchanged clinoptilolite. Zeolites, 4: 129–132.

Lo, C.P., 1986. Applied Remote Sensing. Longman, New York, 393 pp.

Marshall, E., 1985. Selenium poisons refuge, California politics. Science, 229: 144–146.

Ming, D.W., Henninger, D.L. and Brady, K.S. (Editors), 1989. Lunar Base Agriculture: Soils for Plant Growth. American Society of Agronomy, Madison, WI, 276 pp.

Peterson, D.A., Jones, W.E. and Morton, A.G., 1988. Reconnaissance investigation of water quality, bottom sediment, and biota associated with irrigation drainage in the Kendrick Reclamation Project area, Wyoming, 1986–1987. U.S. Geological Survey Water-Resources Investigations Report 87-4255, Washington, DC, 57 pp.

Port, P.S., 1989. Welcoming remarks. In: Selenium and Agricultural Drainage — Implications for San Francisco Bay and the California Environment. Proceedings of the Third Selenium Conference, Berkeley, CA, 15 March, 1986. The Bay Institute, Tiburon, CA, pp. 10–12.

Presser, T.S., Swain, W.C., Tidball, R.R. and Severson, R.C., 1990. Geologic sources, mobilization, and transport of selenium from the California Coast Ranges to the western San Joaquin valley — a reconnaissance study. US Geological Survey Water-Resources Investigations Report 90-4070, Washington, DC, 66 pp.

Prince, S.D. and Justice, C.O., 1991. Editorial. In: Special Issue, Coarse Resolution Remote Sensing of the Sahelian Environment. Int. J. Remote Sensing, 12: 1137–1146.

Rao, B.R.M., Dwivedi, R.S., Venkataratnam, L., Ravishankar, T. and Thammappa, S.S., 1991. Mapping the magnitude of sodicity in part of the Indo-Gangetic plains of Uttar Pradesh, northern India using Landsat-TM data. Int. J. Remote Sensing, 12: 419–425.

Ringrose, S. and Matheson, W., 1991. A Landsat analysis of range conditions in the Borswana Kalahari drought. Int. J. Remote Sensing. 12: 1023–1051.

See, R.B. (Editor), 1991. Assessment of selenium in soils, representative plants, water, bottom sediments, and biota in the Kendrick Reclamation Project area, Wyoming, 1988–990. US Geological Survey Water-Resources Investigations Report 91-394, Washington, DC, 216 pp.

Severson, R.C. and Shacklette, H.T., 1988. Sources and use of fertilizers and other soil amendments in agriculture. US Geological Survey Circular 1017, Washington, DC, 48 pp.

Severson, R.C., Wilson, S.A. and McNeal, J.M., 1987. Analyses of bottom material collected at nine areas in the western United States for the DOI irrigation drainage task group. US Geological Survey Open-File Report 87-490, Washington, DC, 24 pp.

Shacklette, H.T., 1977. Major nutritional elements in soils and plants — a balance sheet. In: O.B. Raup (Editor), Proceedings of the Geology and Food Conference, with Related U.S. Geological Survey Projects and a Bibliography. U.S. Geological Survey Circular 768, Washington, DC, pp. 17–19.

Sheppard, R.A., 1984. Characterization of zeolitic materials in agricultural research. In: W.G. Pond and F.A. Mumpton (Editors), Zeo–Agriculture — Use of Naural Zeolites in Agriculture and Aquaculture. Westview Press, Boulder, CO, pp. 79–87.

Shiva Prasad, C.R., Thayalan, S., Reddy, R.S. and Reddy, P.S.A., 1990. Use of Landsat imagery for mapping soil and land resources for development planning in parts of Northern Karnataka, India. Int. J. Remote Sensing, 11: 1889–1900.

Stowasser, W.F., 1991. Phosphate rock, analysis of the phosphate rock situation in the United States: 1990–2040. Eng. Min. J., 192: 16CC–16II.

Sylvester, M.A., Deason, J.P., Feltz, H.R. and Engberg, R.A., 1991. Preliminary results of the Department of the Interior's irrigation drainage studies. In: R.C. Severson, S.E. Fisher Jr. and L.P. Gough (Editors), Proceedings of the 1990 Billings Land Reclamation Symposium on Selenium in Arid and Semiarid Environments, Western United States. US Geological Survey Circular 1064, Washington, DC, pp. 115–122.

Tidball, R.R., Severson, R.C., McNeal, J.M. and Wilson, S.A., 1989. Distribution of selenium, mercury, and other elements in soils of the San Joaquin valley and parts of the San Luis Drain service area, California. In: Selenium and Agricultural Drainage — Implications for San Francisco Bay and the California Environment. Proceedings of the Third Selenium Conference, Berkeley, CA, 15 March, 1986. The Bay Institute, Tiburon, CA, pp. 71–82.

US Dryland Farming Team Report, 1987. Combined Report of the US Dryland Farming Team and the Economics Team Visits to India, March/April and June 1987, Under the Auspices of the Indo–US Subcommission on Agriculture and the Far Eastern Regional Research Office, Office of International Cooperation and Development, US Department of Agriculture, Washington, DC, 95 pp.

Agriculture, Ecosystems and Environment, 46 (1993) 69–79
Elsevier Science Publishers B.V., Amsterdam

Land degradation and sustainability of agricultural growth: some economic concepts and evidence from selected developing countries

Fred J. Hitzhusen

Department of Agricultural Economics and Rural Sociology, The Ohio State University, 2120 Fyffe Road, Columbus, OH 43210, USA

Abstract

In many developing countries, a large proportion of the population resides and works in rural areas. Agriculture is the dominant sector in rural areas and has the greatest concentration of poverty: landless workers, small tenant farmers, and small farm owners. Thus, any development strategy that is directed towards increasing employment and alleviating a country's hunger must concentrate on sustainable agricultural growth. Historically, economic development, in most countries, has been based on exploitation of natural resources, particularly land resources — in large part because these resources have been undervalued by private markets and political systems. Soil erosion and land degradation have been serious worldwide. Due to reasons such as high population pressure on land, and limited fossil energy supplies, land degradation is more serious in the developing world. Empirical studies show that soil erosion and degradation of agricultural land not only decrease land productivity but that they can also result in major downstream or off-site damage (e.g. reduction of hydroelectric production) which may be several times that of on-site damage. An earlier analysis identified the factors that determine the agricultural production growth rate in 28 developing countries. This study involved statistical estimation of an aggregate agricultural growth function based on cross-country data. The overall results showed that price distortions in the economy and land degradation had statistically significant negative impacts while the change in arable and permanent land was positively related to the growth of agricultural production and food production in 28 developing countries from 1971 to 1980. These results emphasize the importance of 'getting prices right' and implementation of sustainable land and water management practices if future growth in food and agricultural output is to be realized and sustained in developing countries.

Introduction

The sustainability of agriculture, and other natural resource based systems, is a very popular concept. It seems like a common-sense idea, but the concept has many ambiguities and alternative points of view. Dixon and Fallon (1989) suggest that these viewpoints can be grouped into three distinct categories including (1) a purely physical concept for a single resource such as a fishery, where the rule is to use no more than the annual increase in the resource, i.e. maximum sustainable yield, (2) a physical concept for a group of resources or an ecosystem where a variety of system outputs involve trade-offs, i.e. individual resources may be enhanced, maintained or degraded to maintain system integrity, and (3) an environmental economic concept with emphasis on economic rationality and some minimal level of environmentally sustainable economic growth. This latter concept, with emphasis on sustaining some level of agricultural growth, is the conceptual starting point for this paper. A static or steady state notion of agricultural sustainability does not address projected global increases in population and demand for food.

Disparity in agricultural growth among developing countries has been attributed to many factors and a large body of literature exists on the interrelationships between agricultural growth and these factors. Many studies are based on conceptual analysis, or general observations, with few regression analyses. This has resulted in overemphasis of a few factors and neglect of others. Although conceptual analysis and general observations have suggested that land degradation and price distortions greatly influence agricultural production, little effort has been made to statistically evaluate the effect of these factors.

Historically, economic development, in most countries, has been based on exploitation of natural resources, particularly on exploitation of land resources. Soil erosion and land degradation have been serious worldwide. High population pressure on land and limited fossil energy supplies cause land degradation to be generally more serious in the developing world. Empirical studies show that soil erosion and degradation of agricultural land not only decrease land productivity but that they can also result in major downstream off-farm or off-site damage (Crosson, 1985; Hauck, 1985; Clark et al., 1985; Warford, 1987b). Furthermore, the off-site damage may be several times that of on-site damage (Crosson, 1985; Clark et al., 1985).

The exploitation of natural resources, particularly in developing countries, continues in large part because these resources are not priced at their marginal social values. This underpricing in turn occurs because many centrally planned as well as private market economies, with imperfectly defined and enforced property rights, fail to fully internalize the external costs of environmental service benefits related to the use of these natural resources. Examples of environmental services include raw material supply, assimilative capacity of the residuals of economics production and consumption activities, esthetic values, human and wildlife habitat and genetic biodiversity.

This suggests the need for more comprehensive measures of the social benefits and costs of resources or their related environmental services. Environmental economists usually talk about four categories of marginal social costs including, (1) direct costs to current users, (2) external costs borne by others now and in the future, (3) foregone benefits of future users from a depleted resource, and (4) existence values for the sustainable maintenance of a given resource. An estimation of external costs at the watershed level, was reported by Hitzhusen et al. (1984) regarding the foregone hydropower from agricultural erosion and sediment deposition in Valdesia reservoir in The Dominican Republic. The discounted present value of the foregone hydropower was several times larger than the net returns to farmers from erosive practices in the watershed upstream from the dam.

The planned economy or private market imperfections at the micro-economic or watershed level, in the case of soil erosion, also manifest themselves as imperfections in national income accounting at the macroeconomic level. Repetto (1989) argues that by ignoring natural resources (or the broader notion of environmental services), statistics such as the gross national product (GNP) can record illusory gains in income and mask permanent losses in wealth. As a result, a nation could exhaust its minerals, erode its soils, pollute its aquifers and hunt its wildlife to extinction — all without affecting measured national income. For example, Indonesia's high 7.1% economic growth rate as measured by gross domestic product (GDP) from 1971 to 1984 is only 4.0% when GDP is adjusted for unsustainable soil erosion, forest harvest in excess of annual growth, and oil reserve depletion.

Detailed micro or watershed level evaluations of the marginal social costs of natural resource use are possible, but difficult and costly to do. Likewise, major revisions in the national income accounts regarding pricing, depreciation, etc. of natural resources are under way in a few countries, but depend in part on better micro-economic evidence and will take considerable time to complete. As an intermediate step, it would be useful to determine the role of natural resources or specifically the importance of land degradation in agricultural growth, particularly in developing countries.

Land degradation and agricultural growth model

A recent study by Zhao et al. (1991) was primarily concerned with identifying the factors that determine the agricultural production growth rate and in testing the effects these factors have on agricultural growth in developing countries. Specifically, this study involved statistical estimation of an aggregate agricultural growth function based on cross-country data for 28 developing countries. Special attention was devoted to environmental degradation, and agricultural pricing policy and to the policy implications resulting from the effects these variables have on agricultural and food production growth.

The methodology used is based on the concept of a metaproduction function hypothesized by Hayami and Ruttan (1985). Following Lau and Yotopoulos (1987), this metaproduction function can be written as:

$$Y_t = f(X_{it},...,X_{mt},t) \tag{1}$$

where Y_t is the quantity of output, X_{it} is the quantity of ith input, $i = i...,m$, and t is time. This production function can be used to represent the input–output relationship of agriculture or food production. As defined by Lau and Yotopoulos, the implicit hypothesis for this metaproduction function is that all producers (or countries) have potential access to the same set of technology options but each may choose a particular one, depending upon its natural endowment and relative prices of inputs. In the Zhao (1988) research, a metaproduction function similar to Eq. (1) was estimated. However, since his focus was not on the estimation of productivity changes, the time variable t was replaced by two variables, land degradation and price distortion. These two variables were hypothesized to have impacts on technology choices among different countries during the study period. (Note that the price distortion level measure used in this study reflects all price distortions in the economy as discussed later. It is not limited to price distortion in the agricultural sector. Therefore, it is appropriate to treat this variable as exogenous in the model.)

This study was concerned with estimating the relative changes in output rather than the absolute levels of output. Thus, the dependent variables were expressed in relative terms as a percent change or average level of output during the study period. Following Hayami and Ruttan (1971), the inputs may be categorized as: (1) internal resource accumulation, including an expansion of the arable and permanent crop land, and the growth of labor force; (2) technical inputs supplied by the non-agricultural sector. Two industrial inputs, fertilizer and machinery, represent proxies for the whole range of inputs that include modern biological and mechanical technologies.

It is widely recognized that government policies may greatly affect agricultural production growth in developing countries (Krishna, 1982; Argawala, 1983; IBRD, 1986). Most of these studies conclude that the general economic policies pursued in developing countries have limited the growth of agricultural production and hampered efforts to reduce rural poverty.

Although price is not a complete measure of the overall incentives in agricultural production, it is one of the most important policy devices. Agricultural price policy has been actively used by virtually all governments to pursue a wide variety of resource allocation and income distribution objectives. The fact that most developing countries discriminate against agriculture is reflected in price distortions.

The agricultural or food production growth function can be established by estimating the coefficients between agricultural/food production growth and the changes in the relevant independent variables. The growth of total agricultural production or food production is affected by changes in the same factors. Based upon the discussions above, the aggregate agricultural or food production growth function can be expressed in the following form:

$$Y_g = f(A_g, L_g, Q, F_g, M_g, G,)$$

where Y_g is growth rate of agricultural or food production, A_g is rate of change in labor input, L_g is rate of change in land cropped, Q is quality of arable land or soil, F_g is rate of change in fertilizer consumption, M_g is rate of change in machinery power utilization, and G is government policies, e.g. price, land use.

Those variables with subscript 'g' are flow variables, and they should be measured in terms of growth rate. Q and G can be seen as stock variables and the average values are used in estimations.

The approach used in this study involves estimating a cross-country agricultural growth function based on a sample of 28 countries in the Third World. Variations in agricultural growth rate are accounted for by differences in the growth rate of agricultural inputs and related factors. All the data used in this study are from the period 1971–1980. The data for flow variables are the average for 1971–1980 and the data for stock variables are the actual 1975 figures. More detailed development of the study methodology is presented in Zhao et al. (1991).

Analysis and general results

Appendix 1 summarizes the main results of the estimation of the aggregate agricultural production and the food production growth function. Considering the aggregate nature of the secondary data, the levels of statistical significance of several of the estimated coefficients are quite good. The six independent variables in the model can explain as much as 82% of the variance in total agricultural production, and about 78% of the variance of food production when growth is measured by the average index. However, the models based on percentage growth measure are less satisfactory, and the R^2s for the TAP and TFP models are 0.66 and 0.68, respectively. The F tests show that one can be 95.00–99.99% confident of rejecting the hypothesis that all the estimated coefficients are zero for the four models. These models, as a whole, are quite well defined and the multicollinearity problem in the statistical analysis does not appear to be serious, based on the SAS collinearity diagnostics procedure.

Land degradation and area change

Even though the data are categorical (three classifications) and thus gross measures, the statistical significance is relatively strong. (There were no countries in this study with a very severe classification, e.g. land denuded of vegetation, severe gullies and blow-out areas from erosion, and crop yields reduced more than 90%. Slight degradation involves less than 10% reduction in crop yields, excellent to good range condition and to slight erosion. See Dregne (1982) for further detail.) Based on the percentage measure, the estimate of the moderate level of land degradation is significant at the 5% level in the total agricultural production growth model, and the estimate for food production growth is significant at the 2% level. If the growth is measured by the average index, the estimate for moderate land degradation is not significant in the total agricultural production growth model. It is, however, still significant at the 15% level in the food production growth model.

When comparing the absolute values of the estimates and their corresponding significance levels, the land degradation variable has higher absolute coefficient values, and the estimates are more significant in the TFP than in the TAP models. This difference indicates that land degradation tends to affect food production more significantly than it affects non-food agricultural production. The result seems to confirm the belief that land degradation threatens food production growth. It also impedes income increases in rural areas because of a direct relationship between farmers' income and food production growth. Notice that the estimated coefficients of the high degradation level have incorrect signs and are insignificant at the 15% level. This may be due to the fact that there are only three countries with a high degradation level, which results in large standard errors of the coefficient estimates.

The positive relationship between the land area change rate and agricultural production growth or food production growth are quite strong with both significant at the 1% level based on the index measure of growth. However, the estimates are insignificant in percentage measured growth models. In contrast to the negative influence of the land degradation factor discussed above, the amount of arable and permanent crop land is strongly correlated with agricultural and food production growth. Reductions in severe land degradation should increase the future availability of arable and permanent land which may increase agricultural output.

Summary and conclusions

The results of the Zhao (1988) study showed that price distortion in the economy and land degradation had statistically significant negative impacts while the change in arable and permanent land was positively related to the growth of agricultural production and food production in 28 developing countries from 1971 to 1980. These results emphasize the importance of 'getting prices right' and implementation of sustainable land and water management practices if future growth in food and agricultural output is to be sustained in developing countries.

After price distortion, the variable of greatest significance is the degree of land degradation. The reduction in overall agricultural (as opposed to food) production growth caused by land degradation is smaller in magnitude and less significant statistically. The regression results confirm the belief that land degradation in developing countries constitutes an immediate as well as a long-term threat to these countries' capacity to produce food. The estimation in this study failed to capture the off-site damage from land degradation, e.g. water pollution and siltation of hydropower reservoirs and harbors. Thus, the actual negative effects are likely to be much larger and more significant than estimated, since off-site impacts of soil erosion are generally much greater than on-site impacts.

Most developing countries are more dependent on their natural resources, notably land and water, and land degradation significantly threatens agricultural growth. Soil and water conservation is of great importance to sustainable economic development. Most past development efforts have been based on the exploitation of natural resources in many developing countries. In the long-term, land protection and agricultural growth are complementary rather than competitive, even though there may be some trade-off between the two in the short-term. Not only conservation projects but also policy reforms are needed to protect the soil base. Policy reforms require that measures such as soil conservation and proper drainage in irrigation projects be incorporated as an integral part of a development program.

The policy reforms should focus on increasing economic incentives for conservation. Since, in the majority of developing countries, most agricultural activities are performed by small operational units, such as households and small farms, appropriate economic incentives for millions of small farmers are vital in the Third World in order to channel development activities into sustainable development patterns. Various studies show that serious environmental degradation is primarily due to the cumulative effects of many small agricultural operations that are not affected by environmental regulations (IIED and World Resources Institute, 1986; Repetto, 1989). The appropriate economic incentives may include increasing agricultural prices to the competitive level, reducing taxation on agricultural production, establishing effective property rights, providing subsidies and assistance for conservation practices, and eliminating input subsidies.

Correcting only farm level disincentive problems is inadequate because soil erosion causes major off-site impacts which are not borne by farmers. In addition, farmer's time horizons are much shorter and their discount rates much higher than those of society at large. Therefore, public interventions and national actions are usually required to ameliorate the effects of soil erosion and degradation including better defined property rights, government regulations on land use and the traditional approach to environmental problems — public authorities investing in reforestation and pollution control projects.

Price distortions result in great losses of agricultural production in the Third World. The statistical results from the Zhao (1988) study confirm the conclusions that many theoretical analyses and empirical observations have reached. Schultz's (1978) perspective that the economic potential for agricultural development has been largely unexploited because of the low incentives caused by the low prices of agricultural products is borne out, but development must also be sustainable from an environmental perspective. To promote industrialization by means of lowering agricultural prices is, at the very least, inefficient. The costs are too high, and evidence shows that the costs are borne not only by agriculture, but by the whole economy. In order to have a high economic growth and at the same time have a high agricultural growth, 'getting prices right' is a necessary condition.

In some sense, 'getting prices right' also applies to the argument for a sustainable pattern of growth. One of the most important causes of environmental degradation is that environmental services are undervalued. Activities that exploit land resources and cause land degradation are not fully priced or taxed, or, at least, not valued at their marginal social valuation. This, in turn, leads to inaccurate measures of national income. The problem of inadequate land and water management or exploitation of land and water resources is similar to the undervaluation of commodities or credit — the natural resources are overused and undervalued. Thus, the arguments for well-defined property rights, pollution taxes, revised national income accounts, and implementation of conservation measures is 'getting prices right on environmental services'.

Acknowledgments

Professor Wen Chern at The Ohio State University made a major contribution to the thesis research of Fenkun Zhao on which the empirical section of this paper is based. A more comprehensive version of this paper has been published by F. Zhao, F. Hitzhusen and W. Chern (1991) in Agric. Econ., 5: 311–324.

References

Argawala, R., 1983. Price distortion and growth in developing Countries. World Bank Staff Working Paper No. 575, World Bank, Washington, DC.
Bale, M.D. and Lutz, E., 1981. Price distortions in agriculture and their effects: An international comparison. Am. J. Agric. Econ., 63: 8–22.
Clark, II, E.H., Havercamp, J. and Capman, W., 1985. Eroding Soils: The Off-farm Impacts. The Conservation Foundation, Washington, DC.
Crosson, P., 1985. Impacts of erosion on land productivity and water quality in the United States. In: S.A. El-Swaify (Editor), Soil Erosion and Conservation.

Dixon, J.A. and Fallon, L., 1989. The concept of sustainability: origins, extensions and usefulness for policy. Soc. Nat. Resour., 2: 73–84.

Dregne, H.E., 1982. Impact of Land Degradation on Future World Food Production. ERS-677. US Department of Agriculture, Washington, DC.

Hauck, F.W., 1985. Soil erosion and its control in developing countries. In: S.A. El-Swaify (Editor), Soil Erosion and Conservation. Soil Conservation Society of America, Ankeny, IA.

Hayami, Y. and Ruttan, V.W., 1971. Agricultural productivity differences among countries. Am. Econ. Rev., 60: 895–911.

Hayami, Y. and Ruttan, V.W., 1985. Agricultural Development: An International Perspective. Johns Hopkins University Press, Baltimore.

IBRD, 1980. World Development Report. Oxford University Press, New York.

IBRD, 1986. World Development Report. Oxford University Press, New York.

IIED and World Resource Institute, 1986. World Resource 1986. Basic Books, New York.

Hitzhusen, F., Macgregor, B. and Southgate, D., 1984. Private and social cost–benefit perspectives and a case application on reservoir sedimentation management. Water Int., 9: 181–189.

Kawagoe, T., Hayami, Y. and Ruttan, V.W., 1985. The intercountry agricultural production function and productivity differences among countries. J. Dev. Econ., 19: 113–132.

Krishna, Raj, 1982. Some aspects of agricultural growth, price policy and equality in developing countries. Food Res. Inst. Studies, 18 (3).

Lau, L.J. and Yotopoulos, P.M., 1987. The metaproduction function approach to technological change in world agriculture. Mcmorandum No. 270, Stanford University, Stanford, CA.

Myers, N., 1987. The environmental basis of sustainable development. Ann. Reg. Sci., 21(3).

Repetto, R., 1989. Wasting assets: The need for national resource accounting. Technol. Rev., 39–44.

Schultz, T.W. (Editor), 1978. Distortions of Agricultural Incentives. Indiana University Press, Bloomington.

Timmer, C.P. and Falcon, W.P., 1975. The political economy of rice production and trade in Asia. In: L.G. Reynolds (Editor), Agriculture in Development Theory, Yale University Press, New Haven, pp. 373–408.

Warford, J.J., 1987. Natural resources and economic policy in developing countries. Ann. Reg. Sci., 21(3).

Zhao, F., 1988. Some new evidence on factors related to agricultural growth in the Third World. M.S. Thesis. Department of Agricultural Economics and Rural Sociology, The Ohio State University, Columbus, OH, unpublished.

Zhao, F., Hitzhusen, F. and Chern, W., 1991. Impact and implications of price policy and land degradation on agricultural growth in developing countries. Agric. Econ., 5: 311–324.

Appendix 1

Estimates of agricultural growth function on cross-country data (1971–1980)

Dependent variable	Independent variable	Coefficient estimations	
		% Growth measure	Average index measure
TAP	(Intercept)	3.433***	80.552***
		(4.14)	(5.28)
	LAND	0.114	0.461***
		(0.54) [0.046]	(3.73)
	LD		
	Moderate	−1.135**	−0.871
		(−2.07)	(−0.34)
	High	0.180	7.049
		(0.22)	(1.53)
	FERT	0.003	0.001
		(0.12) [0.011]	(0.15)
	LABOR	0.229	−0.033
		(0.94) [0.061]	(−0.22)
	TRACT	0.053**	0.009
		(1.89 [0.162]	(1.21)
	PDL		
	Moderate	−0.246	−9.637**
		(−0.41)	(−3.17)
	High	−1.858***	−20.013***
		(−3.29)	(−7.00)
		$R^2=0.655$	$R^2=0.821$
		PROB$>F=0.0051$	PROB$>F=0.0001$
TFP	(Intercept)	3.297***	80.568***
		(4.28)	(4.55)
	LAND	0.136	0.404**
		(0.56) [0.056]	(2.82)
	LD		
	Moderate	−1.628**	−4.795*
		(−2.65)	(−1.62)
	High	0.517	4.424
		(0.56)	(0.83)
	FERT	0.004	0.003
		(0.13) [0.015]	(0.31)
	LABOR	0.273	0.040
		(0.97) [0.075]	(0.23)
	TRACT	0.064**	0.006
		(1.98) [0.203]	(0.68)

Dependent variable	Independent variable	Coefficient estimations	
		% Growth measure	Average index measure
TFP			
	PDL		
	Moderate	0.600	− 5.332*
		(0.87)	(− 1.51)
	High	− 1.677**	− 19.150***
		(− 2.62)	(− 5.77)
		$R^2 = 0.682$	$R^2 = 0.778$
		PROB > $F = 0.0017$	PROB > $F = 0.0001$

Source: Zhao et al. (1991).

Linear equations are estimated by the least squares method. ***Indicates that the coefficient is significant at 1% level, **at 10% level, and *at 15% level. Figures in parentheses are *t*-values and figures in square-brackets are mean value elasticities.

Agriculture, Ecosystems and Environment, 46 (1993) 81–88

Elsevier Science Publishers B.V., Amsterdam

Agriculture and global warming*

David W. Orr

Oberlin College, Oberlin, OH 44074, USA

Abstract

Global warming poses a distinct threat to agricultural systems worldwide and to attempts to place farming on more sustainable foundations. The magnitude and severity of that threat is systematically understated by the use of inappropriate models of risk and economic analysis that underestimate potential costs of global warming. The author proposes that the logic underlying Pascal's wager about the existence of God, a variant of minimax strategies in game theory, may be an instructive alternative.

The future of agriculture here and in the rest of the world is increasingly complicated by the prospect of global warming and the depletion of stratospheric ozone. Changing rainfall and temperature will compound the uncertainties of farming and may increase temptations to adopt short-term technical, genetic, and biochemical practices that farmers might otherwise prefer to avoid. At the same time increased ultraviolet radiation will tend to reduce the productivity of soils and lower resistance of animals to disease. The prospect of steadily (or rapidly) changing conditions of climate and atmospheric chemistry are unlike the variations of weather within some 'normal' range to which farmers have adapted over the past 10 000 years. Farmers now face the prospect of ongoing changes affecting soil chemistry, temperature, moisture, and incidence of weeds, plant pathogens, livestock diseases, and insect predation. In some prime agricultural regions of the world these changes will mean continual efforts and capital expense to maintain soil fertility and water availability, to adapt to new crops and perhaps to new cropping methods, and to restructure operations in what could be worsening conditions. Even in regions that may benefit from global warming (e.g. Scandinavia) the adjustments to 'opportunity' may be expensive and uncertain. In other words, over the next 50–100 years global warming could reverse the gains in agricultural productivity made since 1945 and set back attempts to rebuild agriculture on a more sustainable foundation.

In fact, past gains in agricultural productivity have contributed signifi-

---

*Adapted from: Orr, D.W., 1992. Pascal's Wager and Economics in a Hotter Time. Ecol. Econ., 6 (1): 1–6.

cantly to the problem of global warming. Post World War II agriculture became increasingly dependent on fossil fuels, chemicals and methods that release substantial amounts of the greenhouse gases: CO_2, N_2O, and CH_4. The food system here and elsewhere requires the long-distance transport of food, feed grains, fertilizer, and water — hence the further release of CO_2 into the atmosphere from transportation, processing, and pumping. Moreover the loss of soil and soil carbon contribute further to the problem of global warming (Lovins and Lovins, 1991). In other words, unsustainable agricultural practices and the energy requirements of the present food system are part of the problem. When governments begin to develop comprehensive policies to deal with global warming they are likely to require changes in agricultural practices that will aim to sequester carbon in soils and biomass, reduce emissions of methane and nitric oxides, and reduce dependence on fossil fuels. These changes will test the adaptability of farmers and the food system on a larger scale and in a shorter time-horizon than ever before. Yet it is widely assumed in influential places that agriculture can adapt readily to gradual climate change (Parry, 1990; National Academy of Sciences, 1991). Whether farmers can adapt to ongoing change decade after decade or to more rapid changes they do not say. Given the history of surprises in the depletion of stratospheric ozone, the prospect of similarly unpleasant climatic surprises is not an unlikely possibility. Nevertheless, these experts tend to oppose efforts to limit the emission of greenhouse gases in the near term and do so in large measure because they believe agriculture to be highly adaptable or otherwise unimportant. If it is not sufficiently adaptable then the case for business as usual is far less robust. By failing to control the emission of greenhouse gases which change the conditions under which farmers must work, we run the risk of undermining the very adaptability of agriculture on which such recommendations depend — a paradox of more than academic importance. In this paper I will explore this paradox in greater detail and particularly the positions of Yale economist William Nordhaus and the Panel convened by the National Academy of Sciences to study how we might adapt to climate change.

   In weighing the question concerning the existence of God, seventeenth century philosopher and mathematician, Blaise Pascal, proceeded in a manner perhaps instructive for other and more mundane questions. "Reason," he declared, "can decide nothing here". Nonetheless, "you must wager. It is not optional". We have, he believed,

> "two things to lose, the true and the good; and two things to stake, your reason and your will, your knowledge and your happiness; and your nature has two things to shun, error and misery".

What would one lose by believing that God exists and living a life accordingly? Pascal's answer was: "if you gain, you gain all; if you lose, you lose nothing". By doing so one would become "faithful, honest, humble, grateful, generous, a sincere friend, truthful". The opposite decision was that God did not exist and that a life lived in pursuit of 'poisonous pleasures, glory and luxury,' whatever its short-term gains, would be one of misery. In other words if he chose not to believe and it turned out that God did exist, he would go to hell. On the other hand if God did not exist and he had lived a life of faith he would have sacrificed only a few fleeting pleasures but gained much more. Thus, his argument for faith rested on the sturdy foundation of prudential self-interest aimed at minimizing risk.

The world now faces a somewhat analogous choice. On one side a large number of scientists believe that the planet is warming rapidly. If we continue to emit large amounts of heat-trapping gases, such as CO_2, CH_4, CFCs, and NO_x they say we will warm the planet intolerably within the next century. The consequences of dereliction and procrastination may include killer heat waves, drought, sea-level rise, superstorms, vast changes in forests and biota, considerable economic dislocation, and increases in disease — a passable description of hell. But like Pascal's wager, no one can say with absolute certainty what will happen until the consequences of our choice, whatever they may be, are upon us. Nonetheless, "we must wager. It is not optional".

Others, however, claim to have looked over the brink and have decided that hell may not be so bad after all, or at least that we should research the matter further. Yale University economist William Nordhaus, for example, believes that a hotter climate will affect mostly "those sectors [of the economy] that interact with unmanaged ecosystems" such as agriculture, forestry, and coastal activities (Nordhaus, 1990b). The rest of the economy, including that which operates in what he calls 'a carefully controlled environment', which includes shopping malls, and presumably the activities of economists, will little notice that things are considerably hotter. "The main factor to recognize", Nordhaus asserts, "is that the climate has little economic impact upon advanced industrial societies" (Nordhaus, 1990a).

Professor Nordhaus concludes that "approximately 3 percent of U.S. national output originates in climate-sensitive sectors and another 10 percent in sectors moderately sensitive to climatic change". There may even be, he notes, beneficial side-effects of global warming: "The forest products industry may also benefit from CO_2 fertilization". (It is, I think, no mistake that he did not say 'forests' but rather 'forest products industry'.) Construction, he thinks, will be 'favorably affected' as will 'investments in water skiing'. In sum, Professor Nordhaus' 'best guess' is that the impact of a doubling of CO_2 "is likely to be around one-fourth of 1 percent of national income", an estimate that he confesses has a "large margin of error".

Professor Nordhaus, however, wishes not to be thought to favor climate change. Rather the point he wants to make is that "those who paint a bleak picture of desert Earth devoid of fruitful economic activity may be exaggerating the injuries and neglecting the benefits of climate change". Whether a hotter earth, but one not 'devoid of fruitful economic activity' might, however, be devoid of poetry, laughter, sidewalk cafes, forests, or even economists he does not say. But he does note that there are a number of possible technological responses to our plight including "climatic engineering ... shooting particulate matter [books on economics?] into the stratosphere to cool the earth or changing cultivation patterns in agriculture". Professor Nordhaus, an economist, gives no estimate of the costs, benefits, or even feasibility of these 'options'. He does, however, estimate the cost of reducing CO_2 emissions by 50% as $180 billion per year. Faced with such costs, Nordhaus thinks that "societies may choose to adapt", which in his words means "population migration, capital relocation, land reclamation, and technological change", solutions for which he again gives no cost estimate. And what about those who cannot adapt, migrate, buy expensive remedies, or relocate their capital? Professor Nordhaus does not say, and one suspects that he does not say because he has not thought much about it.

The complications Nordhaus does notice have to do with those pertaining to "how to discount future costs and how to allow for uncertainty". A discount rate of, say, 8% or higher would lead us to do nothing about global warming for a few decades while the problem grows gradually or perhaps rapidly worse. A rate of 4% or less "would give considerable weight today to climate changes in the late twenty-first century". Nordhaus' solution? "The efficient policy", he argues, "would be to invest heavily in high-return capital now and then use the fruits of those investments to slow climate change in the future". He describes this as a "sensible compromise" between what he asserts is a "*need* for economic growth" and "the desire for environmental protection" (emphasis added).

To his credit, Professor Nordhaus does acknowledge that "most climatologists think that the chance of unpleasant surprises rises as the magnitude and pace of climatic change increases". He also notes that the discovery of the ozone hole came as a 'complete surprise' suggesting the possibility of more surprises ahead. But in the end he comes down firmly in favor of what he calls 'modest steps' that "avoid any precipitous and ill-designed actions that [we] may later regret". These actions he does not specify, making it impossible to know whether they would be in fact precipitous, ill-designed, or regrettable. He believes that "reducing the risks of climatic change is a worthwhile objective", but not more important than "factories and equipment, training and education, health and hospitals, transportation and communications, research and development, housing and environmental protection" and so forth. He does not regard the relationship between heat, drought and climate instability on the one hand and the economy, public health, human behavior under stress, or even what he calls 'environmental protection' on the other as particularly noteworthy.

One might dismiss Professor Nordhaus' analysis as an aberration were it not characteristic of the recklessness masquerading as caution in the highest levels of government and business here, and elsewhere, and were he not as influential at these levels as he certainly is. His views on global warming are neither an aberration nor are they without consequence where portentous choices are made.

His opinions about global warming, for example, weighed heavily in the 1991 report issued by the 'Adaptation Panel' of the National Academy of Sciences (National Academy of Sciences, 1991). The Panel, of which he was a member, approached global warming as an investment problem requiring the proper discount rate. But for those whose interests were discounted, such as the poor and future generations, the problem would appear differently as one of power and intergenerational responsibilities. The Panel, moreover, made many assumptions about the adaptability of complex, mass, technological societies under what may be extreme conditions. In citing "the proven adaptability of farmers", for example, are they referring to the four million failed farms in the past 50 years, or to those 1.5 million farms presently at or close to the margin? Or are they referring to the over-dependence of agriculture and food distribution systems on the very fossil energy sources that are now heating the earth? Or are they referring to present rates of soil loss and groundwater depletion due to current farm practices? Can farmers adapt if global warming occurs rapidly? Since people live "in both Riyadh and Barrow", the Panel drew the conclusion that humans are almost infinitely adaptable, while admitting that some cities will have to be abandoned, and people in poorer countries may be substantially harmed. The Panel, however, hedged its bets by admitting that the warming could be sudden and catastrophic, but dismissed these possibilities quickly without further comment. They did not ask what could happen beyond their 50 year horizon, nor did they ask about the effects on American society of making such portentous decisions in the same way as we make investment decisions about building bridges or shopping malls.

It is therefore a matter of general concern that such analysis provides considerable aid and comfort for those with much to gain by ignoring the risks involved in climate change or the benefits of a far-sighted energy policy. Accordingly, we should attempt to understand how such thought arises, whose ends it serves, and what consequences it risks.

For comparison, it is instructive to note that atmospheric physicists, climate experts, and biologists agree, almost without exception, that the theory of global warming is beyond dispute (Intergovernmental Panel on Climate Change, 1991). It is widely agreed that heat-trapping gases in the atmosphere do in fact trap heat. It is agreed further that if we put enough of these gases into the atmosphere we will trap a great deal of heat. And there is unanimity on the belief that if global warming turns out to be rapid the consequences would, in all probability, be widely catastrophic, even though we cannot predict these with absolute certainty. Disagreement focuses mainly on matters having to do with rates, thresholds, and the effects of feedbacks that might enhance or retard rates of global warming. However these issues are decided, there is no doubt whatsoever that by increasing heat-trapping gases to levels higher than any in the past 160 000 years, and at rates far more rapid than characteristic of past climate shifts, we are conducting an unprecedented experiment with the Earth and its biota which includes us (Peters and Lovejoy, 1992). This experiment need not, and should not, be carried out. But like Pascal's wager, certainty about the consequences will come only after all bets are called in.

Given what is at stake, errors of fact and logic committed by Nordhaus and the Adaptation Panel deserve close attention. For example, the belief that a decline in agriculture and forestry would be of little consequence because they are only 3% of the US economy is equivalent to believing that since the heart is only 1–2% of body weight it can be removed or damaged without consequences for one's health. Both Nordhaus and the Adaptation Panel regard the economy as linear and additive without surprises, thresholds of catastrophe, or even places where angels would fear to go. The biological facts underlying the research are also suspect. There are many reasons to believe that 'CO$_2$ fertilization' will not enhance farm and forest productivity as Nordhaus and the Panel believe (Körner and Arnone, 1992). Changes in rainfall, temperature, and biological conditions necessary for propagation would more likely reduce growth (Houghton and Woodwell, 1989). Higher temperatures mean higher rates of plant respiration, hence the release of still more carbon and methane. The rates of climatic changes may well be many times faster than those to which plants and animals can adapt. This would mean at some time in the future a dieback of forests and the release of even more carbon through fire and rapid decay (Environmental Protection Agency (EPA), 1989). It would also mean a sharp reduction in biological diversity (Peters and Lovejoy, 1992).

Economic estimates used by Nordhaus and the Adaptation Panel are also questionable. Both ignore a large and growing body of evidence that the actions necessary to minimize global warming would be good for the economy, human health, and the land (Romm, 1992; Dower and Zimmerman, 1992). Studies by the EPA, the Electric Power Research Institute, and independent researchers all point to the same conclusion: energy efficiency, which reduces the emission of CO$_2$, is not only not costly, but it is in fact a prerequisite of economic vitality. It is also worth noting that other economists, such as William Cline, conclude that "it appears sensible on economic grounds to undertake aggressive abatement to curtail the greenhouse effect" (Cline, 1992). But there are reasons that go beyond risk aversion. The US economy is roughly one-half as energy efficient as that of the Japanese. This translates into a 5% cost disadvantage for comparable US goods and services (Lovins, 1990). Instead of an annual cost which he estimates at $180 billion, more reliable studies have shown a net savings of approximately $200 billion from improvements in energy efficiency. This, in Amory Lovins' words, "is not a free lunch, but a lunch we are paid to eat". Moreover, estimates by Nordhaus and the Academy Panel do not include the costs of relocating millions of people, or those of failing to do so, those entailed in diking coasts, those of international conflicts over water, those of importing food when the plains states become drier, or those of changes in diseases due to climate change. Nor does he say what it might cost if global warming turns out to be rapid and full of even worse surprises.

The practice of discounting the future creates other costs which cannot be quantified but which will be assessed. If they had included the preferences of, say, the third generation hence in the equation their conclusions would have been quite different. Nordhaus and the Academy Panel chose not to do so, however, by assuming that investments in more of the same kinds of activities that have created the problem in the first place are 'worthy goals'. On closer examination, most of these will intensify the problem of global warming while ignoring opportunities to invest in energy efficiency and renewable resources which would reduce the emission of heat-trapping gases in an economically sound manner.

The economic estimates of Nordhaus and the Academy Panel are not valid because their economy is an abstraction independent of biophysical realities, comparable, say, with an airline pilot who regarded the law of gravity as merely an interesting but untested theory. Their economics are not to be trusted because they fail to acknowledge the vast and unknowable complexity of planetary systems which cannot be 'fixed' by any technology without courting other risks. Their economics are not very good economics. They have ignored the relationship between economic prosperity and energy efficiency, as well as that between energy efficiency and the emission of greenhouse gases. Their economic conclusions are not valid because the problem of global warming is not first and foremost one of economics, as they believe, but rather one of judgement and wisdom. Their economics cannot be trusted because they do not include flesh and blood people who will be sorely stressed by a rapidly changing climate and who will not act with the rationality presumed in abstract models concocted in air-conditioned rooms.

Finally, the conclusions of Nordhaus and the Academy Panel cannot be accepted because they would have us risk another decade or two of business as usual, which as we now know does not mean sustainable prosperity or basic fairness. This is a foolish risk for reasons Pascal described well. If it turns out that global warming would have been severe and we forestalled it by becoming more energy efficient and making a successful transition to renewable energy, we will have avoided disaster. If, however, it turns out that factors, as yet unknown, minimized the severity and impact of warming while we became more energy efficient in the belief that it might be otherwise, we will not have saved the planet, but we will have reduced acid rain, improved air quality, decreased oil spills, reduced the amount of strip mining, reduced our dependence on imported oil and thereby improved our balance of payments, become more technologically adept, and improved our economic competitiveness. In either case we will have set an instructive and far-sighted precedent for our descendants and for the future of the earth. If we gain, we gain all; if we 'lose' we still gain a great deal. On the other hand, if we do as Professor Nordhaus and the other members of the Adaptation Panel would have us do and the warming proves to be rapid there will be hell to pay.

References

Cline, W., 1992. Global Warming: The Economic Stakes. Institute for International Economics, Washington, DC, p. 6.

Dower, R. and Zimmerman, M.B., 1992. The Right Climate for Carbon Taxes. World Resources Institute, Washington, DC.

Environmental Protection Agency, 1989. Policy Options for Stabilizing Global Climate. Environmental Protection Agency, Washington, DC.

Houghton, R.A. and Woodwell, G., 1989. Global Climatic Change. Sci. Am., April.

Intergovernmental Panel on Climate Change, 1991. Climate Change. Oxford University Press, New York.

Körner, C. and Arnone, J., 1992. Responses to Elevated Carbon Dioxide in Artificial Tropical Ecosystems. Science, 1672–1675.

Lovins, A., 1990. The role of energy efficiency. In: J. Leggett (Editor), Global Warming. Oxford University Press, New York.

Lovins, A. and Lovins, H.L., 1991. Least-Cost Climatic Stabilization. Annual Review of Energy and Environment. Annual Reviews, Palo Alto, CA.

National Academy of Sciences, 1991. Policy Implications of Greenhouse Warming: Report of the Adaptation Panel. National Academy Press, Washington, DC.

Nordhaus, W.D., 1990a. Greenhouse economics: count before you leap. Economist.

Nordhaus, W.D., 1990b. Global warming: slowing the greenhouse express. In: H. Aaron (Editor), Setting National Priorities. Brookings Institution, Washington, DC.

Parry, M., 1990. Climate Change and World Agriculture. Earthscan, London.

Peters, R. and Lovejoy, T. (Editors), 1992. Global Warming and Biological Diversity. Yale University Press, New Haven.

Romm, J., 1992. The Once and Future Superpower. William Morrow, New York.

Agriculture, Ecosystems and Environment, 46 (1993) 89–97
Elsevier Science Publishers B.V., Amsterdam

The concept of agricultural sustainability

Neill Schaller

Institute for Alternative Agriculture, 9200 Edmonston Road, Suite 117, Greenbelt, MD 20770, USA

Abstract

Sustainable agriculture has become a popular code word for an environmentally sound, productive, economically viable, and socially desirable agriculture. This paper reviews reasons for growing interest in agricultural sustainability (mainly the unanticipated, adverse side-effects of conventional farming), examines the proposed ends and means of sustainability, and discusses two issues frequently debated — the profitability of sustainable farming and the adequacy of food production from sustainable systems. The concept of agricultural sustainability does not lend itself to precise definition, partly because it implies a way of thinking as well as of using farming practices, and because the latter cannot be specified as final answers. Consequently, people's beliefs and values will continue to mold public understanding of the concept. Two different views of sustainable agriculture are held. One is that fine-tuning of conventional agriculture — more careful and efficient farming with sensitive technologies — will reduce or eliminate many undesirable effects of conventional agriculture. The other is that fundamental changes in agriculture are needed, requiring a major transformation of societal values. Those who believe that only fine-tuning is needed tend to argue that sustainable farming is inherently unprofitable. If widely adopted, it would not feed the world's expanding population as well as conventional agriculture. Those who see a need for more fundamental changes in conventional systems believe that sustainable farming, on the contrary, can be even more profitable than the conventional, especially when the calculation of profit counts all of the benefits and costs of farming. Further, resource conservation, protection of the environment, and farming in partnership with nature — all requirements of sustainability — will enhance, not reduce, global food production. Other issues, such as the connections between sustainable farming and the rest of the food and fiber system, and the implications of sustainability for rural communities and society as a whole, have yet to be addressed significantly.

Introduction

In less than a decade, agricultural sustainability has become a popular code word at home and abroad. To most people, it seems to mean an agriculture that will continue to conserve natural resources and protect the environment indefinitely, enhance the health and safety of the public, and produce adequate quantities of food at a profit for farmers. Others extend the concept to include goals such as social justice and the safeguarding of animal welfare.

However defined, a sustainable agriculture is generally regarded as an alternative to modern industrialized, or conventional agriculture, an agriculture described as highly specialized and capital intensive, heavily dependent upon synthetic chemicals and other off-farm inputs.

Why the interest in sustainability?

The idea of sustainability, at least with regard to agriculture and natural resources, is not new. Throughout history, people have faced the challenge of balancing food production with protection of the environment. In recent times, interest in sustainability has risen periodically in response to environmental crises and health hazards, such as the Dust Bowl of the 1930s and the ill-effects of pesticides detailed in Rachel Carson's landmark book, *Silent Spring* (Carson, 1962). Highly publicized evidence, or fears, of ill-effects of conventional farming practices on food safety — illustrated recently by the use of Alar on apples — have intensified public interest in sustainability. After all, everyone eats.

Problems associated with conventional farming are now widely recognized as hidden costs of modern industrialized farming, costs that until recently have been all but justified by the spectacular gains in food production during this century. The problems include the following: (1) Contamination of ground and surface water from agricultural chemicals and sediment; (2) Hazards to human and animal health from pesticides and feed additives; (3) Adverse effects of agricultural chemicals on food safety and quality; (4) Loss of the genetic diversity in plants and animals, a key to the sustainability of agriculture; (5) Destruction of wildlife, bees, and beneficial insects by pesticides; (6) Growing pest resistance to pesticides (exacerbating the effects noted above); (7) Reduced soil productivity due to soil erosion, compaction, and loss of soil organic matter; (8) Over-reliance on non-renewable resources; (9) Health and safety risks incurred by farm workers who apply potentially harmful chemicals.

Chemicals are, perhaps, the cause of most of the public concern about conventional farming, due to a pervasive fear that chemical pesticides and fertilizers will cause cancer or other health hazards. Evidence exists to support the reality of that fear (National Research Council and Board on Agriculture, 1989, pp. 89–134). Lack of evidence, when it is reported, can lead people to think that the hazards have been exaggerated. For instance, a national survey of drinking water wells recently showed lower levels of chemical contamination in well water than had been suspected (US Environmental Protection Agency, 1990).

The most serious problem due to chemicals may not be their direct effects, but rather the kind of farming they make possible, which in turn can have harmful consequences seemingly unrelated to chemicals. For example, synthetic chemical pesticides and fertilizer permit farmers to grow corn or other crops on the same land year after year, factory style. While those practices often boost efficiency and output, they also invite environmental and economic risks of specialization, overdependence on off-farm purchased inputs, and the habit of 'formula farming'.

Undesirable economic and social impacts of conventional farming were added to the above list of adverse effects of conventional agriculture in the 1980s — and deserve much of the credit for putting sustainability on the national agenda. In that period, foreign demand for US farm products stagnated, crop prices sagged, land values plunged, and thousands of farmers were financially ruined. Though not due solely to the use of conventional farming practices, the financial crisis of the 1980s illustrated painfully what can happen if farmers rely on a few crops produced mainly for export, borrow too much to pay for the chemicals and other inputs used to produce those crops, and depend on the federal government to protect them when things go wrong. The problem, for many farmers, persists today.

Related psychological and social consequences, though mentioned less often, are among the most insidious impacts of conventional agriculture. Personal stress on farm families due to declining or uncertain farm incomes, the persistent loss of family farms, and a steady deterioration of rural communities continue to take their toll. As a result, many farmers in the US today, rather than watching the sustainability movement from the sidelines, are leading it, forming and joining new organizations — such as Practical Farmers of Iowa — to test better farming ideas and to share their experiences using potentially sustainable practices.

The choice and meaning of the term 'sustainable agriculture'

Sustainable agriculture is a popular term because it is general enough to appeal not only to people interested in an environmentally beneficial and healthful farming but also to those concerned with its economic and social dimensions. It also avoids the ambiguity and controversy that has often accompanied other terms, those used to emphasize different dimensions of sustainability or particular farming practices. Examples are 'organic', 'biological', 'ecological', 'reduced-input', 'low-input', 'regenerative', and the more encompassing term, 'alternative agriculture'.

The concept of sustainable agriculture suggests not only a destination for agriculture but particular farming practices that could move agriculture toward that destination (with both ends and means). But neither characteristic lends itself to precise definition. As a destination, sustainability is like truth and justice — concepts not readily captured in concise definitions. Nor can sustainable farming practices be defined easily, simply because no one can ever know precisely and finally which farming practices may be the most sustainable in every location and circumstance.

True, common sense and experience will generate more and more answers over time. There is already wide agreement that sustainability is enhanced by substituting management, brainpower, and patience for many of the inputs now bought in bottles and bags. It is known that crop rotations can break pest cycles and restore soil nutrients. Growing crops and raising livestock can be equally vital. Livestock contribute manure and harvest feed and forage from rotation crops. Insects, weeds, and diseases can be controlled using biological, mechanical, and other non-chemical methods, or what is called integrated pest management. Modern soil and water conservation techniques are time-tested components of a sustainable system. However, because sustainability means 'forever', and because good science requires it, the door must always be left open to new and better information on what could be the most sustainable combinations of practices.

Many people, in fact, prefer to talk about the technical side of sustainable agriculture — about the different kinds of practices that may conserve soil and water, protect the environment, and provide the farmer with a decent profit. Scientists, in particular, tend to think of the ends of sustainability as 'givens' and the role of science to determine how best to achieve them. For that reason, and because it is both difficult and unwise to try to define sustainable farming practices, people's beliefs and values will continue to play a critical role in determining what sustainability means and how it can and should be achieved.

Current beliefs and values differ markedly. At one extreme is the view that conventional agriculture needs only to be fine-tuned. At the other extreme is the conviction that sustainability will not be attainable until conventional agriculture is essentially redesigned (MacRae et al., 1990).

Those who believe that fine-tuning or modification will suffice give the following arguments:

(1) Conventional agriculture is basically sound. More careful and efficient farming will ensure its sustainability, e.g. farmers should use only the amounts of fertilizer and pesticides actually needed.

(2) No agriculture is sustainable if it is not first and foremost a profitable agriculture. Practices commonly recommended by proponents of sustainability are inherently unprofitable.

(3) Let us not sacrifice the production gains achieved by conventional agriculture by cutting back on modern inputs and relying on the whims of nature.

(4) New and sensitive technologies will be designed to correct whatever environmental, health, and safety problems might arise from the use of conventional farming practices.

This view is still widely held by many people and organizations within the traditional agricultural community, such as the US Department of Agriculture, the land grant colleges of agriculture, farm and commodity organizations, and agribusiness firms. Undoubtedly, together with many other citizens, these people are persuaded by the impressive gains in food production throughout this century that drastic changes in agriculture are not necessary. They have an abiding faith in technology. Continuing evidence of this view is seen in federal food, agricultural, and natural resource policies, as well as in public-supported research and education programs.

In contrast, the view at the other extreme is supported as follows:

(1) We cannot expect to have a sustainable agriculture unless all of us adopt a fundamentally different way of thinking about agriculture, which will require major changes in personal beliefs, values, and life styles (Hill, 1982, 1985; MacRae et al., 1990).

(2) Resource conservation, environmental protection, and health and safety are just as important as profitable production. In the long run, they are not conflicting goals. The future productivity and profitability of farming will depend increasingly on measures taken from now on to conserve resources and protect the environment.

(3) Soil erosion, pesticide pollution, and other adverse effects of conventional farming must be prevented from occurring, not continually covered up with bigger and bigger 'band-aids'. Instead of mixing farm chemicals away from the farm well, in case they spill, consider not using them at all.

(4) The farm is an organism, not a factory. Nature is not something to be conquered. Nature and naturally occurring processes beneath and above the soil are allies contributing not only to agronomic sustainability but to economic profitability, due in large measure to the synergism of plants, animals, soil, and the farmer's stewardship that are too easily forgotten in conventional farming.

(5) More careful and efficient farming must be encouraged, but only as a hopeful starting point on the long journey to sustainability.

(6) Agricultural sustainability, if viewed only as sustainable farm production, is a partial or intermediate goal. The concept and its pursuit should extend beyond the farm gate to the rest of the food and fiber system. Social justice and equity should be added to the more widely recognized environmental and economic goals of sustainability.

I subscribe to this view (although I would not qualify as a true practitioner). Although I cannot prove that attainment of agricultural sustainability is impossible by continually fine-tuning conventional agriculture, history and common sense tell me that fundamental changes, not just more careful and efficient farming, are essential. At the same time, I remain open to new information that could cause me to think differently.

Two critical issues regarding agricultural sustainability

Two issues in particular that continue to invite myths and misunderstandings are the profitability of sustainable farming and the adequacy of food production under a sustainable system.

Profitability of sustainable farming

The profitability of sustainable versus conventional farming is often the most contentious issue encountered when the subject of sustainability is discussed. Conventional agriculturalists point out that highly specialized, capital-intensive, chemical-intensive methods have been widely adopted, not only because they increase production but because they have been more profitable than practices now recommended by proponents of sustainability. To condemn current and more profitable methods would be to go back to hoes, hard labor, lower yields, and lower farm income.

Unfortunately, such an interpretation was probably encouraged a few years ago, and innocently so, with the coining of the term 'low-input sustainable agriculture', or LISA, by the Congress and the US Department of Agriculture (Schaller, 1991). 'Low-input' was chosen partly to head off the interpretation among agricultural groups that sustainable agriculture was just another name for chemical-free, or organic, farming. No one who suggested the term meant that farmers should simply cut back on all their inputs (which of course would lower output), but skeptics, especially farm chemical producers and distributors, were never sure.

Proponents of sustainability respond to the claims, and fears, of conventional agriculturalists by explaining that even if a shift to sustainable practices were to result initially in lower total production from the farm, those practices generally involve lower costs and therefore potentially higher profits.

The way in which profitability is calculated is really the basic problem. As currently defined and counted, profit or farm income automatically favors conventional agriculture and penalizes sustainable agriculture. This is because its calculation excludes many of the benefits of sustainable farming to farmers and the rest of society, as well as the costs associated with conventional farming, such as soil erosion and groundwater contamination. A recent study suggests that when the economic 'costs' of erosion are included in the profit equation, farming systems that conserve soil but which are less profitable than those that cause erosion will often be the more profitable of the two (Faeth et al., 1991).

The basic meaning of profit must also be considered. In an unsustainable world, profit may well become almost an end in itself. Implicitly or not, activities are regarded as noble because they are profitable (Kristol, 1974). In a truly sustainable world, profit is pursued and respected more clearly as a means. Activities, including farming, are profitable because they are noble. It follows that the true economic advantages of sustainable agriculture would be more apparent if the meaning of profit and the way it is calculated were to move toward that concept.

Adequacy of food production

Some observers feel that if agriculture is effectively redesigned to achieve sustainability, it may not be able to produce the amount of food the world needs to feed its rapidly growing population (Ruttan, 1988; York, 1990). They reason that global population, now just over 5 billion, is expected to grow to 8.5 billion by the Year 2025 (World Resources Institute and International Institute for Environment and Development, 1990). Most of the world's arable land is now in use. Farming systems believed to be potentially the most sustainable will require the use of much of the available land to produce inputs that are now manufactured and purchased off the farm. Also, it is not clear that crop yields under otherwise sustainable systems can consistently equal those of chemical-intensive agriculture. This reasoning clearly supports the argument for fine-tuning conventional agriculture.

Proponents of agricultural sustainability disagree, citing evidence that sustainable yields can equal if not exceed, and be less variable than, those achieved using conventional methods (National Research Council and Board on Agriculture, 1989; Lee, 1991). Higher and more stable yields would also be expected to come with farmer experience, the development of cultivars that contribute to sustainability, and the availability of new information and other services for farmers using sustainable practices. Moreover, the future productivity of conventional farming could begin to decline precipitously and permanently just when the number of people to be fed in the world reaches its highest level.

Adherents of sustainable agriculture also point out that those who question the ability of sustainable farmers to feed the world seldom consider the possibility that diets might change in ways that would reduce substantially the mounting pressure on available agricultural resources. Consumption of livestock products in the developed world today requires vast areas of land to produce animal feed and forage. In the US alone, about 4 out of 5 acres of 'agricultural land' are used for that purpose. Not all, but certainly much of that land could be used to produce food for direct human consumption. Fears of inadequate production also ignore the possibility of government policies and programs to reduce population growth rates in many parts of the world.

Future directions

So far, the pursuit of agricultural sustainability in the US, as elsewhere, has been limited almost entirely to the search for potentially more sustainable farming practices. It has also focused on the crop and livestock enterprises that are currently the major enterprises in each producing area, rather than on both current and alternative enterprises. It addresses the partial goal of sustainable production, not the sustainability of the entire food and fiber system.

The sustainable agriculture movement has had a good beginning. It should now recognize and embrace further dimensions of sustainability. For example, what are the connections between sustainable farming and the rest of the food and fiber system? If farmers adopt potentially sustainable practices, but the processing and transportation of food beyond the farm gate are highly industrialized and chemical-intensive in philosophy and fact, what does that say about the sustainability of agriculture as a whole? As one observer has put it, when you consider the energy inputs and costs in the distribution as well as production of food, you must ask harder questions. For instance, if you live on the East Coast, is it better to eat lettuce grown in California without any synthetic chemicals, or lettuce produced locally with those chemicals?

Other challenging issues waiting to be addressed include the following. To what extent does sustainable farming increase the well-being of rural people and communities? Do rural communities and institutions enhance or impair the ability of farmers to adopt sustainable practices? Beyond that, what is the connection between agricultural and rural sustainability and the rest of society? A sustainable agriculture alone, cannot remedy problems such as injustice and inequity, but might it set an example that would help to open new doors to a more just and equitable society? And what does all of this mean for public policies and for research and education? Few other subjects afford so much to ponder.

References

Carson, R., 1962. Silent Spring. Fawcett Publications, Greenwich, CT, 304 pp.

Faeth, P., Repetto, R., Kroll, K., Dai, Q. and Helmers, G., 1991. Paying the Farm Bill: US Agricultural Policy and the Transition to Sustainable Agriculture, World Resources Institute, Washington, DC, 70 pp.

Hill, S.B., 1982. A global food and agricultural policy for western countries: laying the foundation. Nutr. Health, 1: 108–117.

Hill, S.B., 1985. Redesigning the food system for sustainability. Alternatives, 12(3/4): 32–36.

Kristol, I., 1974. Horatio Alger and profits. Wall St. J.

Lee, L.K., 1991. Reducing chemical use: farm level and consumer impacts. Working Paper, University of Connecticut, Storrs, CT, 17 pp.

MacRae. R.J., Hill, S.B., Henning, J. and Bentley, A.J., 1990. Policies, programs, and regulations to support the transition to sustainable agriculture in Canada. Am. J. Alternative Agric., 5(2): 76–92.

National Research Council, Board on Agriculture, 1989. Alternative Agriculture. National Academy Press, Washington, DC, pp. 89–134, 247–417.

Ruttan, V.W., 1988. Commentary: Sustainability is not enough. Am. J. Alternative Agric., 3(2 and 3): 128–130.

Schaller, N., 1991. Background and status of the low-input sustainable agriculture program. Sustainable Agriculture Research and Education in the Field, A Proceedings, National Research Council, Board on Agriculture, National Academy Press, Washington, DC, pp. 22–31.

US Environmental Protection Agency, 1990. National survey of pesticides in drinking water wells. Phase I Report, EPA 570/9-90-015, Environmental Protection Agency, Washington, DC.

World Resources Institute and International Institute for Environment and Development, 1990. World resources 1990–91, Basic Books, New York, pp. 1 and 8.

York, E.T., 1990. Sustainable agriculture systems. Speech presented at Phillip C. Hamm Memorial Lecture, University of Minnesota, April 12, 1990, 14 pp.

Agriculture, Ecosystems and Environment, 46 (1993) 99–121
Elsevier Science Publishers B.V., Amsterdam

The role of agroecology and integrated farming systems in agricultural sustainability

C.A. Edwards[a,*], T.L. Grove[b], R.R. Harwood[c], C.J. Pierce Colfer[d]

[a]*Department of Entomology, Ohio State University, Columbus, OH 43210, USA*
[b]*Office of International Programs, College of Agriculture and Life Sciences, North Carolina State University, Raleigh, NC 27695, USA*
[c]*Department of Crop and Soils, Michigan State University, East Lansing, MI 48824, USA*
[d]*512 SW Maplecrest Drive, Portland, OR 97219, USA*

Abstract

Maintenance of biological diversity and nutrient cycling mechanisms are global principles that are common to all agroecosystems and therefore essential in the design of sustainable agricultural systems. Regional or site-specific factors include climate, soils and socio–economic preferences and conditions. These regional factors differ greatly among agroecosystems and may assume major importance in some. Research and development on global commonalities has potentially the most universal impact across all agroecosystems. Interdisciplinarity, participation of farmers and a whole farm level approach are fundamental to such research and development.

Definitions and concepts

Background and definitions

In developed and developing countries, there is a general perception that many agricultural practices are leading to degradation of natural resources through soil erosion, water contamination, deforestation, desertification, and loss of productivity (Dover and Talbot, 1987). These trends are most evident in developing countries of the tropics, where soils commonly are highly weathered, low in fertility, shallow, or susceptible to erosion (Jacobson, 1988). Additionally, agricultural ecosystems in the tropics are subjected to greater numbers and intensity of pests, diseases, and weeds than in the temperate zones (Dover and Talbot, 1987). There is concern that increased yields from high chemical inputs may not be sustainable over the long-term on many tropical soils. Currently, even on many prime lands, yields are decreasing, and the rural poor who are typically subsisting on marginal natural resources

---

*Corresponding author.

rarely have adequate or secure access to chemical inputs (Francis et al., 1986, Lal, 1990).

The development of the concept of sustainable agriculture is a relatively recent response to concerns about degradation of natural resources. Harwood suggests that the concept of sustainable agriculture was first articulated by Jackson (1980) and by Rodale (1983), who outlined a concept of regenerative agriculture which renewed natural resources. These early discussions emphasized the importance of maintaining the renewal capabilities of agricultural ecosystems and claimed that many conventional agricultural practices were deleterious to renewal. This concept has promoted much discussion and has evolved into a framework for a more sustainable agriculture that integrates the principles of ecology and that emphasizes interactions among and within the biological components of agricultural ecosystems.

About 1987, the phrase 'sustainable agriculture' took on additional meaning. As more and more groups and organizations began to recognize the need for adjustments to conventional agriculture that are environmentally, socially, and economically compatible, the phrase sustainable agriculture was used to connote a global agriculture that could provide for the needs of current and future generations while conserving natural resources (Douglass, 1984). In this new context, the phrase is often used to refer to agriculture and all of its interactions with society, which can be considered as sustainable development.

Within the literature, hundreds of definitions of sustainable agriculture have been offered (e.g. Rodale, 1983; Committee on Agricultural Sustainability for Developing Countries, 1987; Edwards, 1987; Weil, 1990), but virtually all have the following characteristics: adequate economic returns to farmers; indefinite maintenance of natural resources and productivity; minimal adverse environmental impacts; optimal production with minimal external inputs; satisfaction of human needs for food and income; provision for the social needs of farm families. In other words, they promote environmental, ecological, economic, and social stability and sustainability. They provide a framework and agenda for the indefinite evolution of agriculture to meet new needs in changing environments.

Edwards (1987) provided a detailed definition of sustainable agriculture as:

"Integrated systems of agricultural production, with minimum dependence upon high inputs of energy, in the form of synthetic chemicals and cultivation, that substitute cultural and biological techniques for these inputs. They should maintain, or only slightly decrease, overall productivity and maintain or increase the net income for the farmer on a sustainable basis. They must protect the environment in terms of soil and food contamination, maintain ecological diversity and the long-term structure, fertility, and productivity of soils. Finally, they must meet the social needs of farmers and their families and strengthen rural communities in a sustainable manner."

Aims of sustainable agricultural systems

The similarities among various definitions of sustainable agriculture lead us to a set of common goals for farms or agricultural ecosystems. Sustainable agricultural systems should maintain or increase biological and economic productivity. Biological productivity is required to feed individual farm families and the population of non-farm people. Economic productivity is required to provide income for farmers. They should enhance the efficiency of use of inputs. Increased efficiency leads to greater economic productivity because it lowers the input requirements. It also lessens adverse environmental impacts both on and off the farm. Sustainable agricultural systems should be both stable and resilient. Stability reduces risk and leads to continuity in income and food supply by fulfilling the short-term needs of farmers without endangering natural resources. Resilience permits adaptation to changes in the biophysical and socio–economic environments. They should be environmentally compatible to avoid contamination and to minimize adverse environmental impact on adjacent and downstream environments. Finally, they should be socially compatible with local people and political economies.

Integrated farming systems

The components of a farm must be integrated for it to persist. A dynamic conceptual model for a farm (Fig. 1) is a simple mass flow model, which consists of two parallel paths of flow through the farm. One pathway is a socio–economic flow which has inputs of land, labor, capital, culture, and knowledge or information. Land, labor, and capital are the traditional inputs of economic models, and labor and capital-based inputs are often considered interchangeable, e.g. herbicides are substitutes for hand-weeding or cultivation (National Research Council (NRC), 1989). Tradition, information, and knowledge can also be substituted for labor and capital, e.g. appropriate intercropping can minimize weed problems.

Outputs of the socio–economic pathway are fulfillment of livelihood needs and include income, health, knowledge, social stability, and a sense of community. Income is the most commonly recognized output of a farm, but livelihood goes well beyond income because farming is also a lifestyle. The formulation and perpetuation of values and the persistence of a community sense are probably equally important outputs.

The importance of internal socio–economic processes on a farm are often overshadowed by the inputs from which they develop and the outputs they produce. The farm provides collateral for loans to cover capital expenditures

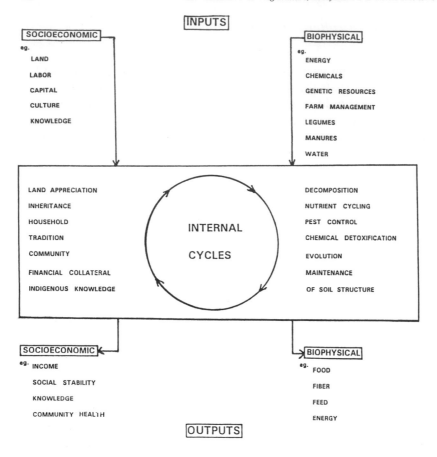

Fig. 1. Conceptual model of a farm.

and operating costs. Land appreciation is a hedge against inflation, provides money in the bank, and represents a retirement account. Farming provides a lifestyle that binds a household together and is a heritage that can be passed to future generations through inheritance of the farm. Indigenous knowledge is typically a product of generations of adaptation of farming practices to local environments. Family, tradition, and formulation of values are all processes that are facilitated on the farm.

The biophysical flows run in parallel with the socio–economic flows. Physico–chemical inputs include energy for operations such as tillage, harvesting and storage, and agrochemicals for fertilization and pest control. Biological inputs include organic matter such as crop residues, animal manure, legume nitrogen, cover crops, rotations, and cropping patterns. Physico–chemical and biological inputs are familiar components of conventional agriculture, but their values have not usually been optimized or integrated. Less obvious, but of great importance is knowledge or information. Use of genetically improved varieties of crops and animals, management of farming practices, and biological pest control are examples of information. These inputs often provide substitutes, that are both cost effective and environmentally benign, for inputs of labor, energy, or chemicals. For example, seeds for a pest-resistant line of a crop costs a farmer little more than for one without resistance but eliminates the need for pesticides and the labor to apply them and stabilize production.

The internal processes occurring within the farm are important in maintaining its natural resource base and its ability to support continued production. Nutrients are recycled and made available for plant growth through the soil decomposer community (Wild, 1973). The combined activities of the soil fauna and flora stabilize soil structure and increase the permeability, long-term fertility, and resilience of the soil (Coleman et al., 1984). Nitrogen-fixing bacteria increase the soil stock of nitrogen (Miller, 1990). Soil organisms assist in control of pests through various mechanisms including competition, antagonism, predation, and allelopathy (Baker and Cook, 1974). Soil microorganisms can metabolize toxic organic compounds (Edwards, 1966). The maintenance of a healthy soil ecosystem provides for evolution and adaptation of the soil biota and community to changing conditions. Sustainable systems add greater biological diversity to nutrient cycles, thereby increasing their longevity and promoting interactions between the decomposer and consumer communities.

The socio–economic and biophysical flows run in parallel and are interdependent. The ability of a farmer to produce food and income is tied directly to internal biological processes. Adoption of the best farm management practices and indigenous knowledge depend upon natural resources and are the results of interactions between farmers and their biophysical and socio–economic environments. Socio–economic and biophysical inputs can often substitute for one another. For example, weeds can be controlled by herbicides or by cover crops, rotations, hand-weeding or cultivations (NRC, 1989). It is the understanding of the pattern of parallel flows and the interdependence between the flows that defines an integrated farm or farming system. For a farm to be sustainable, the flows must be coupled. When either flow or the interaction between them is sufficiently disrupted, the system becomes unsustainable. Consequently, intervention by science and technology in an integrated farm requires consideration of the whole system and the socio–economic and biophysical flows within and through the system.

Such an integrated farm model represents an agroecosystem. Both agricultural sciences and ecology have contributed to our knowledge of agroecosystems. The traditional agricultural sciences have developed a large body of information on the components of agroecosystems, and many of the current ecological theories were developed in agroecosystems. For example, much of the literature on insect ecology (e.g. Andrewartha, 1954) resulted from work done in agricultural systems with a goal of pest control. Agroecology is a relatively new discipline that integrates the techniques and paradigms of ecology with the practices of agricultural sciences for the study of agroecosystems.

We realize that a farm is only one level of an agroecosystem. The farm sits within a hierarchy of levels that includes, for example, fields; farms, households, watersheds and regions (Conway, 1985; Lowrance et al., 1986) In the interest of simplicity, we have restricted our discussion to the farm level, but many of the concepts apply equally to other subsystems and levels of the hierarchy.

Commonalities and differences among agroecosystems

Traditional agroecosystems

There are many inputs into farming systems, all of which can be related to cropping patterns, soils, cultivations, supply of nutrients or pest control. When planning research on integrated sustainable agricultural systems, it is important to differentiate between those processes that are common to all agroecosystems and the inputs, factors, and processes that are of a more regional and specific nature. Commonalities are appropriate for basic research with broad application. Regional and specific differences are appropriate for applied research and development. We suggest that biological diversity and nutrient cycling mechanisms are the key factors or common processes in the function and persistence of traditional agroecosystems (Fig. 2) which are based on relatively low inorganic supplements and depend upon rotations and cultural practices. Biodiversity of plants, animals, and microbes supports production of crops and stock in such traditional agroecosystems. The specific biological components that contribute to biodiversity and the capacities, rates, and patterns of nutrient cycling differ among agroecosystems, but biodiversity and nutrient cycling are common to all. By contrast, soil types, climate, and farming practices are regional in nature and differ considerably among agroecosystems. Production, rates of recycling, and constituents of the biological community vary regionally with climate and edaphic characteristics. Different political economies and cultural histories result in different farming practices. In traditional systems the biological community and the farming practices are generally well adapted to local conditions (Logsdon, 1984). It is this adaptation to local environments that is termed site specificity.

The specific functions of the commonalities of biological diversity and nutrient cycling are further subdivided in Fig. 3. Biodiversity, may be spatial, e.g. the soil biota and cropping patterns, or temporal, e.g. rotations, control of pests, weeds, and diseases through mechanisms such as competition, predation, shading, allelopathy, antagonism, and antibiotics. Biological diversity of soil organisms supports nutrient cycling because the decomposition of or-

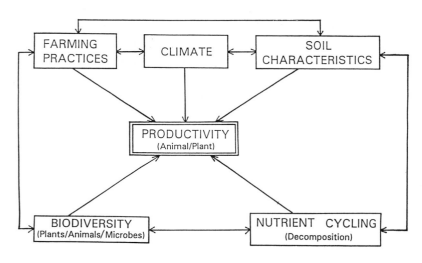

Fig. 2. Key factors and relationships in the management of traditional agroecosystems.

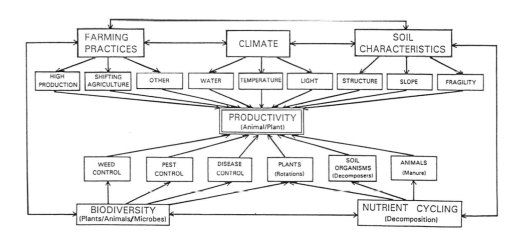

Fig. 3. Key processes and functional relationships in a traditional agroecosystem.

ganic matter is a biotic process. Plants and animals are the sources of organic matter and invertebrate animals, such as earthworms, physically disrupt and mix it, and microbes mineralize nutrients (Brady, 1990). Biodiversity is relatively easily manipulated by farming practices. The regional soil physical characteristics and climate are much more difficult to manipulate other than by cultivation or additions of organic matter or both. Similarly, the only component of climate that can be manipulated is water availability through irrigation. Productive potential and specific farming practices vary regionally and are related to the cultural and economic characteristics of farmers (Logsdon, 1984).

Conventional agroecosystems

In high-input or conventional agriculture (Fig. 4), pesticides typically replace many of the functions of biodiversity as controls of pests, weeds, and diseases. Inorganic fertilizers substitute for biological nutrient cycles that are based on biological decomposition and mineralization of organic matter. These practices may destabilize agroecosystems and create increasing dependence upon chemical inputs (Edwards, 1990). Pesticides often kill beneficial organisms that control pests biologically through antagonism, competition, or predation (Thibodeaux and Field, 1985). Fertilizers can suppress microbial populations and the enzymes that they produce which are key factors in

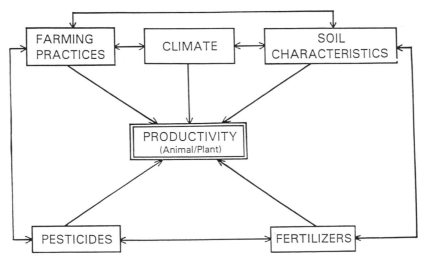

Fig. 4. Key factors and relationships in the management of a conventional monocultural agroecosystem.

nutrient cycling (Fukuaka, 1985); thus the biologically based commonalities are replaced by chemicals. The biological productivity of the agroecosystem thus becomes dependent upon chemical inputs and alternatives for manipulating the system to provide sustainability are reduced.

Sustainable agroecosystems

We propose that a key alternative to this chemical dependence (Fig. 5), which has become common in developed countries, is to maximize the contributions of biodiversity to pest control and nutrient cycling and to supplement this with agrochemicals, only as necessary to attain optimal productivity with minimal inputs (NRC, 1989). Such practices of pest control are termed integrated pest management. Using the same principles, Edwards and Grove (1991) proposed an analogous term for management of nutrients — 'integrated nutrient management'. This approach capitalizes on the adaptive features of traditional systems and incorporates additional advantages of conventional and innovative biologically-based technology.

It is important to recognize that there is a strong link (Fig. 5) between the availability of organic matter and both biodiversity and nutrient cycling (Palm et al., 1987). The practice of removing organic matter from the land for fuel and other purposes is a serious constraint to long-term sustainability in many

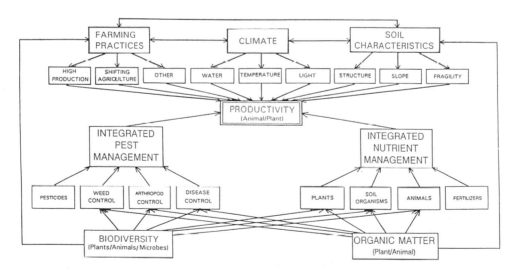

Fig. 5. Key processes and functional relationships in a sustainable agroecosystem.

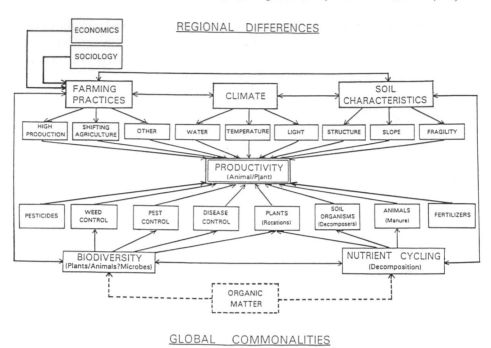

Fig. 6. Social and economic processes and functional relationships in a sustainable agroecosystem.

developing countries (Oram, 1988). The most sustainable farming practices and components of the managed biodiversity can be developed only by understanding the functioning of the agroecosystem and how social and economic conditions of farmers and their climatic and edaphic environments impact upon overall crop and animal productivity (Fig. 6). No matter how well an agroecosystem functions biologically, it will only be sustainable, if it is socially and economically sound (Altieri, 1987).

Simulation model of agroecosystems

There is often confusion about which processes are common to all agroecosystems and which are local or regional in nature. Also confusing is the relationship between the functioning of an agroecosystem and a natural ecosystem. In an attempt to clarify these points, a simulation model is offered (see Figs. 7, 8, 9) to illustrate agroecosystems by relating them to the structure and functions of a barn. A brief discussion of relevant ecological paradigms precedes description of the model to establish a context for the discussion.

Ecologists usually consider three levels of organization, populations, com-

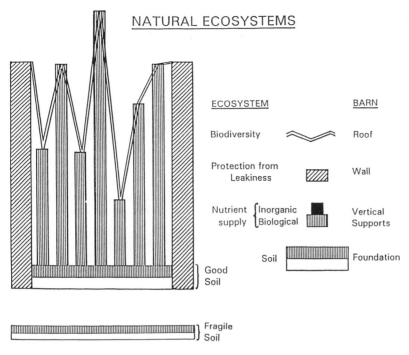

Fig. 7. Cross-section of a barn as an analogy representing a natural ecosystem.

munities, and ecosystems. Populations are groups of organisms belonging to the same species, occupying a contiguous area and defined in terms of reproduction, birth rates, mortality rates, and immigration and emigration rates (Smith, 1990). Species live in complex associations or communities controlled by interactions between their members. A community is linked closely to its environment. Both climate and soil affect a community and a community in turn affects the soil and its own internal climate or microclimate (Whittaker, 1975). Energy and matter are taken from the environment to run dynamic processes, transferred from one organism to another in the community and released back to the environment. A community and its environment, when it is treated as a functional system of complementary relationships and transfer and circulation of energy and matter is called an ecosystem (Whittaker, 1975).

Undisturbed ecosystems mature with time. Biomass, species diversity, and spatial heterogeneity (canopy layers) typically increase (Dover and Talbot, 1987). Trophic interactions are generally more complex in mature systems, and there is a higher degree of organization (Odum and Margalef, 1969).

Ecosystems function through the capture of solar energy and production of biomass by plants (primary production) and consumption of plant material by other organisms (secondary production) to produce secondary biomass. The ratio of productivity to biomass decreases with increasing maturity and

TRADITIONAL AGROECOSYSTEMS

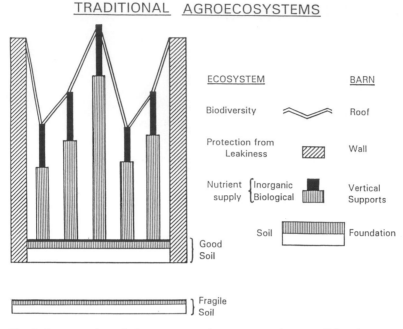

Fig. 8. Cross-section of a barn as an analogy representing a traditional agroecosystem.

the captured energy is more fully used within ecosystems as they mature (Dover and Talbot, 1987). In mature ecosystems, energy is used more for maintenance than for production of additional biomass (Dover and Talbot, 1987). Immature ecosystems have high production to biomass ratios and living material accumulates (Dover and Talbot, 1987). Most agroecosystems are maintained at early successional stages to exploit this production.

A mature ecosystem is relatively closed, i.e. essential minerals and other nutrients stay within it (Odum, 1971; Miller, 1990). Ecosystems become more efficient at trapping and holding nutrients as they mature. In contrast, an agroecosystem does not hold nutrients tightly, and they leak readily from the system (Dover and Talbot, 1987).

Mature natural ecosystems have a diverse flora and fauna, whereas immature ones are less diverse. Agroecosystems, particularly those that do not involve crop rotations, tend to have a much less diverse flora, and the populations and interactions of the invertebrates and soil microorganisms also tend to be much less complex. This has strong implications for pest, disease, and weed control and for the decomposition of organic matter in soil. The maintenance of stable communities depends upon the diversity of competitors and predators that prevent explosions of populations of pest species. When diversity of plants and animals is decreased, pest populations have greater opportunity for rapid growth (NRC, 1989). Decomposition of organic matter de-

HIGH INPUT/CONVENTIONAL AGROECOSYSTEMS

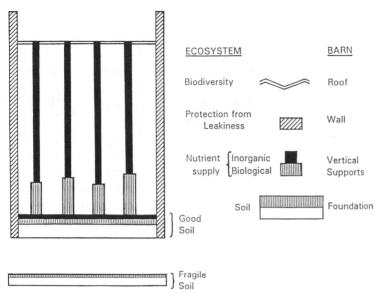

Fig. 9. Cross-section of a barn as an analogy representing a conventional monocultural agroecosystem.

pends upon a diverse community of plants, animals, and microbes (Tivy, 1990). When any component of the community is eliminated, there are effects upon rates of decomposition.

Some ecosystem functions are common to all ecosystems whether natural or agricultural. These are biodiversity of microorganisms and plants, and invertebrate and vertebrate animals, and biological recycling of nutrients from plant and animal organic matter, which is mediated by complex interactions between soil organisms. The functioning of agroecosystems can be understood better by a simulation model that likens ecosystems to barns. All barns are built for storage purposes and their stability, potential lifespan, and carrying capacity can vary with their design and the materials from which they are built. If we consider a natural ecosystem, a traditional rotational agroecosystem and a monocultural agroecosystem as three types of barns (see Figs. 7, 8, 9), it is easy to identify their efficiencies, and shortcomings and to relate them to one another. Thus, biodiversity can be considered as having a similar function to the roof of the barn, the soil as the foundation, the nutrient supply as the vertical support pillars, and the wall as a protection from 'leakiness'.

In a natural ecosystem (Fig. 7), the nutrient supply is ensured by the strength and diversity of the biological recycling of nutrients from the soil surface where organic matter is deposited. This can be considered as strong vertical supports for a barn firmly implanted in a solid foundation of biologically active soil. The diversity of plants and animals results in stability and spatial heterogeneity (Weil, 1990), which can be considered as a roof that protects the contents of the barn from adverse environmental shocks and stresses. The maturity of the ecosystem can be likened to the wall — a thick wall minimizes leakages from the system.

In a traditional agroecosystem (Fig. 8), that involves rotations and other polycultures, there is still a reasonable degree of biological recycling of nutrients although these are often supplemented by inorganic inputs (Tivy, 1990). The supplements reduce the strength of the supports, e.g. dependence upon external inputs exposes the system to risks of their unavailability. The biodiversity is less, the system is thus more susceptible to shocks and stresses than a natural ecosystem, but there is still sufficient diversity for reasonable protection from shocks and stresses (Conway, 1985). Within the barn analogy, a severe storm or heavy snowfall might collapse the roof of the traditional agroecosystem, but little damage would result from typical storms. The walls of the barn are thinner than those of its natural analogues, and, therefore, leakage is greater.

A conventional monocultural agroecosystem (Fig. 9) is much more dependent upon inorganic supplements (Francis et al., 1986), and the vertical supports are therefore thinner, taller, and less stable. They need continual renovation. The biodiversity is less and provides less protection from shocks and stresses. Thus, the roof of the barn can be considered to be flatter and much less able to support storms and snows. There is a much greater tendency for nutrients to leak from the system as the walls are thinner.

The differences among the three ecosystems is greater on the poorer or more fragile soils that have fewer biological resources and reserves. The monocultural agroecosystem often has a greater carrying capacity than its traditional counterparts, but the sustainability and stability of the monoculture is much less because it rests on more fragile supports. It can be likened to a barn that needs continual renovation and that may eventually fall down, if the renovations fail to protect its foundations which are susceptible to erosion.

Practical tools for developing sustainable agricultural systems

Manipulation of inputs

Currently we have limited our discussion to an understanding of the requirements for development of sustainable agricultural systems. The concept of the existence of two global commonalities, biological diversity and nutrient cycling, among agroecosystems, which we present in this paper, is supported by the ecological literature, anecdotal accounts of indigenous practices, and the rapidly emerging literature of agroecology. The basic research challenges are to understand these main commonalities, to focus research responses upon them, and to capitalize upon our knowledge of these principles in designing productive, stable, and equitable sustainable agricultural systems for general application to all regions.

Questions concerning biodiversity are both quantitative and qualitative. We have little understanding of how much and what kinds of biodiversity are required. We have only a limited understanding of the mechanisms by which biodiversity stabilizes ecosystems. While we understand many of the principles of nutrient cycling, there are many gaps in our knowledge. Fundamental questions — such as which fractions of organic matter are labile and active in recycling and which factors control the rates of mineralization of plant organic matter (Edwards, 1987) — are as yet largely unanswered and need resolution by research to provide the basis for better management of nutrient recycling.

Organic matter is the foundation of all nutrient cycles. The labile and available nutrients within an ecosystem are largely contained in organic matter (Tivy, 1990). They depend upon a community of producer and decomposer organisms. The producers are typically well managed, since they are the crops and animals that are produced by the farm and have economic value. The decomposer organisms in soils are less obvious, and their importance is often ignored (Coleman et al., 1984). The fundamental issues concerning efficient use of organic matter are leakage of nutrients from agroecosystems and the rates of decomposition. Organic matter and the nutrients it contains are lost from soils by run-off and mineralization (Tivy, 1990), both of which can be controlled by appropriate tillage practices. Loss of nutrients to mineralization is also controlled by assuring sufficient inputs of plant or animal material to maintain the soil organic matter reserves (Woodmansee, 1984).

Legumes are important in maintaining adequate soil organic matter and increasing the soil nitrogen supply. They are components of virtually all native terrestrial ecosystems and are typical of traditional agroecosystems (Powers, 1987). Legumes can be used as food or forage crops and managed as intercrops or fallows. In addition, they protect the soil from run-off, wind, and water erosion and frequently improve infiltration.

The importance of tillage practices to sustainability is becoming more obvious with time. Conventional mold-board ploughing reduces biodiversity, exposes organic matter to run-off and oxidation, and increases the soil's susceptibility to erosion (NRC, 1989). Alternative tillage practices, such as no-till (direct drilling), various forms of conservation tillage and ridge tillage, offer alternatives that minimize erosion and conserve soil (Altieri, 1987), soil organic matter (Crossley et al., 1984), and biodiversity (Stinner and Blair, 1990) while maintaining crop productivity (Phillips et al., 1980). They also play a role in insect and disease control (Edwards, 1990).

Living and dead mulches and trap crops protect the soil from erosion (Reganold et al., 1987), capture nutrients and hold them within the agroecosystem (Hoyt and Hargrove, 1986), and serve as sources of organic matter (Hoyt and Hargrove, 1986). They can also be utilized as intercrops (where their benefits also include weed control (Edwards, 1987)) or as relay crops that are planted at the end of the crop cycle.

Agroforestry systems use leguminous and other trees to provide alternative crops (Steppler and Lundgren, 1988), produce animal forage and fuel (Spears, 1988), recycle nutrients for crop use (Altieri, 1987), and protect soil from wind and water erosion (Altieri, 1987). Traditional agricultural systems often contain trees (Farrell, 1987). Some of the higher value products of small farmers in developing countries are tree crops, e.g. coconut, papaya, coffee, tea, cloves (Altieri, 1987). Many staples are also tree crops, e.g. bananas. Leguminous trees provide nitrogen and recycle other nutrients when grown as alley crops with field crops between (Wilson et al., 1986). There is a need to design appropriate mixes and patterns of trees and to integrate animals into agroforestry systems to optimize productivity and sustainability (Baker et al., 1989).

Plant biodiversity plays an important role in pest, disease, and weed management. Rotations and various forms of polyculture are effective in controlling pests, diseases, and weeds (Altieri, 1987). The control of weeds by rotation of crops in traditional agroecosystems is well known in principle, but practical recommendations are poorly documented. Living mulches control weeds and minimize the need for herbicides (Regnier and Janke, 1990). Invertebrate animal pests such as insects are also controlled by rotations (Altieri, 1987). For instance, in the USA the life-cycle of the corn borer is sufficiently disrupted by rotations to eliminate the need for insecticides (Pimentel, 1993).

Polycultures and other management patterns such as strip cropping (Edwards et al., 1992) that increase biological diversity control pests (Altieri, 1987). Increases in structural diversity within the crop canopy leads to greater diversity in insects and less damage from insect pests (Stinner and Blair, 1990).

Integration of animals into agroecosystems offers further diversity and stability. Animals can utilize plant products that are not useful to humans. Since their harvest is not seasonal, they serve as a 'bank account' and can be harvested when cash needs arise.

Development of integrated systems

The challenge is to design practical integrated systems of crops and animals that can be adapted to regions, minimize energy-based inputs and have long-term sustainability. The tools described above are examples of alternatives which can be integrated into systems to meet the challenges. With suitable integration of such tools, the low productivity of traditional systems can be greatly improved.

Socio–economic inputs into the development of sustainable agricultural systems

In addition to the ecological aspects of sustainability, farmers of both sexes must perceive benefits to themselves and their children, if improved agricultural systems are to be sustained (World Bank, 1990). This requires an ongoing involvement of farmers in the planning, experimentation, and extension of agricultural innovations.

A good start in involving farmers in agricultural research and development was made in many of the farming systems research and development (FSRD) efforts during the past 10–15 years (Bunch, 1982; Chambers et al., 1989). Building on the successes of FSRD should be a prime goal of the new concern for sustainability.

There are several entry points for involving farmers meaningfully and profitably in efforts to develop sustainable agriculture. Indeed, in our view, there can be no truly sustainable agriculture without a genuine commitment on the part of the rural people who ultimately implement whatever scientists and developers propose. In the following sections, we specify examples from one country (Indonesia). Our goal is to provide some concrete examples. These kinds of examples, however, are available from all over the world.

Indigenous knowledge

A first step in developing sustainable agricultural systems is to draw on the existing body of knowledge and practices. For instance, Clay (1988) has practical suggestions on how to implement this approach. Look at what rural people are doing and identify problems, but also identify opportunities that existing systems present. Build on what local people know and do because the experience, knowledge, and capabilities of rural people which are grossly under recognized and under utilized at present, represent a priceless resource.

Studies of several indigenous farming systems in Indonesia indicated that agronomic and ecological research into their practices and approaches represented a potentially more useful approach to agriculture in the humid tropics than did techniques that were imported from temperate areas (Colfer, 1983; Colfer et al., 1988). In Sumatra, Sitiung's Minangkabau communities had an agricultural system which was sustainable under conditions of low population density. It combined shifting cultivation of upland rice and rubber with permanent paddy rice. Their approach was to choose an appropriate piece of land for the particular crops they wished to grow on the basis of soil, water, plant cover, and topographical features. This practice differed markedly from the common western approach which involves overcoming constraints rather than recognizing alternatives which could avoid or lessen the impacts of constraints.

Collaboration with farmers

Scientists and rural people each have knowledge and experience of great value, but both can be more successful in their respective endeavors, if they work together. Working with farmers throughout the research process and modifying experiments as indicated by observations and responses of farmers will ensure easier acceptance of resultant products. Scientists mus be willing to test their findings in the real world of farmer's fields.

Farmers often experiment (Chambers et al., 1989), and their results can provide useful foundations for further scientific exploration. Regular interaction with farmers alerts scientists to the problems that farmers face both in their own experiments and in implementation of recommendations by researchers.

During the implementation phase of a TropSoils collaborative trial, 20 Indonesian farmers complained that hoe cultivation, to which all had agreed, was too difficult. The first inclination of the project staff was to malign them as lazy and unreliable. However, daily exposure to the fields allowed the staff to discover that below the soil surface was a tangled mass of nearly impenetrable roots, remnants of the former forest. The research design was subsequently altered. On the same experiment, three farmers refused to hoe because of personal labor constraints. The tillage factor, which was included in the analysis because of the farmers' intransigence, was statistically more significant than the fertility factors which were the original objects of study (Wade et al., 1985). Farmers also contribute positively. When the project staff proposed a green manure of *Callopogonium* spp., the farmers suggested an alternative that enriched the soil equally well and also produced an edible bean.

Biodiversity in the human context

The cultural aspect of biodiversity is rarely addressed. Currently the world contains thousands of distinct cultural groups. Yet there are powerful social, political, and educational forces at work which encourage increasing standardization of cultures. Just as the biological diversity that characterizes humid tropical forests is now recognized as worthy of protection and preservation, so is the diversity of cultures that characterize the Earth. This diversity of cultures should be seen as a risk reducing mechanism and a resource to be conserved.

Implementation of sustainable agricultural systems research

A major objective of sustainable agricultural systems research is the integration of available information to solve complex problems of agricultural development. The lack of systems research has been identified (NRC, 1989) as a major constraint to adoption of alternative farming practices and as a necessity for development of an alternative and more sustainable agriculture. While the value of systems approaches has been recognized increasingly over the past decade, few crop and livestock production systems have been studied in detail. Because of the extreme variability among agroecosystems, it is important to identify the major components of any agroecosystem and the regional factors that are constraints.

We propose a simple conceptual framework for the conduct of integrated agricultural systems research. The steps include:

(1) Description of the target agroecosystem including its goals, boundaries, components, functioning, interactions among components, and interactions across its boundaries.

(2) Detailed analysis of the agroecosystem to determine factors which limit or could contribute to attainment of productive and sociological goals.

(3) Design of interventions and identification of actions to overcome the constraints.

(4) On-farm experimental evaluation of interventions.

(5) Review effectiveness of newly designed systems.

(6) Redesign as necessary.

We believe that all steps should be conducted on farms by an interdisciplinary team of agricultural, social, and ecological scientists and with full participation of farmers.

The description of an agroecosystem must be based upon discussions with farmers and upon recommendations of the disciplinary specialists. Techniques for describing agroecosystems have been reported in the literature (e.g. Conway, 1985; Altieri, 1987; Harwood, 1987). Understanding the farmer's goals is especially important as the role of the proposed interventions is to help the farmer attain these goals. Description of the boundaries and limitations of the agroecosystem is essential in providing focus for study, but should not limit understanding of its interactions with adjacent ecosystems or with local, regional, national, and international political economies. Description of the components of the system is the traditional occupation of many agricultural scientists, but a description and analysis of interactions among components requires farmer participation as well as interdisciplinarity.

Although the descriptive phase is largely qualitative, the analytical stage takes maximal advantage of quantitative information. The proposed descriptions may lead to development of hypotheses that require experimental study for resolution and quantification. For example, if nitrogen is suspected to be a limiting factor, then nutrient response studies may be required. If losses to pests are hypothesized as a key factor, they can be quantified experimentally and integrated management measures recommended for the pests identified. The result of the analytical phase is to approach a more detailed understanding of the limitations to the attainment of the farmer's goals.

The design phase involves forming hypotheses about appropriate interventions that will contribute to farmers' goals. It is a deductive process based upon the description and analysis of the system and the final design represents the best collective judgments of the study team and the participating farmers.

The evaluation phase tests the proposed interventions empirically. Effects must be measured in terms of the goals of the system and trade-offs among goals must be determined for any proposed intervention. Interdisciplinary involvement and participation are essential in a successful evaluation phase.

We hypothesize that if a similar descriptive and analytical process is employed for the study of different agroecosystems in a number of agroecological zones, the commonalities among them will emerge. Furthermore, if the commonalities identified are verified by further experimentation in farmer's fields, they become, in essence, global principles. We contend that there is currently sufficient evidence to suggest that maintenance of biological diversity and nutrient cycling mechanisms are likely to be global principles and worthy of hypothesis status in the design of sustainable agricultural systems. The regional influences may differ greatly among agroecosystems and may assume major importance in some, but, action on the commonalities will be of value in most agroecosystems. Interdisciplinarity is fundamental to such action.

Acknowledgments

We would like to express our gratitude to John E. Bater for preparing the diagrams in this paper so well and to Margaret Fredericks for her long hours in the library and her editorial assistance. We thank the National Research Council of the National Academy of Sciences and the US Agency for International Development who contributed financial support for the preparation of this paper.

References

Altieri, M., 1987. Agroecology. Westview Press, Boulder, CO, 185 pp.
Altieri, M. and Merrick, L., 1988. Agroecology and in-situ conservation of native crop diversity

in the Third World. In: E.O. Wilson (Editor), Biodiversity. National Academy Press, Washington, DC, pp. 361–368.

Andrewartha, H., 1954. The Distribution and Abundance of Animals. University of Chicago Press, Chicago, 580 pp.

Baker, K.F. and Cook, R.J., 1974. Biological Control of Plant Pathogens. W.H. Freeman, San Francisco, CA, 433 pp.

Brady, N., 1990. The Nature and Property of Soils. MacMillan, New York, 750 pp.

Bunch, R., 1982. Two Ears of Corn: A Guide to People-Centered Agricultural Improvement. World Neighbors, Oklahoma City, 250 pp.

Chambers, R., Pacy, A. and Thrupp, L.A., 1989. Farmer First: Farmer Innovation and Agricultural Research. Intermediate Technology Publications, London, 218 pp.

Clay, J.W., 1988. Indigenous Peoples and Tropical Forests: Models of Land-use and Management from Latin America. Cultural Survival, Cambridge, MA, 116 pp.

Coleman, D.C., Cole, C.V. and Elliott, E.T., 1984. Decomposition, organic matter turnover, and nutrient dynamics in agroecosystems. In: R. Lowrance, B. Stinner and G. House (Editors), Agricultural Ecosystems: Unifying Concepts. John Wiley, New York, pp. 83–104.

Colfer, C.J.P., 1983. Change and indigenous agroforestry in East Kalimantan, Borneo. Res. Bull., 15(1,2): 70–87.

Colfer, C.J.P., Gill, D. and Agus, F., 1988. An indigenous agroforestry model from West Sumatra: A source of insight for scientists. Agric. Sys., 26: 191–209.

Committee on Agricultural Sustainability for Developing Countries, 1987. The transition to sustainable agriculture: an agenda for AID, 29 pp.

Committee on Agricultural Sustainability for Developing Countries, 1989. The transition to sustainable agriculture: a two-year review of AID's agriculture and rural development programs and Agenda for the 1990s.

Conway, G.R., 1985. Agroecosystem analysis. Agric. Adm., 20: 31–55.

Crossley, D.A., House, G., Snider, R., Snider, R. and Stinner, B., 1984. The positive interactions in agroecosystems. In: R. Lowrance, B. Stinner and G. House (Editors), Agricultural Ecosystems: Unifying Concepts. Wiley, New York, 310 pp.

Douglass, 1984. The meaning of agricultural sustainability. In: Douglass, Agricultural Sustainability in a Changing World Order. Westview Press, Boulder, CO, pp. 3–29.

Dover, M. and Talbot, L., 1987. To Feed the Earth — Agro-Ecology for Sustainability in a Changing World Order. World Resources Institute, 122 pp.

Edwards, C.A., 1966. Pesticide residues in soils. Residue Rev., 13: 83–132.

Edwards, C.A., 1987. The concept of integrated systems in lower input/sustainable agriculture. Am. J. Alternative Agric., 2(4): 148–152.

Edwards, C.A., 1990. The importance of integration in sustainable agriculture systems. In: C.A. Edwards, R. Lal, P. Madden, R.H. Miller and G. Hause (Editors), Sustainable Agriculture Systems. Soil and Water Conservation Society, Ankeny, IA, pp. 249–264.

Edwards, C.A., Brust, G.E., Stinner, B.R. and McCartney, D.A., 1992. Work in the United States on the use of cropping patterns to promote natural enemies of pests. Aspects Appl. Biol., 31: 139–148.

Edwards, C.A. and Grove, T.L., 1991. Integrated nutrient management for crop production. Toward sustainability. Natural Research Council, Washington, DC.

Farrell, J.G., 1987. Agroforestry systems. In: M. Altieri, Agroecology, Westview Press, Boulder, CO, 185 pp.

Francis, C., Harwood, R. and Par, J., 1986. Potential for regenerative agriculture in developing world. Am. J. Alternative Agric., 1(2): 65–73.

Fukuoka, M., 1985. The Natural Way of Farming. Japan Publications, Tokyo, Japan, 284 pp.

Hart, R.D., 1987. Ecological framework for multiple cropping research. In: C.A. Francis (Editor), Multiple Cropping Systems. MacMillan, New York, 383 pp.

Harwood, R.R., 1987. Agroforestry and mixed farming systems. In: A.E. Lugo, J.R. Clark and R.D. Child (Editors), Ecological Development in the Humid Tropics: Guidelines for Planners. Winrock International, Morrilton, AR, 362 pp.

Hoyt, G.D. and Hargrove, W.H., 1986. Legume cover crops for improving crop and soil management. Hortic. Sci., 21: 397–402.

Jackson, W., 1980. New Roots for Agriculture. Friends of the Earth, San Francisco, CA, 294 pp.

Jacobson, J.L., 1988. Environmental Refugees Yardstick of Habitability, Worldwatch Paper No. 86. Worldwatch Institute, Washington, DC, 88 pp.

Logsdon, G., 1984. The importance of traditional farming practices for a sustainable modern agriculture. In: W. Jackson et al. (Editors), Meeting the Expectations of the Land, North Point Press, San Francisco, pp. 3–18.

Lowrance, R., Hendrix, P.F. and Odum, E.P., 1986. A hierarchical approach to sustainable agriculture. Am. J. Alternative Agric., 1: 169–173.

Miller, G.T., Resource Conservation and Management. Wadsworth Publishing Company, Belmont, CA, 546 pp.

Miller, R.H., 1990. Soil microbiological inputs. In: C.A. Edwards, R. Lal, P. Madden, R.H. Miller and G. Hause (Editors), Sustainable Agricultural Systems. Soil and Water Conservation Society, Ankeny, IA, pp. 614–623.

National Research Council (NRC), 1989. Alternative Agriculture. National Academy Press, Washington, DC, 490 pp.

Odum, E., 1971. Fundamentals of Ecology. W.B. Saunders, Philadelphia, PA, 535 pp.

Odum, E. and Margalef, 1969. The strategy of ecosystem development. Science, 164: 262–270.

Oram, P., 1988. Building the agro-ecological framework. Environment, 14–34.

Palm, O., de Silva, A. and Guildford, M., 1987. Nutrient cycling in paddy rice of a traditional farming system. Trop. Agric., 87: 129–133.

Phillips, R.E., Blevins, R.L., Thomas, G.W., Frye, W.W. and Phillips, S.H., 1980. No tillage agriculture. Science, 208: 1180–1113.

Pimentel, D., 1993. Environmental and economic benefits of sustainable agriculture. In: M.G. Paoletti et al. (Editors), Proceedings of the International Conference on Agroecology and Conservation Issues in Temperate and Tropical Regions. Elsevier, Amsterdam, in press.

Powers, J., 1987. Legumes: their potential role in agricultural production. Am. J. Alternative Agric., 2(2).

Reganold, J.P., Elliott, L.F. and Unger, Y.L., 1987. Long-term effects of organic and conventional farming on soil erosion. Nature, 330: 370–372.

Regnier, E. and Janke, R., 1990. Evolving strategies for managing weeds. In: C.A. Edwards, R. Lal, P. Madden, R.H. Miller and G. Hause (Editors), Sustainable Agriculture Systems. Soil and Conservation Society, Ankeny, IA, pp. 174–202.

Rodale, R., 1983. Breaking new ground: the search for sustainable agriculture. Futurist, 17(1): 15–20.

Russell, E.W., 1983. Soil Conditions and Plant Growth. Longmans, London, 360 pp.

Smith, R.L., 1990. Ecology and Field Biology. Harper & Row, New York, 85 pp.

Spears, J., 1988. Preserving biological diversity in the tropical rain forests of the Asian region. In: E.O. Wilson (Editor), Biodiversity. National Academy Press, Washington, DC, pp. 393–402.

Steppler, H. and Lundgren, B., 1988. Agro-forestry: now and in the future. Outlook Agric., 17(4): 146–151.

Stinner, B. and Blair, J., 1990. Ecological and agronomic characteristics of innovative cropping systems. In: C.A. Edwards, R. Lal, P. Madden, R.H. Miller and G. Hause (Editors), Sustainable Agriculture Systems. Soil and Water Conservation Society, Ankeny, IA, pp. 123–140.

Thibodeaux, F. and Field, H.F. (Editors), 1985. Sustaining Tomorrow. University Press of New England, Hanover, 186 pp.

Tivy, J., 1990. Agricultural Ecology. Longman Scientific and Technical, Essex, UK, 280 pp.

Weil, R., 1990. Defining and using the concept of sustainable agriculture. J. Agron. Educ., 19(2): 126–130.

Whittaker, R.H., 1975. Communities and Ecosystems. Second Edition. MacMillan, New York, 162 pp.

Wilson, G.F., King, B.T. and Mulongoy, K., 1986. Alley cropping: trees as source of green manure and mulch in the tropics. In: J.M. Lopez-Real et al. (Editors), The Role of Microorganisms in a Sustainable Agriculture. AB Academic Publishers, UK, 292 pp.

Woodmansee, R., 1984. Comparative nutrient cycles of natural and agricultural ecosystems. In: R. Lowrance et al. (Editors), Agricultural Ecosystems. Wiley, New York, pp. 152–156.

Agriculture, Ecosystems and Environment, 46 (1993) 123–134
Elsevier Science Publishers B.V., Amsterdam

Designing the future: sustainable agriculture in the US

Charles A. Francis[a,*], J. Patrick Madden[b]

[a]*University of Nebraska, Lincoln, NE 68583-0910, USA*
[b]*Madden Associates Consulting, Glendale, CA 91209, USA*

Abstract

Global agriculture is entering a challenging and difficult period with an increasing human population and an accelerating need for food, fiber, feed, and raw materials for other industries. This challenge will need to be met on fewer hectares of available land and a reduced supply of the fossil fuel inputs that have catalyzed the increased productivity of the past five decades. Agriculture in some forms has negative and lasting effects on the environment. The research and education community is seeking a more resource-efficient, sustainable system of food production that has less negative impact on the environment. This system is characterized by increased resource use efficiency, greater reliance on internal or renewable resources, increased short- and long-term profitability, enhancement of soil productivity, minimal negative environmental impact, and social viability for families and communities. Agricultural research over the last half century has contributed many components to sustainable productivity, but its focus in the future will be more on systems, interactions among components, and the impact of the activity on the broader environment and community. Education in agriculture is moving from a concentration on memorizing detail and cook-book approaches to a development of creative thinking and problem solving skills. We are building the capacity to access and apply a wide range of information resources. There is a growing congruence of classroom teaching and adult education in extension, an evolution that will lead to better curriculum planning for a life-long educational and learning experience. All the key players in US agriculture will take greater responsibility for their own learning in this system, being empowered to conduct both on-station and on-farm research, design learning activities, and evaluate progress and applications of information to real world challenges. This paper describes what is happening in the US in research, in teaching, and in extension. We also envision a new paradigm for education in the future. Instead of preparing to react or adjust to a predictable future, an empowered rural populace can begin to design a more desirable future. With increased focus on scarce resources, fragility of the environment, and the lessons of nature, we can take creative approaches to systems design and begin to make decisions today to create a more sustainable future for tomorrow.

Introduction

"Avoiding the wholesale breakdown of natural systems requires a shift from the pursuit of growth to that of sustainable progress."

---

*Corresponding author.

"No economy can be called successful if prosperity comes at the expense of future generations and if the ranks of the poor continue to grow."

(Postel and Flavin, 1991)

We operate today in a global economy. The success of a farmer in the state
of Nebraska, USA or in Arusha Province, Tanzania is dependent not only on
rainfall and crop prices in that region, but on climatic, economic, and political events around the world. New crop varieties and technologies move across
borders, oceans, and continents. We no longer live in an isolated information
environment. We live in an increasingly interdependent world in which many
climatic changes and economic decisions in one country can ultimately affect
humans and other species around the globe. In today's world, long-term sustainable progress is possible only if we take into account a multiplicity of interacting resource, environmental, economic, social, and political issues.

For many individual farmers and ranchers in the US, there has been economic success in crop and livestock production. Others have not been able to
sustain their agricultural operations, or can do so only with an outside job.
There is an awareness of the fragility of an industry that is highly dependent
on government subsidies for success. Pressure is growing from other elements
of the population who perceive an adverse impact of some agricultural practices on food quality and the environment. What is needed is a move toward
more sustainable production systems that reduce the risk to farmers, both in
the short- and long-term, and that can prosper in a future with greater competition for scarce fossil fuel and land resources. Systems need to enhance soil
productivity, while minimizing the impact on the off-farm environment. The
development of these systems is both a challenge and an opportunity. There
is a strong research and educational commitment to agriculture in the US,
and this is increasingly focused on sustainable practices in crop and animal
production.

There is a rich history of component research into crop rotations, insect
and pathogen resistance in crops, grazing patterns and forage species improvement, efficient water use by crops, and other sustainable practices. This
is expanding into areas of system performance, resource use efficiency, and
minimizing environmental impacts of crop and livestock production. Schaller
et al. (1986) evaluated 6413 CRIS projects in the US that were "designed
and conducted to investigate problem areas and ideas to aid and improve on-
farm production of crops and livestock". He concluded that 88% of the projects so selected from this base were considered 'neutral' in their application,
and thus useful to a wide range of producers with different philosophies. Of
the total number of projects, 24 (or 4%) were considered appropriate to organic systems that avoid chemical products and encourage crop rotations, animal and green manures, and some biological pest control. Which of these

categories could be considered to include 'sustainable farming systems' depends on definition.

The land grant university system in the US has a long tradition and commitment to classroom teaching and extension programs that promote increased productivity and profitability of farming and ranching systems. The close integration of research, teaching, and extension has been a model for other national agricultural systems around the globe. In this university and federal system, there is a growing emphasis on challenges of efficient resource use, more reliance on renewable inputs, reducing negative environmental impacts, and reducing economic risks in the production system. Those in the system are beginning to explore the social and ethical dimensions of technologies that are developed and introduced. Finally, there is a new focus on greater participation by all those involved in agriculture, on empowerment of clients, and on tapping a wider range of information resources to help farmers and ranchers make decisions. It is this new research and educational environment that will lead to a more sustainable agriculture in the US, and may provide examples that will be useful elsewhere. In this paper, we explore the challenges and opportunities in designing the future of US agriculture.

Challenges in research

Past agricultural research in the US

The majority of agricultural research in the US has focused on improving components of farming systems. This has been a highly successful endeavor over the past half century, with increased crop yields per acre, improved rates of gain in livestock, and an impressive productivity per unit of labor. The labor efficiency of this system is due, in part, to high levels of capital investment and large inputs of fertilizers, chemicals, and irrigation in some areas. Most systems have been profitable, at least in the short-term, although recent success in most major crops has been due to government subsidies. Specific research contributions to the productivity and profitability of agricultural systems with reduced fossil fuel inputs include: selection of insect and pathogen resistance in crops; soil testing, precise yield goals, and careful budgeting of nitrogen and other nutrients; reduced or zero tillage and planting into crop residues; crop rotations to improve soil physical structure, nutrient status, and pest management; rotational grazing and greater reliance on forage-based animal nutrition; genetic selection for greater nitrogen use efficiency in crops; selection for drought tolerance and resistance to extremes in temperature.

These advances and others have contributed to greater resource use efficiency in agriculture. There is a growing movement in the research community to place more emphasis on research of total systems, on how these components fit together and interact, and how specific crop rotations and systems impact the larger environment. One of the most significant new initiatives with federal support is the low-input, sustainable agriculture (LISA) program.

Research with LISA funding

The low-input, sustainable agriculture program is a federally funded initiative that began in 1988 as a partial implementation of the 1985 Food Security Act, better known as 'the Farm Bill'. This research and education program of the US Department of Agriculture (USDA) "responds to an emerging interest by many farmers for a more cost-effective and environmentally benign agriculture" (Madden et al., 1990). Although the funding has been modest (less than $9 million for the first 2 years) there has been a strong response from the university and private community with people submitting proposals. For example, in the first 2 years there were 802 proposals submitted and 105 funded. Examples of these projects from the four regions of the US include the following:

(1) Biological and cultural control of root diseases in cotton as an alternative to chemical pesticides, a project in California and Arizona that included plant pathologists, extension specialists, and farmers. Beneficial bacteria and fungi that are antagonists to root pathogens were introduced using a low-volume watering system, and this was found effective in controlling post-emergence damping off in cotton. Treated plants were almost twice as tall as plants in non-treated plots early in the season (Madden et al., 1990).

(2) The transition to alternative farming systems was studied by Smolik and Dobbs (1991) in South Dakota. Over a 5 year transition period, three crop rotation systems were studied in both row crops and small grains, including alternative systems with no commercial fertilizers or pesticides. Average net incomes for the 5 years over all costs except management were highest for the alternative systems, and these have both agronomic and economic promise in the Great Plains.

(3) Weed control with greatly reduced rates of herbicide was studied in Arkansas in a number of crops and locations (Madden et al., 1990). Dr. Ford Baldwin and farmer cooperators found that "herbicide inputs can be greatly reduced by substituting mechanical weed control, spraying herbicide in narrow bands, targeting herbicide to most susceptible weed species, and making very early application". For example, they were able to reduce herbicide costs in cotton from $21 to only $2.30 per acre. In another example, about one-third of soybean farmers in the state have adopted similar low-rate practices at a cost saving of $7 million annually.

(4) Cover crops for New England vegetable growers have been studied by university and non-government researchers in a three-state area in order to

use legumes as both a source of nitrogen and a cultural practice to suppress weeds (Madden et al., 1990). Both biological and economic data are being collected on farms to assess the resulting enterprise budgets for conventional and alternative cover crop systems on New England vegetable farms.

Characteristics of future research

Building on a foundation of historical research on resource use efficiency, along with the new initiatives under LISA funding, we anticipate a number of research directions that will become more prevalent as we focus on developing a sustainable agriculture. Some of these relate to the types of technologies that will be needed, others to the methods that will be used to find and test new technologies, and a small number to the evaluation of the research process itself and its impact on society. This list should be considered as an example of research priorities, and is not exclusive of other ideas.

Types of technologies to promote sustainability

(1) Efficient use of all production inputs, with substitution where possible of renewable resource-based technologies for those that depend on fossil fuels.

(2) Increased reliance on biological components in production systems, through understanding of biological structuring in the soil/plant/pest environment.

(3) Efficient exploitation of nutrient and water cycling in production systems, based on information gleaned from the functioning of natural systems.

(4) Reliance on genetic manipulation of crops and livestock species to adapt to their environments, rather than high-cost domination of the environment to fit them.

(5) Creation of more genetically diverse production systems that capture and use nutrients and water through a greater part of the total growing season and year.

(6) Design of diverse agricultural landscapes within fields, both in space and time, through multiple species of crops and trees on contour patterns.

(7) Increased use of crop rotations and crop/animal integration within each farming operation to enhance nutrient cycling and increase income diversity.

(8) Greater focus on the integration of technologies into production systems and the interactions among components, as well as the impact of systems on the off-farm environment.

Methods to find and test new technologies

(1) Greater reliance on interdisciplinary teams in research, with team members representing other agencies outside the university.

(2) Increased use of on-farm research, with the farmer (or rancher) play-

ing an active role as a full member of the team that chooses treatments, collects data, and interprets results.

(3) Development of practical on-farm designs and statistical procedures for analyzing and summarizing results.

(4) Enhanced integration of research, teaching, and extension with information moving rapidly among specialists in each of these areas.

(5) Increased recognition of team research and rewards for solutions to practical challenges facing farmers and ranchers in US agriculture.

Evaluation of the impact of research on society

(1) Determination of how research priorities are decided, how funding is allocated, and how results of university research impact society.

(2) Evaluation of production efficiency measures other than yield per acre, net income per acre, and yield per unit of labor; these may include yield or return per unit of capital, per farm, per unit of renewable or fossil fuel energy.

(3) Measurement of research contributions to quality of rural life, to availability of food in the US, to the security of the global food system.

(4) Exploring the potentials for value-added products on farm and in the rural community as a route to increased and more stable returns to basic agricultural production.

(5) Evaluation of social impacts of agriculture and research, including potentials for owner/operator farms and for entry level farmers or ranches.

Much of our research under way in the US contributes to these directions. There is likely to be an increase in team research and an expansion in the concept of who belongs to those teams. Closer working relationships among farmers, ranchers, scientists in industry and university, and extension specialists will help the industry achieve the goal of long-term resource use efficiency and sustainability of food production. A number of these options have been described in more detail in the National Research Council (1989) report on alternative agriculture and in recent books by Edwards et al. (1990) and Francis et al. (1990).

New directions in education

Past educational programs in agriculture

In a similar way to the resource efficiency and profitability dimensions of past research in US university programs, classroom teaching and extension meetings have focused on appropriate use of inputs for profitable crop and livestock production. The emphasis has evolved over several decades from maximum yields to maximum economic yields to the current recommendations for best management practices. Not only formal classes in agriculture but also adult extension activities have included topics such as

(1) Choice of crop hybrids and varieties according to the total length of frost-free growing season and available rainfall or irrigation.

(2) Fertilizer decisions on nitrogen and other nutrients according to soil tests, appropriate yield goals, and careful budgeting of nutrients from all sources.

(3) Reduced primary tillage and no-till/reduced tillage planting systems to minimize moisture loss and cut down on energy use.

(4) Rotations of herbicides to prevent selection of resistant weeds, and band application of fertilizers and pesticides to reduce costs of chemical weed control.

(5) Careful adjustment of planters, cultivators, and combines to establish and maintain precise plant densities and harvest the crop as completely as possible.

(6) Precise crop and livestock cost accounting, separation of expenses and receipts by enterprises, and overall financial management of farms and ranches.

(7) Conscious and deliberate marketing of crops and livestock, use of futures and other ways of forward marketing, as compared to selling to the closest buyer.

These educational programs, whether in the classroom or the conventional extension meeting, have followed the classical land grant university format of lecture setting, use of some visuals, and time for discussion and questions by the audience or class. Visual aids have improved immensely over the past two decades, and some practical exercises in laboratory or field trips have used innovative approaches and participation. The entire system generally is constrained by lack of funding for up-to-date projection equipment and facilities for computer-aided instruction.

Educational projects under LISA funding

A number of demonstration and educational activities have been funded by the LISA program described above. Most of these involve both university extension specialists and county agents, working together with farmers and farmer organizations in a given state or several-state area. Some examples include the following.

(1) A northeast Dairy Farm Forage Demonstration Project was designed in New York to "teach farmers, feed dealers, and consultants how to implement a total year round forage management system utilizing intensive rotational grazing techniques" and dry hay and other forages, and at the same time reducing tillage on highly erodible soils. The economic analysis of these demonstrations further promotes the educational goals of the project.

(2) An overview Sustainable/Low-Input Agriculture videotape was produced by the Agronomy Department and the Educational Television Network in Nebraska that described a number of resource-efficient practices for the north–central region (Madden et al., 1990). There was a definition of sustainable agriculture, video footage of a number of farmers describing their practices in fertility, tillage, and alternative crops. The video was distributed to the 12-state region, and has been used extensively in classroom and extension meetings in Nebraska.

(3) Enhancing Farmer Adoption and Refining a Low-input Intercropping Soybean–Wheat System was a project implemented in Mississippi and Arkansas with university, USDA, SCS, and farmer participation (Madden et al., 1990). The project was designed to enhance adoption of a low-input, reduced tillage intercropping system for wheat–soybean, to test and demonstrate alternative cultural practices such as row spacing, and show the enhanced profitability of this system compared with conventional monocropping.

(4) Options for Reducing Production Inputs in the Cereal and Legume Growing Regions of the northwest were studied on station and on farm in a six-state area where these systems are important (Madden et al., 1990). Long-term plots from Pendleton, Oregon with a 50 year history were evaluated and summarized in a farmer-friendly report. Another ten locations throughout the region had legume trials for at least 8 years. In addition, farm surveys were used to acquire practical results from farmers using legumes and cereals. Results were presented in conferences, a newsletter, farm tours, and other presentations around the region.

The emergence of a farming system approach

About a decade has passed since the Farming System Research/Extension approach began to have an impact on international programs in agriculture (see Gilbert et al., 1980). This emphasis on total farm systems rather than individual enterprises and on recognition of the farmer as more than a passive receiver of information has done much to revitalize our thinking in agricultural education. The approach identifies the farmer as a key actor on the stage, from identification of research priorities to choice of experimental treatments to design of trials and collection of data to interpretation and demonstration of results. A new awareness of the importance of total systems is apparent in a renewed classroom emphasis on crop rotations, use of alternative fertility approaches, and integrated pest management. There are many capstone courses at the senior level where concepts and information from lower division and specialized classes are brought together in a focus on the total system. The same change in emphasis is seen in land grant extension programs, and one private initiative that has been highly successful is holistic resource management (Savory, 1988). Some of the characteristics of this approach include:

(1) Emphasis on the integration of component technologies into production systems, as activity that farmers and ranchers must pursue in the design of an enterprise.

(2) Setting long-term goals for family and farm, and translating these into short-term activities designed to reach those goals.

(3) Reduction of variable costs through eliminating or curtailing use of purchased inputs when they are not absolutely needed for crop or livestock production.

(4) Careful examination of each enterprise and its place in the entire crop or crop/animal array of enterprises, distribution of labor and equipment use, efficient management.

Agroecology and sustainable development

A new discipline that is emerging simultaneously in the US and in Latin America is agroecology. This combination of agriculture with ecology brings the science of crop/livestock production together with the study of natural systems to see what can be done in designing 'managed ecosystems'. One long-term approach is to use nature as guide and standard, and to design mixtures of economic species with others that supply nutrients or assist in pest management, in some emulation of natural ecosystems (Jackson, 1980). A number of textbooks or edited volumes have been published in the past few years that provide a conceptual and a practical base for this work (Poincelot, 1986; Altieri, 1987; Gliessman, 1990; Carroll et al., 1990). These books emphasize the relationship between ecology and managed cropping systems, including dimensions of component integration:

(1) Study of natural ecosystems and their spatial and temporal resource use as a guide to potential economic cropping systems in the same ecoregion (Jackson, 1990).

(2) Adjustment of stocking rates and time of grazing in specific pastures according to rainfall, season, and growth of forage (Savory, 1988; Murphy, 1990).

(3) Careful evaluation of nutrient cycling in soils between the biological and the mineral phases and the dynamic interactions among nutrient pools (Stevenson, 1986).

(4) Evaluation of multiple cropping systems in tropical regions and their potential applications in the temperate zone (Francis, 1986; Gliessman, 1990).

(5) Development of classes in agroecology that examine the resources of a

region, the human population, and the potentials for agricultural systems to meet human needs.

(6) Development of integrated curricula that bring together the principles of ecology with the science of crop and animal production in a series of classes or programs in extension.

(7) Expanding the concept of ecosystem to include the human population and its social structures and needs as vital components of the system (Altieri, 1987).

These perspectives and methods of looking at agricultural systems are just beginning to permeate the planning and programming of agricultural classroom teaching programs in the US. They are also found in a very limited number of extension activities to date. The strides made toward integrating agroecology with sustainable development in Latin American universities and non-governmental programs can serve as a model for similar programs in other parts of the world. The potentials and future applications of this approach are reviewed in the final section.

Designing the future

The perspectives on research and education outlined above represent some of the current thinking and a few ongoing programs in sustainable agriculture in the US. What will we find in the future, as we further develop these concepts of agroecology, sustainable development, and participatory education in the classroom and extension activities? The following list of philosophies, approaches, programs, and networks includes ideas from other sectors of the economy, integration of thoughts and discussions, and some 'blue sky projections' toward the future. In general, they focus on highly participatory activities and broad ownership of the agenda for research and education. The ideas include a proactive empowerment of all those in agriculture to design and develop the future. Some examples follow:

(1) Broadening the intellectual and practical base for research, including a balance of on-station and on-farm research with farmers and ranchers as key players on each team.

(2) Developing new criteria for system evaluation, including yield or net return per unit energy, per unit capital, per unit of renewable/non-renewable resources.

(3) Developing indices for farm level, regional, or national productivity such as the 'gross sustainable product' used by Indonesia.

(4) Evaluating systems based on ecological impact as well as direct human benefit in the short-term, with careful assessment of the long-term costs of off-farm impact.

(5) Collecting and evaluating a broad range of information resources such as farmer trials and industry tests, and including a rigorous review before including these in a comprehensive database for agriculture.

(6) Including risk analysis in evaluation of system success, both risk to producer and risk to those who work in the field and to those who live nearby.

(7) Developing systems based on integrated biological and farm structuring, nutrient and water cycling, crop and animal integration, and long-term biological planning (Francis et al., 1986).

(8) Incorporating social perspectives into the analysis of cropping and livestock systems, including long-term farm stability and rural community viability.

(9) Focus of universities on catalyzing the learning process rather than on conventional teaching approaches; move toward development of critical thinking and communication skills, problem solving processes, and information access and evaluation.

(10) Evaluating new thinking on topics such as Gaia, deep ecology, and ecofeminism (Lovelock, 1979) and seeking potential applications in sustainable agroecosystems.

These are among the ideas that will influence the directions taken in research and education in the future. Systems will be highly participatory, and people will be empowered to direct their own research and interpret results. The location specificity of information can be determined by careful observation or scouting of individual fields and farms. The efficient sharing of information will provide the basis for a biologically sustainable and economically viable owner/operator farming sector. Most important, people in agriculture will have the perspectives and tools needed to determine their future in this industry. To be able to envision the future, decide on a more desirable structure and make decisions today to make that future a reality tomorrow is a positive and empowering alternative. This is one possible direction for a sustainable agriculture in the US, a direction that is highly dependent on a positive vision and a willingness to change and pursue opportunities as they arise. Our farming and ranching populations have shown the capacity to grow and adapt, and they will be challenged to accelerate positive change in the future.

References

Altieri, M.A., 1987. Agroecology: the Scientific Basis of Alternative Agriculture. Westview Press, Boulder, CO, 227 pp.

Carroll, C.R., Vandermeer, J.H. and Rosset, P.M. (Editors), 1990. Agroecology. McGraw Hill, New York, 641 pp.

Edwards, C.A., Lal, R., Madden, P., Miller, R.H. and House, G., 1990. Sustainable Agricultural Systems. Soil and Water Cons. Soc., Ankeny, IA, 696 pp.

Francis, C.A. (Editor), 1986. Multiple Cropping Systems. Macmillan, New York, 383 pp.

Francis, C.A., Harwood, R.R. and Parr, J.F., 1986. The potential for regenerative agriculture in the developing world. Am. J. Alternative Agric., 1: 65–74.

Francis, C.A., Flora, C.B. and King, L.D. (Editors), 1990. Sustainable Agriculture in Temperate Zones. John Wiley, New York, 487 pp.

Gilbert, E.H., Norman, D.W. and Winch, F.E., 1980. Farming systems research: a critical appraisal. MSU Rural Dev. Paper No. 6, Dep. Agric. Econ., Michigan State Univ., East Lansing, Michigan, 135 pp.

Gliessman, S.R. (Editor), 1990. Agroecology: Researching the Ecological Basis for Sustainable Agriculture. Springer, New York, 380 pp.

Jackson, W., 1980. New Roots for Agriculture. Univ. Nebraska Press, Lincoln, Nebraska, 150 pp.

Jackson, W., 1990. Agriculture with nature as analogy. In: C.A. Francis, C.B. Flora and L.D. King (Editors), Sustainable Agriculture in Temperate Zones. Wiley, New York, pp. 381–422.

Lovelock, J.E., 1979. Gaia: a New Look at Life on Earth. Oxford Univ. Press, Oxford and New York, 157 pp.

Madden, J.P., Deshazer, J.A., Magdoff, F.R., Pelsue, N., Loughlin, C.W. and Schlegel, D.E., 1990. LISA 88–89: Low-Input Sustainable Agriculture Research and Education Projects Funded in 1988 and 1989. US Dep. Agric., CSRS, Office of Special Projects and Program Systems, Washington, DC, 133 pp.

Murphy, B., 1990. Pasture management. In: C.A. Francis, C.B. Flora and L.D. King (Editors), Sustainable Agriculture in Temperate Zones. Wiley, New York, pp. 231–262.

National Research Council, 1989. Alternative Agriculture. National Academy of Sciences, Washington, DC, 448 pp.

Poincelot, R.J., 1986. Toward a more sustainable agriculture. AVI, Westport, CT, 241 pp.

Postel, S. and Flavin, C., 1991. Reshaping the global economy. In: L.R. Brown et al. (Editors), State of the World 1991. W.W. Norton, New York, pp. 170–188.

Savory, A., 1988. Holistic Resource Management. Island Press, Covelo, CA, 564 pp.

Schaller, F.W., Thompson, H.E. and Smith, C.M., 1986. Conventional and organic related farming systems research: an assessment of USDA and state research projects. Special Rep. 91. Agric. Home Econ. Exp. Stn., Iowa State Univ., Ames, IA, 74 pp.

Smolik, J.D. and Dobbs, T.L., 1991. Crop yields and economic returns accompanying the transition to alternative farming systems. J. Prod. Agric., 4: 153–161.

Stevenson, F.J., 1986. Cycles of Soil: Carbon, Nitrogen, Phosphorus, and Sulfur, Micronutrients. Wiley, New York, 380 pp.

Agriculture, Ecosystems and Environment, 46 (1993) 135–145
Elsevier Science Publishers B.V., Amsterdam

Do we need a new developmental paradigm?

Thurman L. Grove[a,*], Clive A. Edwards[b]

[a]*College of Agriculture and Life Sciences, North Carolina State University, Raleigh, NC 27695, USA*
[b]*Department of Entomology, Ohio State University, Columbus, OH 43210, USA*

Abstract

Many question the applicability of past developmental approaches or paradigms to the current issues of agricultural development and suggest that new paradigms are required. In this paper we characterize the common agricultural science-led development paradigm of the past and a newly emerging approach that we have titled the socio–ecological paradigm. We examine the assumptions, successes and failures of these paradigms and propose changes in approach that seem appropriate to address current developmental issues. These changes include a move towards systems based research and development and greater involvement of rural people who are the targets of developmental efforts. A systems protocol is proposed that attempts to reconcile the appropriate interface between systems and component based research and development.

Introduction

Perhaps the most significant result of the report of the World Commission on Environment and Development (World Bank, 1987), is its suggestion that our past developmental paradigms may require modification to allow further development in the future. The reasons for the World Commission's suggestions are straightforward and include principally the increasing incidence of poverty and environmental degradation.

In his foreword to the World Development Report (1978) Robert McNamara stated: "The past quarter century has been a period of unprecedented change and progress in the developing world. And yet despite this impressive record, some 800 million individuals continue to be trapped in what I have termed absolute poverty: a condition of life so characterized by malnutrition, illiteracy, disease, squalid surrounding, high infant mortality and low life expectancy as to be beneath any reasonable definition of human decency". Today, 15 years later, the essence of McNamara's statement continues to be relevant. The numbers of poor in developing countries have continued to increase and in many countries the productivity per capita is declining. In addition,

---

*Corresponding author.

numerous publications (e.g. the World Commission on Environment and Development, 1987; Brown, 1990; World Resources Institute, 1990) have documented the continuing degradation of natural resources that is particularly prevalent in many areas of the tropics. Such degradation further erodes the capabilities of the poor to improve their livelihoods and threatens the life-support systems of the entire planet.

The challenges of changing our approach to development include identification of the successful elements of development that we should continue and formulation of new hypotheses about how we might improve and accelerate development while conserving renewable natural resources. In the following discussions we examine the assumptions underlying our current science-led development paradigm, comment on its successes and failures and explore some alternatives that seem worthy of further examination.

Science-led development

Since the Second World War most developmental programs in developing countries have relied heavily on a science-led developmental paradigm. This paradigm holds that the first step in development is intensification of agriculture. The resulting increased productivity supplies labor and capital for initiation of other industries. It also lowers the price of food, which effectively increases the incomes of the poor and frees resources for improvements in their living standards. This progression is familiar to many because it reflects the history of Europe and North America. The Green Revolution of the 1960s was based on this model. Most scientists, especially agricultural scientists, and developmentalists subscribe to it at least to some degree. When the model fails, the typical response is a call for even greater investments in agricultural research and development. However, rarely are the assumptions and practices that underlie the paradigm examined and articulated. We shall discuss selected assumptions and practices.

Assumptions and practices

In the practice of science-led development it is typically assumed that humankind is food limited — that human carrying capacity is determined by food availability. This assumption has persisted at least since the time of Malthus. We are unaware of conclusive evidence that humans are food limited and are not convinced that it is worthwhile to debate the ultimate limiting factors for humankind. However, when this assumption is embraced by, and coupled with, the power and self-interest of the agricultural community, it leads to investments in agricultural science-led development to the exclusion of other models. Consequently, there is little information on alternatives.

Agricultural scientists typically assume that the role of agriculture is to feed all the inhabitants of the globe. Virtually every report (e.g. Technical Advisory Committee of the Consultative Group on International Agricultural Research, 1988; York et al., 1989) that discusses continued investments in agricultural research and development, argues that the challenge is to meet the food needs of an expanding population. While this may be a worthy goal, it has consequences on the focus of agricultural research and development. This assumption centers attention on yield increases of staple grains, especially rice, wheat and maize. Much of the world's population, for example those in tropical humid Africa, does not depend on these grains and thus does not benefit from such research and development. A focus on staple grains is not a good strategy for lifting rural small-holders from poverty. Their holdings are small and the value of grains is low. Increased production may contribute to their food security, but usually is too small to produce significant additional income. A better strategy might involve stabilization of their food supplies with grain and production of income with market crops of higher value. The assumption that agriculture's role is to feed the inhabitants of the globe also directs research and development efforts to the better lands and to the farmers with access to resources to exploit such lands. It is implied that benefits will trickle down to poorer people as food costs decrease. However, the productivity of the poor rarely increases because the technologies developed for better-endowed lands and more prosperous farmers are rarely appropriate to the poorer land and the impoverished farmers who are typically found upon them.

Agricultural science-led development seems to assume that rural community inhabitants should be farmers. This assumption directs research and development towards production of commodities rather than to the welfare of people. It leads to technologies that require a farmer to invest all of his or her resources in the farm. It fails to recognize that there are few, if any, full-time farmers. Poor farmers in developing countries typically spend less time and attention on their farming activities as their incomes increase (Kusterer, 1989).

Fundamental to the science-led paradigm is the assumption that agriculture is the engine of economic growth. This implies that development occurs only through economic growth. These beliefs are so deeply ingrained in the practice of development that they have inhibited the examination of other options.

The current practice of agricultural science is reductionist. For nearly four centuries Cartesian philosophy has dominated science and led to a belief that humans can always overcome ignorance. The method of overcoming ignorance is based on simplification through inductive reasoning and dissection of natural systems. When this philosophy is applied to development it often leads to a belief that technological solutions are all that is required. Development is considered much as a seed that is placed in the soil and faces constraints to germination and growth. The objective of such an approach then becomes one of overcoming the constraints, usually through technology. Such an approach is usually a gross simplification of a very complex phenomenon that includes social, economic and political factors in addition to complex biophysical factors.

The practice of science-led development is typically driven by the 'first biases of professionals' (Chambers, 1987). Scientists and other professionals seek the answers to questions which are professionally satisfying within the norms of their professions and disciplines. In agricultural sciences, success is defined typically either as yield increases or as anecdotal explanations of how things are ordered, i.e. a pattern or component analysis.

Successes and failures

Science-led development has had tremendous impacts. The success of the Green Revolution is legendary. Since 1950, global economic productivity has quintupled. During the same period grain yields in North America and western Europe quadrupled and increased globally by a factor of 2.6 (Brown, 1990). Many consider (e.g. Ruttan, 1990) agricultural development in most of Asia to be a success. During the 1970s and 1980s, real agricultural gross domestic product increased annually by an average of more than 3% (Vyas and James, 1988). On the better soils yields from farmers' fields are nearly equal to those obtained on agricultural experiment stations (Byerlee, 1989). Yields of rice, wheat and maize, especially in east Asia, are equal to or greater than those obtained in many developed countries (Ruttan, 1990).

Despite the success of the Green Revolution, in many areas of North America, western Europe and Asia, it has failed in many locations and has led to inequities within and among countries. The rate of growth in per capita gross national product is falling, especially in Africa. During the latter half of the 1980s the average global per capita grain production fell by 7% (Brown, 1990). Differentials between yields on experiment stations and farmers' fields are much higher in Africa than in Latin America or Europe (Lyman and Herdt, 1988). Even in India, where the success of the Green Revolution is often cited, much of the population lives below the poverty standard and much of the land is degraded beyond productive use.

The intensive technology that characterizes the Green Revolution has resulted in considerable environmental degradation. Its cropping practices cause excessive erosion, the pesticides and fertilizers used contaminate underground aquifers and surface waters and pests are consuming an increasing proportion of production as they develop resistances to pesticides.

Socio–ecological developmental paradigm

During the past decade, many scientists who work in development have recognized the limitations of the science-led developmental paradigm

(Chambers, 1983: Dover and Talbot, 1987; Lugo et al., 1987; Conway and Barbier, 1990; National Research Council, 1991a,b). In the face of persistent poverty, malnutrition and environmental degradation they formulated new hypotheses and applied different methodologies in attempts to improve the livelihoods of the poor and to conserve the environment. The new methods used were typically employed by anthropologists, sociologists and ecologists and considered the full range of activities of rural people, not just their agricultural pursuits. The experiences of the last decade are sufficient to encourage us to look at characteristics and successes of these new experiments which we have titled the socio–ecological paradigm.

Assumptions and practices

The socio–ecological paradigm is founded on an assumption that human development must be adaptive. Development processes and components can be successful only if they are compatible with local biophysical and socio-cultural environments. Success is measured best in terms of improved livelihood and extends beyond food and income to include, for example, good health and the freedom of choice. The role of science in this paradigm is to enhance the rate of adaptation. Delegation to a supportive role is often unsatisfying to scientists who believe that science should provide leadership in development.

This assumption can be frightening because it implies that there are a nearly infinite number of adaptive scenarios. Many feel that development of technological packages for each of these situations will exceed the capacity of the global agricultural support system. However, such concerns are a result of thought processes that assume that technological packages are the driving forces of development. The concept is less frightening, when considered in terms of process. Process can be adapted relatively easily, if local people are involved, since they have historically coped with their environments by adaptation. Their processes for coping with change are probably the closest to what is required for further development. If processes are adaptable and compatible with the environment, there will be a choice of components and the ease of their adaptation will be a criterion of selection. The processes that are adapted will lead to components that require less development before implementation.

There is a danger in this paradigm as it assumes that technology is on the shelf and well ahead of political and economic development. Such an assumption has led in the past to complacency about the need for support for scientific research. During the past decade, the US Agency for International Development and the World Bank have significantly reduced their support for agricultural research and technology (Paarlberg and Lipton, 1991). However, development is dynamic in nature. Future development requires present and continuing development of its technical bases and foundations.

The socio–ecological pattern differs from the science-led developmental paradigm most strikingly, in its assumption that the role of agriculture is a means of rural livelihood. This is a sharp contrast to the assumption that agriculture's main role is to feed the globe. It focuses on a diversity of crops and products, off-farm sources of income and other sectors of activity such as forestry and gathering of natural products by farmers. Such an approach more easily accommodates the development of marginal lands that require alternate species and management practices.

Practitioners of the socio–ecological development paradigm have shown that full-time farmers are rare and that most small farmers will reduce their investments in agriculture, if alternative sources of income are available (e.g. Kusterer, 1989). Such a finding has tremendous implications for how a farmer uses his or her resources and on the research that supports the farmer's development. Adaptation of agricultural technologies to cultural environments requires a recognition of this behavior pattern. Technologies that reduce a farmer's options for investments or income off the farm will probably face obstacles to implementation.

The socio–ecological paradigm is holistic and typically views a broad range of farmers' livelihood needs and the activities and infrastructure that support them. Included in these are the whole farm, external markets and employment, health and social needs. This approach is concerned more with processes and linkages among components than with the components themselves. Experiments are evaluated in terms of systems properties such as stability, equity, productivity and sustainability (Conway, 1985, 1987). Success is measured as improvement in livelihood, which is complex and much more difficult to measure than yield.

Most practitioners of the socio–ecological approach subscribe to the participatory process in which rural people have a significant voice in the design and implementation of projects and programs that are intended to affect their lives (Chambers, 1983: Altieri, 1987; Clay, 1988). This practice is predicated upon beliefs that indigenous people are a valuable source of information and will most readily change their behaviors, if they have a sense of ownership in the anticipated changes and products.

Successes and failures

Generalizations about the successes and failures of the socio–ecological developmental paradigm are difficult. Such a paradigm is relatively new and there are few practical examples. The details of the examples that exist vary a great deal. The underlying sciences of anthropology, social science and ecology are young relative to agriculture and science-led development. The state-of-the-art has also been limited by resources and technology. Holistic approaches require large amounts of data and are, therefore, costly. Analysis of data has often been limited by technology. This situation is changing rapidly with advances in information and computer sciences.

However, the results are promising despite the limitations. Numerous examples of successful programs exist and are discussed in several publications (e.g. Tull et al., 1987; Gradwohl and Greenberg, 1988; Reid et al., 1988). Common themes in these examples of success include the empowerment of local people, use of novel and simple technology, innovations that are adapted to local biophysical and cultural environments and multi-sectoral approaches to development.

Issues

Approaches to development are varied and could undoubtedly be associated with numerous other paradigms. The two paradigms discussed above are not intended to represent the full spectrum of possibilities, nor are they presented as competitors. Rather they are offered as examples to provide a framework for discussion. The fundamental issue that emerges from the juxtaposition of these two paradigms is that we have little understanding of why development interventions succeed or fail in any particular situation. Each of these paradigms and others have succeeded and failed in a number of diverse locations and situations. Generalizations about the causes of successes and failures are difficult. However, it seems essential that developmentalists increase understanding of what works so that we can increase the efficiency of developmental investments.

As research and development advance towards an increased understanding, we suggest that changes in approach are required. These changes may not constitute a new paradigm, but rather require a new balance and integration among the successful elements of current paradigms. This seems appropriate since development is an applied science. Its role is to integrate the best of its components — not to advance the state of the individual component disciplines. This is the essence of applied science. For example, engineering, natural resource management and medicine have relied historically upon their component disciplines for advancement by choosing and assembling appropriate parts that have resulted from their fundamental sciences. Such assembly involves deductive reasoning and a systems approach to problems. It is a significant change in philosophy towards synthesis and integration and away from reductionism and towards the multiple goals that are held by people, the objects of development. If peoples' goals are to be fulfilled we should explicitly recognize them and understand the roles and limitations of the component disciplines in their attainment. This is not a new paradigm, but simply the best use of current knowledge.

We propose that there are at least two fundamental reasons for adopting integrated systems approaches in agricultural research and development. First integration through systems approaches can increase scientific rigor. Inductive and reductionist science frequently advances through increased analytical rigor. For example, the sciences of ecosystem ecology and economics typically advance by building mathematical or statistical models. This increased mathematical rigor is considered progress since it offers detailed descriptions of components, but does so at the expense of descriptions of the interactions among components. As a consequence, the models frequently have little relationship to real problems (Kerr, 1976; Hall, 1991). For example, Pavlov's classic work with dogs resulted in detailed descriptions of the relationships between physiology and behavior, which is perhaps a worthy objective, but it offered little aid to the practical problems of feeding a dog. Development is much the same. Detailed descriptions of the components of disadvantaged people are perhaps useful, but when considered in isolation of the complexity of their livelihoods, they often fall short of providing alternatives for improved livelihoods. A deductive and systems approach helps overcome this situation by focusing on the whole. A deductive and iterative process of describing components and interactions among them can lead to identification of constraints to development and facilitate formulation of the breadth of interventions that are required to allow development.

Second, and related, the systems approach overcomes the limitations of linear reasoning. Linear models that ignore feedback processes are the norm in formulation of economic policy and agricultural development. Policy tools such as fertilizer subsidies often ignore limitations of infrastructure and fail when inputs cannot be delivered on a timely basis. Research and development on costly and intensive technologies that are focused on yield increases ignore evidence that small farmers frequently value risk aversion above productivity (Conway and Barbier, 1990) or that farmers will de-emphasize their farms, if other sources of income are available (Kusterer, 1989). Environmental degradation and inequity resulting from agricultural technology occur typically because these issues were not considered in the research and development protocol (National Research Council, 1990). The process of iterative evaluation, that characterizes systems approaches, helps to overcome such deficiencies of linear reasoning.

Proposed systems protocol

We have proposed (Grove and Edwards, 1992; Grove et al., 1993) a simple conceptual framework for the conduct of integrated agricultural systems research. The steps include:

(1) description of the target agroecosystem including its goals, boundaries,

components, functioning, interactions among components, and interactions across its boundaries;

(2) detailed analysis of the agroecosystem to determine factors which limit or could contribute to attainment of productive and sociological goals;

(3) design of interventions and identification of actions to overcome the constraints;

(4) on-farm experimental evaluation of interventions;

(5) review effectiveness of newly designed system;

(6) redesign as necessary.

We believe that all steps should be conducted on farms by an interdisciplinary team of agricultural, social, and ecological scientists and with the full participation of farmers.

The description of the agroecosystem must be based upon discussions with farmers and upon recommendations of the disciplinary specialists. Techniques for describing agroecosystems have been reported in the literature (e.g. Conway, 1985; Altieri, 1987; Harwood, 1987). Understanding the farmer's goals is especially important because the role of the proposed interventions is to help the farmer attain these goals. Description of the boundaries and limitations of the agroecosystem is essential in providing focus for study but should not limit understanding of its interactions with adjacent ecosystems or with local, regional, national, and international political economies. Description of the components of the system is the traditional occupation of many agricultural scientists, but a description and analysis of interactions among components requires farmer participation as well as interdisciplinarity.

Although the descriptive phase is largely qualitative, the analytical stage takes maximal advantage of quantitative information. The proposed descriptions may lead to development of hypotheses that require experimental study for resolution and quantification. For example, if nitrogen is suspected to be a limiting factor, then nutrient response studies may be required. If losses to pests are hypothesized as a key factor, they can be quantified experimentally and integrated management measures recommended for the pests identified. The result of the analytical phase is to approach a more detailed understanding of the limitations to the attainment of the farmer's goals.

The design phase involves forming hypotheses about appropriate interventions that will contribute to farmers' goals. It is a deductive process based upon the description and analysis of the system and the final design represents the best collective judgments of the study team and the participating farmers.

The evaluation phase tests the proposed interventions empirically. Effects must be measured in terms of the goals of the system and trade-offs among goals must be determined for any proposed intervention. Interdisciplinary involvement and participation are essential in a successful evaluation phase.

Conclusion

We believe that traditional approaches to development have typically involved the intensification of agriculture as a first step. While this approach has had numerous successes, it has failed to increase the livelihoods of the rural poor in many locations and has often been accompanied by environmental degradation. This approach seems to fail most frequently when applied to rural poor who live on marginal lands with limited resources. We propose that a new protocol is required to address these intractable environments. The protocol is systems based and involves an iterative process of on-farm description, identification of constraints and hypothesis testing. The involvement of multiple disciplines and farmers is fundamental to this process.

References

Altieri, M., 1987. Agroecology. Westview Press, Boulder, CO, 227 pp.

Brown, L.R., 1990. The illusion of Progress. In: L.R. Brown (Editor), State of the World. Worldwatch Institute Report. W.W. Norton, New York, pp. 3–16.

Byerlee, D., 1989. Food for thought: technological challenges in Asian agriculture in the 1990s. Paper prepared for Conference of USAID Asian and Near East Bureau's Agricultural and Rural Development Officers, Rabat, Morroco, 19–24 February 1989, Agency for International Development, Washington, DC, pp. 1–23.

Chambers, R., 1983. Rural Development: Putting the First Last. Longman Scientific and Technical, London, pp. 1–32.

Chambers, R., 1987. Sustainable rural livelihoods: a strategy for people, environment and development. Commissioned paper for Only One Earth, Conference on Sustainable Development. 28–30 April 1987, International Institute for Environment and Development, London.

Clay, J.W., 1988. Indigenous Peoples and Tropical Forests: Models of Land-use and Management from Latin America. Cultural Survival, Cambridge, MA, 116 pp.

Conway, G.R., 1985. Agroecosystem analysis. Agric. Adm., 20: 31–55.

Conway, G.R., 1987. The properties of agroecosystems. Agric. Syst., 24: 95–117.

Conway, G.R. and Barbier, E.B., 1990. After the Green Revolution: Sustainable Agriculture for Development. Earthscan Publications, London, 205 pp.

Dover, M. and Talbot, L.M., 1987. To Feed the Earth: Agroecology for Sustainable Development. World Resources Institute, Washington, DC.

Gradwohl, J. and Greenberg, R., 1988. Saving the Tropical Forests. Earthscan Publications, London, 207 pp.

Grove, T.L. and Edwards, C.A., 1992. Agroecological perspectives on agricultural research and development. In: C. Valdivia (Editor), Sustainable Crop-Livestock Systems for the Bolivian Highlands. Proceeding of an Small Ruminant CRSP Workshop. University of Missouri. Columbia, MO.

Grove, T.L., Edwards, C.A., Harwood, R.R. and Colfer, C.J., 1993. The role of agroecology and integrated farming systems in agricultural sustainability. Agric. Ecosystems Environ.

Hall, C.A.S., 1991. An idiosyncratic assessment of the role of mathematical models in environmental science. Environ. Int., 17: 507–517.

Harwood, R.R., 1987. Agroforestry and mixed farming systems. In: A.E. Lugo, J.R. Clark and

R.D. Childs (Editors), Ecological Development in the Humid Tropics: Guidelines for Planners. Winrock International, Morrilton, AR, pp. 271–301.

Kerr, S.R., 1976. Ecological analysis and the Fry paradigm. J. Fish. Res. Board Can., 33: 329–332.

Kusterer, K., 1989. Small farmer attitudes and aspirations. AID Program Evaluation Discussion Paper Number 26, USAID, Washington, DC, 25 pp.

Lugo, A.E., Clark, J.R. and Childs, R.D., 1987. Ecological Development in the Humid Tropics: Guidelines for Planners. Winrock International, Morrilton, AR, 362 pp.

Lyman, J.K. and Herdt, R.W., 1988. Sense and sustainability: sustainability as an objective in international agricultural research. CIP-Rockefeller Foundation Conference on Farmers and Food Systems, Lima, Peru, September 26–30, 1988, 27 pp.

National Research Council, 1990. Alternative Agriculture. National Academy Press, Washington, DC, 448 pp.

National Research Council, 1991a. Toward Sustainability: Soil and Water Research Priorities for Developing Countries. National Academy Press, Washington, DC, 65 pp.

National Research Council, 1991b. Toward Sustainability: A Plan for Collaborative Research on Agriculture and Natural Resource Management. National Academy Press., Washington, DC, 147 pp.

Paarlberg, R. and Lipton, M., 1991. Changing missions at the World Bank. World Policy J., 8: 475–493.

Reid, W.R., Barnes, J.N. and Blackwelder, B., 1988. Bankrolling Successes: A Portfolio of Sustainable Development Projects. Environmental Policy Institute and National Wildlife Federation, Washington, DC, 48 pp.

Ruttan, V.W., 1990. Constraints on agricultural production in Asia: into the 21st century. In: CIMMYT 1989 Annual Report, Centro Internacional de Mejoramiento de Maiz y Trigo, Mexico City, pp. 1–2.

Technical Advisory Committee of the Consultative Group on International Agricultural Research, 1988. Sustainable agricultural production: implications for international research. TAC Secretariat, Food and Agriculture Organization of the United Nations, Rome, 14 pp.

Tull, K., Sands, M. and Altieri, M., 1987. Experiences in Success: Case Studies in Growing Enough Food Through Regenerative Agriculture. Rodale Institute, Emmaus, PA, 52 pp.

Vyas, V.S. and James, W.E., 1988. Agricultural development in Asia: performance, issues and policy. In: S. Ichimura (Editor), Challenge of Asian Developing Countries: Issues and Analysis. Asian Productivity Organization, Tokyo, pp. 119–131.

World Bank, 1978. World Development Report 1978. Oxford University Press, New York, 192 pp.

World Commission on Environment and Development, 1987. Our Common Future. Oxford University Press, Oxford, 400 pp.

World Resources Institute, 1990. World Resources 1990–91. Basic Books, New York, 383 pp.

York, E.T., Fox, R. and Smith, N., 1989. S&T/AGR Research Strategy for the 1990s. Report prepared for USAID. Consortium for International Development, Tucson, AZ, 155 pp.

Agriculture, Ecosystems and Environment, 46 (1993) 147–160
Elsevier Science Publishers B.V., Amsterdam

The need for a systems approach to sustainable agriculture

John E. Ikerd

Center for Sustainable Agriculture, University of Missouri–Columbia, Columbia, MO 65211, USA

Abstract

Differences between conventional and sustainable paradigms of agriculture are much more a matter of differences in farming philosophy than of farming practices or methods. The conventional model of agriculture is fundamentally an industrial development model which views farms as factories and considers fields, plants, and animals as production units. The goal of industrial development is to increase human well-being by increasing production of material goods and services and simultaneously increasing aggregate employment and incomes. The underlying assumption of the industrial model is that a higher quality of life can be derived from increases in income and consumption of goods and services. A fundamental strategy for industrial development has been to specialize, routinize, and mechanize agricultural production in order to achieve the economic efficiencies that are inherent in large-scale industrial production. New technologies are designed to remove physical and biological constraints to production and, thus, make unlimited progress possible. Sustainable agriculture, on the other hand, is based on a holistic paradigm or model of development which views production units as organisms that consist of many complex interrelated suborganisms, all of which have distinct physical, biological, and social limits. People are viewed as part of the organisms or systems from which they derive their well-being. Quality of life is considered to be a consequence of interrelationships among people and between people and the other physical and biological elements of their environment. Fundamental strategies for sustainable development include diversification, integration, and synthesis. Whole systems have qualities and characteristics that are not contained in their individual parts or components. The same set of components or parts may be rearranged spatially or sequentially resulting in a unique system or whole for each new arrangement. People increase their well-being by using information and knowledge to manage or rearrange the components of systems, resources, processes, and technologies in ways that enhance the productivity or 'well-being' of those systems. Human progress is limited only by our ability to enhance the social, biological, and physical systems of which we are a part. Sustainable agriculture requires a holistic systems approach to farm resource management. A component approach focusing on individual farming practices, methods, and enterprises may have been appropriate for the era of agricultural industrialization. However, a systems approach which focuses on knowledge-based development of whole farms and communities will be required to address the environmental, economic, and social challenges of the post-industrial era of agricultural sustainability.

Introduction

The primary public mandate or social agenda for US agriculture throughout this century has been to increase production efficiency in order to reduce consumer food costs, to free farm families for more rewarding occupations, and to free rural resources for other uses. Government programs for agriculture, including publicly funded research and education, have been focused on development and implementation of new technologies that would enhance agricultural productivity and reduce the need for farm labor.

Agricultural development strategies of the past have been highly successful in reducing the proportion of the nation's resources devoted to food production. For example, the 1895 Yearbook of Agriculture indicated that 42% of people in the US were employed on farms in 1890 (US Department of Agriculture (USDA), 1895). This compares with less than 2% of the total US population living on farms a century later. In addition, those living on farms today earn more than half of their income from non-farm sources (USDA,1990). US consumers now spend less than 12% of their income on food. Farmers get only about 25 cents out of each dollar spent for food, and about half of that goes to pay for purchased inputs.

Agriculture has fulfilled its public mandate of the past, but public priorities are changing. Until now, the negative side-effects of the twentieth-century agricultural policies and technologies on the ecological and social environment of rural areas have been largely ignored (National Research Council, 1989), but growing environmental and social equity concerns of the general public are raising questions that must be addressed by all sectors of the US economy, including agriculture. People are increasingly concerned about the negative environmental impacts of modern farming methods and the deteriorating quality of life in rural communities. These concerns have already resulted in changes in agricultural policies and programs at the national level. Additional fundamental changes could well be forthcoming.

Society appears to be giving agriculture a new, much broader public mandate. This new mandate is to develop a food and fiber system that is still efficient and productive, but in addition is ecologically sound, economically viable, and socially acceptable. This new mandate may well dictate a new paradigm or model for US agriculture.

Growing concerns with industrialization of agriculture

The industrial model of agriculture "treats the farm like a factory, with inputs and outputs, and considers fields and animals to be production units" (Kirschenmann, 1991). Industrial farming systems have relied primarily on specialization and mechanization to achieve physical and economic efficiencies through large-scale production. However, these same strategies that have increased past agricultural productivity have now begun to raise significant environmental and economic concerns. There is no general agreement among scientists at present regarding the extent to which these concerns reflect actual threats to long-term agricultural sustainability. None the less, there can be no doubt that the public is in fact concerned.

First, there are growing concerns regarding the continued effectiveness of the inputs and technologies which support these large-scale, specialized systems. Increased concentration of a single crop within a geographic region increases pest pressures on that crop. In addition, insects and weeds are becoming resistant to pesticides and require higher rates of application or new, more costly pesticides for control. Previously fertile soils in some areas have lost organic matter and natural fertility through monocropping, conventional tillage, and removal of crop aftermath year after year. Lower organic matter has meant less microbial activity, less ability to hold water, and less availability of nutrients in root zones, meaning lower yields from a given level of water and fertilization or higher fertilizer and irrigation costs to maintain yields.

Questions of natural resource stewardship are also confronting modern agriculture. Water tables in some of the major irrigated areas are declining as rates of irrigation surpass rates of natural regeneration of aquifers. Irrigation supports some of the largest of the large farming operations. Salinization of soils has also become a major concern in some irrigated farming regions. Soil conservation rose to the top of the political agenda in 1985 primarily because of rising soil erosion rates. Soil losses went up as farmers abandoned forage grass and legume-based crop rotations in the 1960s and rose still further as farmers intensified row crop production for growing export markets during the 1970s.

Other costs of specialization are beginning to show up in the environment of farm families, farm workers, and rural residents. Health risks in handling pesticides, for example, have become a major issue in farm safety. Risks of chemical contamination of drinking water and risks of pesticide residues in food are important public perceptions, regardless of the facts concerning actual risk levels. Nitrate leaching into ground water may be attributed to organic sources, such as livestock waste and crop residues, as well as the use of commercial fertilizer. However, this issue, as much as any other, has increased awareness in rural areas of the potential environmental hazards of chemically dependent farming.

The industrialization of agriculture has also changed rural landscapes. Farmers planted 'fence row to fence row' during the 1970s and many tore down the fences and plowed out the fence rows. Farming areas were no longer patchworks of fields, meadows, grassy hills, and valleys separated by rows of trees. Rural landscapes became field after field of corn, soybeans, wheat, and cotton across the hills and valleys. Timber was cleared to make room for cow herds. Livestock feeding and poultry production became concentrated in large feed lots in animal producing factories.

Larger, more specialized farming operations have meant fewer farming families. Fewer people are needed on farms with industrial farming technologies. Not only have purchased inputs been substituted for land and climate, but machines have been substituted for labor and technology has been substituted for management. The unneeded human resources have been squeezed out of agriculture as a natural economic consequence of the substitution of technology-based inputs for resources.

Technological advances reduced the costs of production and provided incentives for expanded production which, in turn, reduced market prices and ultimately reduced farm incomes. Attempts to mitigate the effects of surplus production through export expansion have instead created a system that is even more dependent on new technologies to remain competitive in global markets. Only those farmers who were among the first to adopt new technologies have realized profits. Those who lagged behind were forced to adopt to the new technologies in order to survive. Those who could not adopt or adapt quickly enough were forced to sell out to their more 'progressive' neighbors.

The continual repetition of this process over time has ensured that the economic returns to those remaining in agriculture were kept well below those of growing economic sectors. This process was an economic necessity to move unwilling people and resources out of agriculture and into other uses within the economy. But there were costs associated with this migration out of agriculture. These costs have included "social disorganization, shrinking rural economic bases, declining rural communities and institutions, and the specter of a permanent underclass in the cities" (Glover, 1988).

Rural communities are as much the victims of a more productive agriculture as are displaced farm families. Rural America has lower levels of income, education, employment, health, nutrition, and community service than urban America, and the agricultural technology treadmill has been a major contributor to this. Today, about one-quarter of the US population lives in rural areas, but only about 8% of all rural residents are farmers. About 75% of total agricultural income comes from the largest 25% of all farming units. Only about one-third of all farmers consider farming to be their principal occupation. About 75% of farmers' incomes come from off-farm sources (Hyman, 1990). Thus, most people who live in rural areas are not farmers, most farmers are not large producers or even full-time farmers, and most farm families depend more on non-farm income than on farming for a living.

These statistics reflect the indirect effects of the forced migration of families out of farming. Farmers have been reluctant to leave their local communities even though they no longer have an economic future in farming. Many have been willing to work in their home town for lower wages than they would be willing to accept elsewhere. Other farm families have been able to continue to farm, but they can no longer make an adequate living from farming alone. In these cases, one or more family members have taken a part-time or full-time job off the farm at lower wages than they might be willing to accept if they had to give up farming. Off-farm employment has allowed many to remain on the land that otherwise have been forced to leave their communities. However, the need for off-farm employment of farmers has added to the oversupply of rural labor and thus has contributed to chronically depressed rural labor markets.

However, the problem is not with agriculture in general but with an industrial, input-dependent agriculture. Today, the sustainability of industrial farming systems is being questioned. The ability of industrial systems to compete in the information age of the twenty-first century is also being questioned. The answers to these questions could well have significant implications both for the future of agriculture and of rural communities.

A new paradigm for agriculture

A sustainable agriculture must be capable of maintaining its productivity and usefulness to society indefinitely (Ikerd, 1990). Such an agriculture must use farming systems that conserve resources, protect the environment, produce efficiently, compete commercially, and enhance the quality of life for farmers and society overall.

Systems which fail to conserve and protect their resource base degrade its productivity and eventually lose their ability to produce. Such systems are not physically sustainable. Systems which fail to protect their environment eventually produce more harm than good; they lose their usefulness to society and, thus, are not socially sustainable. Farming systems which fail to provide the people with adequate supplies of safe and healthful food at reasonable costs and otherwise enhance the quality of life are not politically sustainable. Agricultural systems of Communist Europe were prime examples of such systems. Systems that are not commercially competitive will not generate the profits necessary for financial survival or an acceptable quality of life for producers and, thus, are not economically sustainable.

The industrial model of farming may or may not be sustainable over time. No one can possibly know the future with certainty. It may be possible to fine-tune, refine, or redesign the industrial model, resulting in a new model that will meet the ecological and social standards required to sustain long-term productivity without changing the fundamental approach to farming. On the other hand, an approach to or philosophy of farming that is fundamentally different from the industrial model may be required. Such an alternative agricultural model has evolved, and is being further developed, to address the economic, ecologic, and social balances required for long-term sustainability. This alternative to the industrial model of agriculture is commonly called sustainable agriculture. However, only time will tell which, if any, of our current models of farming are actually sustainable.

The new sustainable paradigm "treats the farm like an organism consisting of many complex, interrelated suborganisms, all of which have distinct biological limits" (Kirschenmann, 1991). Economic performance is dependent on the achievement of the total organism and, thus, requires a holistic systems approach to farm resource management. The sustainable agriculture model relies more on management of the internal resources of the farm and less on purchased commercial inputs in attempting to reduce negative ecological impacts while maintaining economically viable farms. Such an approach was characterized by the National Research Council (1989) in outlining the goals for alternative agriculture as involving:

(1) greater reliance on natural processes such as nutrient cycling, nitrogen fixation, and pest–predator relationships in the production process;

(2) reduction in use of off-farm inputs with the greatest potential to harm the environment or the health of farmers and consumers;

(3) greater productive use of the biological and genetic potential of plant and animal species;

(4) improved matching of cropping patterns and the productive potential and physical limitations of agricultural lands to ensure long-term sustainability of production levels;

(5) profitable and efficient production with greater emphasis on farm resource management and conservation of soil, water, energy, and biological resources.

The sustainable model implies greater reliance on human resources, in terms of the quality and quantity of labor and management, and relatively less reliance on land and capital. Thus, sustainable farming systems may require more farm operators, more farm laborers, and more farm families than do conventional farming systems. In addition, operators of sustainable farmers are motivated by a new mix of objectives. Social considerations are balanced with environmental and economic concerns. Social concerns may cause such farmers to show a preference for local markets and local input supply sources if this preference does not threaten their economic survival.

A fundamental shift in the balance of returns to people versus land and capital must be brought about, by one means or another, if more families are to be able to farm successfully. Disproportionately large quantities of land and capital are often controlled by a few individuals. Thus returns to human resources, to people, tend to be more egalitarian in nature. Government farm policies of the past have undervalued rural resources by failing to recognize the long-term social cost of resource depletion and degradation and even by subsidizing their short-term exploitation. Changes in government farm policies will be required to remove past government subsidies of large-scale, specialized farming. However, smaller diversified farms will become commercially competitive with larger specialized farms, only if human resources can be economically substituted for other resources and commercial inputs.

By implication, diversified sustainable farmers of tomorrow must be more knowledgeable, more creative, and more skilled workers. They cannot expect to earn a larger return for their labor and management, reducing their reliance on land and commercial inputs, unless they possess superior labor and management skills.

Sustainability: a systems concept

Differences between sustainable and industrial models of farming reflect differences in farming philosophy. In fact many individual farming practices, methods, or components used in industrial systems may be useful in sustainable systems. Farmers with different philosophies, however, may choose to integrate the same basic components quite differently in the process of developing farming systems. Differences in philosophy cannot be subjected to scientific analysis. Thus, some scientists have concluded that comparisons of conventional and sustainable systems fall outside the realm of science (Council of Agricultural Science and Technology, 1988). However, all scientific inquiry begins with at least two basic value judgments. How does the world work? What is the basic purpose of human activity? Science is incapable of providing definitive answers to either of these questions.

Agroecology provides a philosophical foundation for the sustainable agriculture concept. Agroecology is a synthesis of agriculture and ecology (Altieri, 1983). Agriculture, by its very nature, represents an attempt to enhance the productivity of nature in ways that favor humans relative to other species. However, the discipline of ecology views humanity as only one component of essentially interrelated ecosystems that include all people as well as the other biological species and physical elements of the biosphere. The concept of agroecology implies a right of humans to shift the ecological balance in favor of humans relative to other elements of the ecosystem. However, attempts to shift the balance too far or too fast in favor of humans relative to other species, in favor of some people relative to others, or in favor of the current generation relative to later generations, may destroy the critical ecological balance upon which the survival of humanity ultimately depends. Quality of human life is a product of relationships among humans and between humans and non-human elements of the biosphere. The humanistic and egocentristic elements of agroecology are constrained in that human well-being is dependent upon the well-being of other species and individual well-being is dependent on the well-being of society.

Actions taken in any part of an ecosystem have consequences for all other parts of the ecosystem, both now and in the future. Agroecologists contend that agricultural technologies must ultimately enhance nature rather than replace nature and must work with nature rather than attempt to conquer nature. The constraints of nature on humankind can be moved but not removed.

The techno–industrial philosophy of agriculture views humans as having dominion over all other species and over the biosphere in general. The quality of human life is a product of bringing this dominion under human control. The purpose of agriculture is to serve humanity, and in this philosophy, any constraints to productivity imposed by nature can be removed by future technology and, thus, are viewed as temporary obstacles to overcome. The purpose of technological development is to replace limited natural resources and limited natural production processes with technology-based industrial alternatives. The implicit assumption is that technology can ultimately remove all constraints to human progress.

Science can neither prove nor disprove the correctness of any philosophy. Intelligent people, including scientists, differ with respect to their philosophies regarding the relationships between people, agriculture, nature, and the fundamental purpose of developing new agricultural technologies. Those concerned with the sustainability of agriculture tend to lean more toward an agroecological viewpoint while those who see little relevance of the sustainability issue tend to lean toward the techno–industrial viewpoint. Lacking the ability to prove that one view is right and the other is wrong, scientists should be willing to pursue knowledge and to develop technologies that are consistent with both.

Agroecology implies a systems approach to farming, integrating technology, and natural processes to develop productive systems. All systems are in fact components of still larger systems and all components of systems are in fact systems made up of still smaller components. The first step in a systems approach to management is to identify the boundaries of the system to be managed (Bird et al., 1984). The purpose in establishing boundaries is to separate those things which can be managed and, thus, are a part of the management system from those things which cannot be managed and are thus a part of the external environment. Those things in the external environment may affect, and be affected by, those things within the system. Thus, systems boundaries do not imply mutual independence between things inside and outside them. On the contrary, the boundaries serve to sharpen the perception of interdependence between systems and their external environment.

The system under the management control of a farmer is the economic unit typically called a farm. A farm may include owned or rented land in several different geographic locations and all related buildings, equipment, and financial assets that are managed. In the case of a part-time farm, the economic unit may include off-farm as well as on-farm enterprises. A farming system also includes people: the principal farm operator, hired farm workers, and in the case of farming families, all members of the family who are considered part of the farming operation. The key concept in defining the boundaries of a farming system is to separate those things that can be managed, or directly influenced, from those things that cannot. A systems approach to farm management then implies that each decision will be evaluated in terms of its impact on the performance of the system as a whole.

Sustainable agriculture is fundamentally a systems approach to farm planning and decision making. The National Research Council (1989) defines a farming practice as a way of carrying out a discrete farming task such as preparing a seed bed, applying fertilizer, or spraying pesticides. A farming method is defined as a systematic way of accomplishing a basic farming function such as establishing, protecting, or feeding a crop that is achieved by integrating a number of complementary farming practices. A farming system, however, must be defined in terms of an overall approach to farming derived from a farmer's goals, values, knowledge, available technologies and opportunities, and is constructed by integrating a number of complementary farming methods.

A given set of farming practices or methods is not inherently more or less sustainable that any other set of practices or methods. Sustainability depends on the nature of whole farming systems. The goals and values of long-term sustainability must be reflected in combinations of practices and methods that are consistent with an individual farmer's unique set of resources, including his or her knowledge base, technical know-how, and farming opportunities. Sustainable farming systems are very much individual farmer and farmsite specific. Sustainability is determined by the system, considered as a whole, not by its individual components.

Synergism: the key to sustainability

Farming for sustainability requires a holistic approach to farm planning and management. Whole systems have qualities and characteristics not present in any of their constituent parts; therefore, one must seek to understand the greater whole in order to understand its parts, not vice versa (Savory, 1988). Systems take on values in and of themselves through the process of synergism. The essence of the whole of something is the arrangement of its parts. Arrangement is not a characteristic of the parts arranged but rather of the whole. Time, place, form, and possession are the fundamental sources of utility or economic value. Thus, creation of value is not a simple matter of changing the form of things through the physical processes of production. Value can also be produced by rearranging the various forms of things so as to affect their dimensions of time, place, and individual possession. Synergism then is the process by which resources and inputs are rearranged spatially, temporally, physically, and individually in order to create more valuable wholes.

Some simple examples may serve to illustrate the basic nature of potential synergistic gains from holistic management of farming systems in general. The time, space, form, and possession characteristics of production systems are obviously interrelated and are treated separately in these examples only for purposes of illustration.

A crop rotation represents a temporal sequence of farming methods and practices within a given spatial context. A particular sequence of crops may result in increased yields, reduced commercial pesticide and fertilizer requirements, and reduced soil erosion. A cropping sequence may break biological pest cycles, fix nitrogen from the air, and keep the ground covered during periods of heavy rainfall. In other words, crops grown continuously in separate fields may result in higher total costs, greater environmental risks, lower production, and less profit than would the same crops grown in a logical rotation or cropping sequence. The added benefits come from the temporal arrangement.

The spatial matching of crop and livestock enterprises to particular climate and soil characteristics is a critical factor in determining both economic and ecologic results. Most crop and livestock species have natural comparative advantages in production in particular regions of a country. Cotton, peanuts, rice, and tobacco, for example, are more common in the southern US because historically they have had comparative advantages under warm, moist growing conditions. When crops and livestock are grown in regions for which they are not particularly well adapted, the natural environment must be modified.

Relatively cheap and effective commercial pesticides, fertilizers, fossil fuels, and irrigation water have allowed commercially competitive production of many commodities outside their range of previous comparative advantage. However, the increased use of these particular inputs and resources is now a primary source of concern regarding environmental risks and resource depletion. Environmental risks are not inherent characteristics of plants or animals, nor even inherent to particular chemicals. Risks and returns, in many cases, are determined by the location of production, or spatial arrangement, among regions of production or even among fields on a farm.

The basic function of agriculture is to convert solar energy into an energy form that will provide human food, clothing, and shelter. Over time, however, US farmers have changed from being basic producers of food and fiber to being primarily converters of purchased inputs into raw materials. However, some farmers now have begun to try to reverse this trend. They are expanding their operation vertically rather than horizontally. They are producing some of their own inputs and substituting resource management for others. They are adding value to their products by integrating some or all of the traditional form-changing processing functions into their farming operations. Many successful niche farmers also tailor the form of output of each production process to fit the input requirement of the next process. In addition, they may utilize wastes from one stage of production as inputs in another in order to reduce costs and environmental risks. Their success may depend more on gains from their unique vertical arrangement of form changing processes than from either their market niche or the individual processes considered separately.

The utility or value of possession or ownership is an individualistic concept. Different individuals have different tastes and preferences which determine the value they place on given goods or services. Thus, the value of a given kind of good or service, at a given place and time, may not be the same for any two individuals. Different individuals also have different sets of skills, knowledge, and other resources that they can use to produce something of value to other people. Thus, the efficiency with which any two people are able to produce a given good or service may be quite different. Consequently, the cost of providing a given kind of good or service may be different depending on the individual(s) involved in the production process. The key to creating value through individualization is to match people, as producers, with other people, as consumers, so that people are producing the things they can produce best for people who value those things most.

In reality, the dimensions of time, space, form, and ownership are inseparable. Thus, a holistic systems approach to farming is a matter of managing the temporal, spatial, physical, and individual arrangements of interrelated sets of markets, resources, inputs, products, people, and processes. Holistic management is complex, but within this complexity lies the potential for synergistic gains. Such gains come from management, the process of choosing arrangements, and not from a given endowment of land, labor, or capital resources.

Knowledge: the key to systems management

Many of the past gains in productivity of agriculture have resulted from applying industrial production and business principles to farming. Likewise, many of the current threats to agricultural sustainability are associated with these same industrial systems of farming, but the industrial era may be coming to an end in the general economy. Thus, the industrialization of agriculture could be nearing an end as well. Further attempts to apply the industrial model in farming may result in declining economic benefits at increasing economic costs. Thus, the era of input-intensive farming, and its associated environmental and social costs, may be coming to an end, with or without a new environmental and social agenda for agriculture. The new public policy agenda will still have an important role in ensuring that agriculture in the post-industrial era is environmentally sound and socially acceptable as well as productive and profitable.

Toffler (1990) contends that knowledge will be the key to economic and political power in the future. He argues that the smoke-stack era in which power was associated with control of capital and the physical means of production is passing. He suggests that power in the future will belong to those who know how to access and synthesize data and information into value-added knowledge. Information and knowledge must be translated into value through the act of decision making. Value from knowledge results from decisions concerning arrangements of things, not from the things available to be arranged. Value created on farms in the future may result much more from the application of knowledge than from the possession of either resources, capital, or production technology. Knowledge is not a characteristic of the components or parts of a system. Knowledge is embodied in the arrangements which are characteristics of wholes.

He summarizes his hypotheses concerning the new system of wealth creation with twelve basic characteristics of a future knowledge-based system.

(1) The new system for wealth creation is increasingly dependent on the data, information, and knowledge.

(2) The new system of flexible, customized, 'de-massified' production will turn out products at costs approaching those of mass production.

(3) Conventional factors of production — land, labor, raw materials, and capital — become less important as knowledge is substituted for them.

(4) Capital becomes extremely fluid and the number of sources of capital multiplies.

(5) Goods and services are modular and configured into systems.

(6) Slow-moving bureaucracies are replaced by 'ad-hocratic', free-flowing information systems.

(7) The number and variety of organizational units multiply.

(8) The most powerful wealth-amplifying tools are inside workers' heads, giving them a critical share of the 'means of production'.

(9) The new heroes are the innovators who combine imagination with action.

(10) Wealth creation is recognized to be a circular process, with wastes recycled into inputs for the next cycle of production.

(11) Producer and consumer, divorced by the industrial revolution, are reunited in cycles of wealth creation.

(12) The new wealth creation system is both local and global, doing things economically on a local basis but with functions which spill over geographic boundaries.

These same basic characteristics can be associated with sustainable farming systems. Sustainable farming systems:

(1) are management-intensive and knowledge-dependent;

(2) are individualistic and site specific;

(3) substitute knowledge and information for inputs;

(4) may require capital from non-traditional sources;

(5) may produce composite products for specific niche markets;

(6) depend on free-flowing information from multiple sources;

 (7) tend to be smaller and more varied in size and character;
 (8) combine functions of thinking and doing in family operations;
 (9) rely on innovative arrangements of parts within whole systems;
 (10) utilize wastes and on-farm inputs in production processes;
 (11) connect production with consumption, producing for market niches;
 (12) rely more on local resources but may produce for global market niches.

Toffler (1990) contends that the smoke-stack industries lack the necessary flexibility to adapt to accelerated changes in needs and desires of society in the twenty-first century. Power in the future will accrue to those who have the knowledge needed to translate resources, inputs, and raw data into goods, services, and information tailored to narrowly segmented markets. He suggests that pursuit of economic power, rather than environmental protection or conservation, may be the primary motivation for adopting these knowledge-based systems of production.

In the post-industrial era, egocentric motives may support more sustainable systems of economic development in much the same sense that profit motives supported more productive development strategies during the industrial era. The industrial model, with its inherent environmental and social costs, may well become economically obsolete. However, public policies will still be required to protect the long-term interests of society in cases where they inevitably conflict with short-term interests of businesses and individuals.

Knowledge-based systems of farming could reduce, if not eliminate, many of the existing capital constraints to future agricultural productivity while conserving the natural resource base and protecting the environment. Knowledge is infinitely expandable since there is essentially no limit to how much we can create or use once it is created. The same knowledge can be used by many people at the same time and is more likely to be expanded than expended through simultaneous use. Knowledge can be created, in principle at least, just as effectively by the weak and poor as by the strong and rich. An economy based on knowledge rather than capital can provide greater equity of opportunity among all people.

Sustainable farming systems are fundamentally knowledge-based systems of farming. Holistic management of the physical, biological, and financial components of farming systems, oriented toward a goal of long-term sustainability, may be a classic example of knowledge-based systems of resource development. The ability of farmers to participate successfully in the era of knowledge-based development may well depend on the ability of the public research and education community to move from an industrial agriculture paradigm, designed to increase productivity, to a knowledge-based paradigm designed for long-term economic, environmental, and social sustainability.

References

Altieri, M.A., 1983. Agroecology — The Scientific Basis for Alternative Agriculture. Westview Press, Boulder, CO, 227 pp.

Bird, G.W., Edens, T., Drummond, F. and Groden, E., 1984. Design of pest management systems for sustainable agriculture. In: C.A. Francis, C.B. Flora and L.D. King (Editors), Sustainable Agriculture in Temperate Zones. Wiley, New York, 487 pp.

Council of Agriculture, Science and Technology, 1988. Alternative Agriculture, Scientist's Review, CAST, Ames, IA, pp. 1–19.

Glover, R.S., 1988. Farmers Pay the Price for Advances in Biotech. Atlanta Constitution, Atlanta, GA, 17 December.

Hyman, D., 1990. Rural Development is Challenge of the 1990s. Farm Economics, January/February, College of Agriculture, Pennsylvania State University.

Ikerd, J.E., 1990. Agriculture's search for sustainability and profitability. J. Soil Water Conserv., 45(1): 18–23.

Kirschenmann, F., 1991. Fundamental fallacies of building agricultural sustainability. J. Soil Water Conserv., 46(3): 165–168.

National Research Council, 1989. Alternative Agriculture. National Academy Press, Washington, DC, 426 pp.

Savory, A., 1989. Holistic Resource Management. Island Press, Covelo, CA, 564 pp.

Toffler, A., 1990. Power Shifts. Bantam Books, New York, 553 pp.

US Department of Agriculture, 1895. Yearbook of the United States Department of Agriculture, 1895, US Government Printing Office, Washington, DC, 653 pp.

US Department of Agriculture, 1990. Agricultural Chart Book, Agricultural Handbook No. 689, Washington, DC, 116 pp.

Agriculture, Ecosystems and Environment, 46 (1993) 161–173
Elsevier Science Publishers B.V., Amsterdam

An economic framework for evaluating agricultural policy and the sustainability of production systems

Paul Faeth

World Resources Institute, 1709 New York Avenue, Washington, DC 20006, USA

Abstract

Rhetorical definitions of agricultural sustainability abound. However, definitions that can be put to operational use in economic analysis have not been adequately developed. An economic definition of sustainability using natural resource accounting techniques is put forward. Two case studies are examined and physical and economic measures of natural resource impacts are developed. Five policy scenarios are tested to determine the public and private gains possible from different approaches to agricultural support and economic incentives.

Introduction

In recent years, researchers and organizations have struggled to define 'sustainable agriculture'. Almost all definitions include maintaining productivity and farm profitability, while minimizing environmental impacts. However, these definitions have been qualitative, not quantitative, in nature. The productivity of the natural resource base, which is fundamental to sustainability, has not been successfully incorporated into definitions of agricultural productivity. The notion of agricultural sustainability has therefore been of only limited operational use to policy makers and researchers attempting to determine the effects of various policies and technologies.

Sustainability means that economic activity should meet current needs without foreclosing future options; the resources required to provide the needs of the future must not be depleted to satisfy today's consumption (World Commission on Environment and Development, 1987). The standard definition of income found in economic and accounting textbooks encompasses this notion of sustainability (Hicks, 1946; Edwards and Bell, 1961). Income is defined as the maximum amount that can be consumed this year without reducing potential consumption in future years, i.e. without consuming capital assets.

Accounting systems for both businesses and nations have included a capital consumption allowance, representing the depreciation of capital during the current year, which is subtracted from net revenues in calculating annual income. Historically, however, changes in the productivity of the natural resource base which, like other forms of capital provides a flow of economic benefits over time have not been included in these accounts. Changes in manmade capital have become pre-eminent in accounting systems, implying that natural resource capital is of negligible value in current production systems. Nations, businesses, and farmers account for the depreciation of assets such as buildings and tractors as they wear out or become obsolete, but ignore changes in the productive capacity of natural resources such as soil or water. Thus, soil can be eroded, ground water contaminated, wildlife poisoned, and reservoirs filled with sediment, all in order to support current agricultural practices and income. No depreciation allowance is applied against current income for the degradation of these resources, even though the loss of asset productivity jeopardizes future income. Current accounting practices can mask a decline in wealth as an increase in income.

For agriculture and other economic sectors that are fundamentally dependent upon the health of the natural resource base, the accounting of natural resource capital is extraordinarily important. When changes in natural resource assets are ignored, resource degradation is encouraged, if not guaranteed. The methods of natural resource accounting provide a relatively simple way to arrive at quantitative measures of sustainability. Soil productivity, farm profitability, regional environmental impacts, and government fiscal costs can all be included within a natural resource accounting framework. In the study reported here (Faeth et al., 1991), a natural resource accounting approach was used to rectify key omissions in past comparisons of conventional and alternative practices. Few previous studies compare profitability under alternative policy scenarios, and none compares the economics of conventional and alternative production systems when natural resources are accounted for (see, for example, Lockeretz et al., 1984; Cacek and Langner, 1986; Domanico et al., 1986; Helmers et al., 1986; Goldstein and Young, 1987; Dobbs et al., 1988). These are critical omissions; if natural resource impacts are ignored, the primary justification for sustainable agriculture will have been overlooked. Additionally, any biases in current agricultural policy will also be reflected in the analysis.

The analytical methodology was designed to quantify economic, fiscal, and environmental costs and benefits of agricultural policy options and can be used to analyze the consequences of a wide range of policy interventions. Moreover, it can be used to analyze the environmental costs of farm policies in both physical and monetary terms, so that the benefits and costs of alternative policies can be compared.

At the core of two case studies in Pennsylvania and Nebraska, which are presented here, are economic comparisons between commonly used conventional farming systems, which rely on heavy inputs of fertilizers and pesticides, and alternative systems, which rely on crop rotations and tillage practices for soil fertility and moisture and pest management. These comparisons cover not only farmers' receipts and production costs but also selected on- and off-farm resource and environmental costs.

Estimates of environmental costs are based on detailed physical, agronomic, and economic modeling of soil, water, and chemical transport from the field and the implications of these processes for water quality and soil fertility. Data from 9 years of field experiments at the Rodale Research Center in Kutztown, Pennsylvania, and at the University of Nebraska at Mead were analyzed using the US Department of Agriculture's (USDA) Erosion-Productivity Impact Calculator (EPIC) Model (Williams, et al., 1989). Output from this model was used to estimate the on- and off-farm soil costs associated with conventional and alternative crop rotations. Other problems associated with agricultural production, such as groundwater contamination, loss of wildlife habitat, soil salinization or build-up of toxic chemicals, and human health problems due to the use of these chemicals were not addressed in this study. Hydrological models, for example, were inadequate, so economic losses associated with groundwater contamination could not be determined. The nature of the case study approach ruled out exploration of large-scale trade-offs in surface water quality, soil erosion, and groundwater quality, in which benefits in one area may be offset by costs in another due to widespread land use changes (Hrubovcak et al., 1990). Future extensions of the method should begin to include these issues.

In both case studies, five policy options were modeled to represent their constraints on, and incentives to, farmers. For example, the implications of different cropping patterns on farmers' base acreage and government support payment receipts are built into each analysis. Each policy option's financial and economic value is analyzed. The financial value (net farm income) of a production program to farmers takes into account current and future transfer receipts but ignores environmental costs borne by others. Net farm income is defined to include the value of changes in soil productivity, the farmer's principal natural asset. This definition is consistent with business and economic accounting practices, which incorporate asset formation and depreciation in their measures of income. By contrast, the same program's economic value to society (net economic value) includes environmental costs that farmers' activities impose on others, such as damages related to water pollution, but ignores transfer payments. Because the most financially rewarding production system to farmers may not generate the greatest economic value, some policy options may induce significant economic losses.

A natural resource accounting framework

Net farm income and net economic value per acre for Pennsylvania's best conventional corn–soybean rotation over 5 years, with and without allowances for natural resource depreciation are compared in Tables 1 and 2. Table 1 (Column 1) shows a conventional financial analysis of net farm income. The gross operating margin (crop sales less variable production costs) is shown in the first row ($45). Because conventional analyses make no allowance for natural resource depletion, the gross margin and net farm operating income are the same. Government subsidies ($35) are added to obtain net income ($80).

When natural resource accounts are included, the gross operating margin is reduced by a soil depreciation allowance ($25) to obtain net farm income ($20) (see Table 1, Column 2 and Table 3, Column 3). The depreciation allowance is an estimate of the present value of future income losses due to the impact of crop production on soil quality. The same government payment is added to determine net farm income ($55).

Net economic value (Table 2, Column 2) subtracts $46 as an adjustment for off-site environmental costs (such as sedimentation, impacts on recrea-

Table 1
Conventional and natural resource accounting economic frameworks compared: net farm income ($ acre^{-1} year^{-1})

	Without natural resource accounting	With natural resource accounting
Gross operating margin	45	45
Minus soil depreciation	–	25
Net farm operating income	45	20
Plus government commodity subsidy	35	35
Net farm income	80	55

Table 2
Conventional and natural resource accounting economic frameworks compared: net economic value ($ acre^{-1} year^{-1})

	Without natural resource accounting	With natural resource accounting
Gross operating margin	45	45
Minus soil depreciation	–	25
Net farm operating income	45	20
Minus off-site costs	–	46
Net economic value	–	(26)

tion and fisheries, and impacts on downstream water users). Net economic value also includes the on-site soil depreciation allowance, but excludes income support payments. Farmers do not bear the off-site costs directly, but nevertheless they are real economic costs attributable to agricultural production and should be considered in calculating net economic value to society. Subsidy payments, by contrast, are a transfer from taxpayers to farmers, not income generated by agricultural production, and are therefore excluded from net economic value calculations. In this example, when these adjustments are made, an $80 profit under conventional financial accounting becomes a $26 loss under more complete economic accounting.

Measuring sustainability

The economic and resource-accounting models used in this study integrate information at four levels, corresponding to the four-fold hierarchy of sustainability defined by Lowrance et al. (1986): field, farm, region, and nation. They represent in a consistent framework the farmer's financial perspective and wider environmental and economic perspectives.

At the field level, the USDA's EPIC model was used to simulate the physical changes in the soil that would result from different agronomic practices. EPIC, a comprehensive model developed to analyze the soil erosion productivity problem, simulates erosion, plant growth, nutrient cycling, and related processes by modeling the underlying physical processes.

A simple farm-level programming model was developed to assess the impact of commodity programs (operating through changes in input and output prices, acreage constraints, and deficiency payments) on net farm income and net economic value. The EPIC and programming models were linked to calculate not only crop sales, production expenses, government deficiency payments, and net farm incomes for each cropping pattern, but also soil erosion, off-site damages, and a soil depreciation allowance.

At the farm level, EPIC's estimates of soil productivity changes were used to calculate the economic depreciation of the soil resource. These estimates were combined with agronomic production data to determine the full on-farm production costs for each rotation and treatment. The farm level information on soil erosion was coupled with regional estimates of off-site damage per ton of eroded soil (Ribaudo, 1989) to derive estimates of off-farm damages resulting from each agronomic practice.

At the national level, agricultural sector models developed by the Food and Agricultural Policy Research Institute (FAPRI) generated estimates of changes in crop prices under the various policies (FAPRI, 1988, 1990). These prices were used in farm programming models to determine net farm income and net economic value. The farm-level model also generated estimates of government payments for the different crop production alternatives under the policy scenarios, which could then be generalized to compare the relative federal budgetary costs of different policy options.

Estimating a soil depreciation allowance

Estimates of long-term soil productivity changes taken from the EPIC model for different farming practices were used in present-value calculations to compute the economic impacts of soil productivity changes due to soil erosion and changes in soil structure.

The prices used to calculate the value of the productivity changes were those projected by FAPRI for each policy scenario tested. The yield change for each rotation period was taken to be the total yield change for the 30 year simulation divided by the number of rotations in 30 years, thereby assuming a linear change in yields. In this way the productivity change for each rotation included only the change attributable to the rotation over one rotation period. Since input costs were invariant to yields, this change in yields was then multiplied by the crop price to determine the loss in net farm income for the period. The present value of all income losses over the next 30 years, using a 5% real (excluding inflation) discount rate, represents the loss in soil asset value.

The formula used to determine the soil depreciation allowance is as follows:

Soil depreciation allowance

$$= [(Y_0 - Y_n)/(n/RL)] \times P_c \times \{[1 - 1/(1+i)^n]/i\},$$

where Y_0 is initial yield, Y_n is final yield, RL is rotation length, n is period under consideration, P_c is crop price, and i is real interest rate. For rotations that include more than one crop, each crop was weighted according to its acreage in the rotation, and these weighted crop depreciation allowances were added to determine the allowance for the rotation as a whole. When comparing rotations of different length, the rotation with the longest period was used to calculate the depreciation allowance for all rotations. The soil depreciation values are shown in Column 3 of Tables 3 and 4.

The off-farm costs of soil erosion

Ribaudo (1989), has presented a comprehensive estimate of the widely varying off-site costs of soil erosion for different areas in the US. In the northeast, where many rivers drain into the densely populated seaboard and the economic value of water is high, damage per ton of erosion is $8.16 (1990 dollars). At the other extreme, in the sparsely populated, dry Northern Plains where the economic value of water is low, damage per ton of erosion is $0.66. These estimates were combined in this study with EPIC erosion estimates to

Table 3
Rotation characteristics, Pennsylvania

Tillage/rotation	Soil erosion (ton acre^{-1} year^{-1})	Off-farm erosion cost[1] ($ acre^{-1} year^{-1})	Soil depreciation ($ acre^{-1} year^{-1})
Conventional tillage			
Continuous corn	9.26	69	24.8
Corn–beans	6.07	47	24.6
Alt. cash grain (ACG)[2]	4.25	32	(2.8)[3]
ACG with fodder[d]	3.29	26	(8.4)
All hay	0.66	5	(4.8)
Reduced tillage			
Continuous corn	7.15	53	24.4
Corn–beans	5.29	41	23.8
Alt. cash grain[2]	3.49	27	(3.6)
ACG with fodder[4]	2.49	20	(10.2)

[1]Estimated using a damage cost of $8.16 per ton. Calculations weighted by crops and set-aside acreages.
[2]Alternative cash grain — organic corn–corn–beans–wheat/clover–barley.
[3]Parentheses indicate appreciation in soil asset values due to increased productivity.
[4]Alternative cash grain with fodder — organic corn–beans–wheat/clover–clover–corn silage.

Table 4
Rotation characteristics — Nebraska

Rotation	Soil erosion (ton acre^{-1} year^{-1})	Off-farm erosion cost ($ acre^{-1} year^{-1})	Soil depreciation ($ acre^{-1} year^{-1})
Continuous corn–corn–beans	6.5	4.0	7.8
With inorganic inputs	3.7	2.3	3.0
With fertilizer only	3.7	2.3	2.8
With organic treatment	3.1	2.0	(2.0)
Corn–beans–corn–oats/clover			
With inorganic inputs	3.1	2.0	(1.3)
With fertilizer only	3.1	2.0	(1.0)
With organic treatment	2.2	1.5	(4.0)

calculate off-farm damages from soil erosion with the various rotations. The erosion rates were weighted by the crop set-aside requirements, where applicable, and multiplied by the regional per ton damage estimates. These values are shown as the off-farm costs in Tables 3, and 4, Column 2.

Policy analysis

The relative profitability and economics of the conventional and alternative practices were tested under five policy scenarios (Table 5). For these scenarios crop prices predicted by FAPRI (1988, 1990) were used. Tables 6 through 9 show the results of the analysis for each of the five scenarios, for each of the case studies, and for both net economic value and net farm income.

Table 5
Principal features of policy options tested

Baseline policy	Continuation of 1985 Food Security Act
Sustainable Agriculture Adjustment Act of 1989 (H.R. 3552) (SAAA)	Farmers using resource-conserving rotations allowed 100% planting flexibility, maintenance of crop base acreages, waiver of crop set-aside acreage and payments on forage crops (US House of Representatives, 1989)
Normal crop acreage	100% planting flexibility, but no deficiency payment for non-program crops if harvested
Input tax of 25%	25% tax applied to fertilizers and pesticides
Multilateral decoupling	Elimination by US, EC, Japan, etc. of all commodity programs tied to production (US Department of Agriculture, 1990)

Accounting for the soil-related resource costs of agricultural production in these two areas makes policy's effect on agricultural sustainability clear: US term income support programs discourage the adoption and generation of resource-conserving practices. Base acreage constraints and commodity programs that cover only seven crops put resource-conserving crop rotations at a financial disadvantage, despite the fact that farming practices that reduce soil erosion and improve soil productivity can provide a significant net economic gain to society. Morever, conventional practices can result in large net economic losses to society by eroding soil or damaging recreation, fisheries and navigation although the taxpayer subsidy dictated by current policy may make these practices most profitable to farmers. If distortions of current policy were corrected, as would happen under a policy of multilateral free trade, net farm income and net economic value could increase considerably for both conventional and alternative practices. In areas with high resource costs, such as Pennsylvania, resource-conserving practices would become most profitable options, greatly encouraging their adoption. In low resource-cost areas such as Nebraska, while farmers would not necessarily change practices, they would enjoy the benefit of higher crop prices, which would result from worldwide free trade. As a result, society would probably bear fewer negative environmental impacts from agriculture, government expenditures for income support could be dramatically reduced, and farm income could be maintained or improved.

The utility of natural resource accounting methods has implications for research funding, as well as for the structure of federal commodity programs. Presumably, public monies fund the research with the greatest net social payback. It is clear that excluding environmental costs and benefits from funding decisions could mean that from society's point of view, the wrong research is funded. In the past, research dollars have been spent on labor- and land-saving technologies. It is time now to evaluate and fund 'environment-saving' technologies as well.

Table 6
Summary results for Pennsylvania — transition period plus present value of the normal period: net farm income ($ acre⁻¹ for 10 years)

	Policy[1]	Conventional tillage[2]				Reduced tillage[2]				
		CC	CCBCB	ACG	ACGF	CC	CCBCB	ACG	ACGF	All hay
Gross operating margin	Baseline	(47)	607	486	508	(75)	581	480	501	247
	SAAA	(118)	461	461	492	(146)	437	455	485	247
	NCA	(118)	461	235	294	(146)	437	229	287	247
	MLDC	631	959	734	639	603	934	728	632	247
	25% Tax	(168)	523	486	508	(196)	497	480	501	247
Minus soil depreciation	Baseline	231	230	(26)	(78)	228	222	(34)	(95)	(45)
	SAAA	222	215	(24)	(73)	218	207	(32)	(90)	(45)
	NCA	222	215	(24)	(73)	218	207	(32)	(90)	(45)
	MLDC	285	246	(28)	(77)	282	241	(37)	(93)	(45)
	25% Tax	231	230	(26)	(78)	228	222	(34)	(95)	(45)
Net farm operating income	Baseline	(278)	377	512	587	(302)	359	514	596	292
	SAAA	(340)	247	485	565	(364)	230	487	574	292
	NCA	(340)	247	259	367	(364)	230	261	376	292
	MLDC	346	712	762	716	322	694	766	725	292
	25% Tax	(399)	293	512	587	(424)	275	514	596	292
Plus government commodity subsidy	Baseline	547	328	315	192	547	328	315	192	0
	SAAA	608	366	407	541	608	366	407	541	0
	NCA	608	366	608	312	608	366	608	312	0
	MLDC	–	–	–	–	–	–	–	–	–
	25% Tax	547	328	315	192	547	328	315	192	0
Net farm income	Baseline	269	706	827	779	244	687	829	788	292
	SAAA	268	612	892	1106	245	596	894	1115	292
	NCA	268	612	867	679	244	596	869	688	292
	MLDC	346	712	762	716	322	694	766	725	292
	25% Tax	147	622	827	779	123	603	829	788	292

[1]SAAA, sustainable agriculture adjustment act; NCA, normal crop acreage; MLDC, multilateral decoupling.
[2]CC, conventional continuous corn; CCBCB, conventional corn–beans; ACG, alternative cash grain — organic corn–corn–beans–wheat/clover–barley; ACGF, alternative cash grain with fodder — organic corn–beans–wheat/clover–clover–corn silage.

Table 7
Summary results for Pennsylvania — transition period plus present value of the normal period: net economic value ($ acre^{-1} for 10 years)

	Policy[1]	Conventional tillage[2]				Reduced tillage[2]				
		CC	CCBCB	ACG	ACGF	CC	CCBCB	ACG	ACGF	All hay
Gross operating margin	Baseline	(47)	607	486	508	(75)	581	480	501	247
	SAAA	(118)	461	461	492	(146)	437	455	485	247
	NCA	(118)	461	235	294	(146)	437	229	287	247
	MLDC	631	959	734	639	603	934	728	632	247
	25% Tax	(168)	523	486	508	(196)	497	480	501	247
Minus soil depreciation	Baseline	231	230	(26)	(78)	228	222	(34)	(95)	(45)
	SAAA	222	215	(24)	(73)	218	207	(32)	(90)	(45)
	NCA	222	215	(24)	(73)	218	207	(32)	(90)	(45)
	MLDC	285	246	(28)	(77)	282	241	(37)	(93)	(45)
	25% Tax	231	230	(26)	(78)	228	222	(34)	(95)	(45)
Net farm operating income	Baseline	(278)	377	512	587	(302)	359	514	596	292
	SAAA	(340)	247	485	565	(364)	230	487	574	292
	NCA	(340)	247	259	367	(364)	230	261	376	292
	MLDC	346	712	762	716	322	694	766	725	292
	25% Tax	(399)	293	512	587	(424)	275	514	596	292
Minus off-site costs	Baseline	641	438	304	242	494	382	250	183	50
	SAAA	641	438	323	250	494	382	265	190	50
	NCA	641	438	295	231	494	382	244	175	50
	MLDC	705	462	323	250	543	403	265	190	50
	25% Tax	641	438	304	242	494	382	250	183	50
Net economic value	Baseline	(919)	(61)	208	345	(796)	(23)	264	413	243
	SAAA	(981)	(191)	162	315	(858)	(152)	222	384	243
	NCA	(981)	(191)	(37)	136	(858)	(152)	17	202	243
	MLDC	(359)	251	438	466	(222)	290	500	536	243
	25% Tax[3]	(919)	(61)	208	345	(796)	(23)	264	413	243

[1]SAAA, sustainable agriculture adjustment act; NCA, normal crop acreage; MLDC, multilateral decoupling.
[2]CC, conventional continuous corn'; CCBCB, conventional corn–beans; ACG, alternative cash grain — organic corn–corn–beans–wheat/clover–barley; ACGF, alternative cash grain with fodder — organic corn–beans–wheat/clover–clover–corn silage.
[3]Columns will not add for the input tax, as the amount of tax has been added back to determine the net economic value.

Table 8
Summary results for Nebraska: net farm income ($ acre⁻¹ for 4 years)

	Policy[1]	Rotation[2]						
		CC	HFCB	FOCB	ORGCB	HFROT	FOROT	ORGROT
Gross operating margin	Baseline	119	501	503	473	351	348	334
	SAAA	99	445	449	422	341	338	325
	NCA	99	445	449	422	299	298	286
	MLDC	305	583	582	551	461	452	436
	25% Tax	83	485	495	473	332	336	334
Minus soil depreciation	Baseline	31	12	11	(8)	(5)	(4)	(12)
	SAAA	30	11	10	(8)	(5)	(4)	(11)
	NCA	30	11	10	(8)	(5)	(4)	(11)
	MLDC	38	13	12	(10)	(6)	(5)	(15)
	25% Tax	31	12	11	(8)	(5)	(4)	(12)
Net farm operating income	Baseline	88	489	492	482	356	352	346
	SAAA	70	434	439	430	246	342	337
	NCA	70	434	439	430	304	302	297
	MLDC	267	571	570	561	467	457	451
	25% Tax	53	473	485	482	338	340	346
Plus government commodity subsidy	Baseline	199	100	100	100	100	100	100
	SAAA	222	111	111	111	185	185	185
	NCA	222	111	111	111	222	222	222
	MLDC	–	–	–	–	–	–	–
	25% Tax	199	100	100	100	100	100	100
Net farm income	Baseline	287	589	592	581	455	451	445
	SAAA	291	545	550	541	531	527	521
	NCA	291	545	550	541	526	524	519
	MLDC	267	571	570	561	467	457	451
	25% Tax	252	572	584	581	437	440	445

[1]SAAA, sustainable agriculture adjustment act; NCA, normal crop acreage; MLDC, multilateral decoupling.
[2]CC, conventional continuous corn; HFCB, conventional corn–beans, with herbicides and fertilizer; FOCB, corn–beans with fertilizer but no herbicide; ORGCB, organic corn–beans; HFROT, corn–beans–corn–oats/clover with herbicides and fertilizer; FOROT, corn–beans–corn–oats/clover with fertilizer but no herbicide; ORGROT, organic corn–beans–corn–oats/clover.

Table 9
Summary results for Nebraska: net economic value ($ acre⁻¹ for 4 years)

	Policy[1]	Rotation[2]						
		CC	HFCB	FOCB	ORGCB	HFROT	FOROT	ORGROT
Gross operating margin	Baseline	119	501	503	473	351	348	334
	SAAA	99	445	449	422	341	338	325
	NCA	99	445	449	422	299	298	286
	MLDC	305	583	582	551	461	452	436
	25% Tax	83	485	495	473	332	336	334
Minus soil depreciation	Baseline	31	12	11	(8)	(5)	(4)	(12)
	SAAA	30	11	10	(8)	(5)	(4)	(11)
	NCA	30	11	10	(8)	(5)	(4)	(11)
	MLDC	38	13	12	(10)	(6)	(5)	(15)
	25% Tax	31	12	11	(8)	(5)	(4)	(12)
Net farm operating income	Baseline	88	489	492	482	356	352	346
	SAAA	70	434	439	430	346	342	337
	NCA	70	434	439	430	304	302	297
	MLDC	267	571	570	561	467	457	451
	25% Tax	53	473	485	482	338	340	346
Minus off-site costs	Baseline	16	9	9	8	8	8	6
	SAAA	16	9	9	8	8	8	6
	NCA	16	9	9	8	8	8	6
	MLDC	17	10	10	8	8	8	6
	25% Tax	16	9	9	8	8	8	6
Net economic value	Baseline	72	480	483	474	348	344	340
	SAAA	54	425	430	422	338	334	331
	NCA	54	425	430	422	296	294	292
	MLDC	250	561	561	553	458	449	445
	25% Tax[3]	72	480	483	474	348	344	340

[1]SAAA, sustainable agriculture adjustment act; NCA, normal crop acreage; MLDC, multilateral decoupling.
[2]CC, conventional continuous corn; HFCB, conventional corn–beans, with herbicides and fertilizer; FOCB, corn–beans with fertilizer but no herbicide; ORGCB, organic corn–beans; HFROT, corn–beans–corn–oats/clover with herbicides and fertilizer; FOROT, corn–beans–corn–oats/clover with fertilizer but no herbicide; ORGROT, organic corn–beans–corn–oats/clover.
[3]Columns will not add for the input tax, as the amount of tax has been added back to determine the net economic value.

References

Cacek, T. and Langner, L.L., 1986. The economic implications of organic farming. Am. J. Alternative Agric., I(1): 25–29.

Dobbs, T.L., Leddy, M.G. and Smolik, J.D., 1988. Factors influencing the economic potential for alternative farming systems: case analyses in South Dakota. Am. J. Alternative Agric., 3(1): 26–34.

Domanico, J.L., Madden, P. and Partenheimer, E.J., 1986. Income effects of limiting soil erosion under organic, conventional and no-till systems in eastern Pennsylvania. Am. J. Alternative Agric., 1(2): 75–82.

Edwards, E.O. and Bell, P.W., 1961. The Theory and Measurement of Business Income. University of California Press, Berkeley.

Faeth, P., Repetto, R. Kroll, K., Dai, Q. and Helmers, G., 1991. Paying the Farm Bill: U.S. Agricultural Policy and the Transition to Sustainable Agriculture. World Resources Institute, Washington, DC.

Food and Agricultural Policy Research Institute (FAPRI), 1988. Policy Scenarios with the FAPRI Commodity Models. Working Paper 88-WP 41. Center for Agricultural and Rural Development, Iowa State University, Ames, IA, p. 42.

FAPRI, 1990. Draft Report: An Evaluation of Price Support Equilibration Options for the 1990 Farm Bill. Ames, Iowa: Center for Agricultural and Rural Development, Iowa State University, Ames, IA, p. 35.

Goldstein, W.A. and Young, D.C., 1987. An agronomic and economic comparison of a conventional and a low-input cropping system in the Palouse. Am. J. Alternative Agric., 2(2): 51–56.

Helmers, G.A., Langemeier, M.R. and Atwood, J., 1986. An economic analysis of alternative cropping systems for east–central Nebraska. Am. J. Alternative Agric., 1(4): 153–158.

Hicks, J.R., 1946. Value and Capital: An Inquiry into Some Fundamental Principles of Economic Theory. Oxford University Press, Oxford.

Hrubovcak, J., LeBlanc, M. and Miranowski, J., 1990. Limitations in evaluating environmental and agrichemical policy coordination benefits. In: AEA Papers and Proceedings: Environmental and Agricultural Policies, 80(2): 208–212.

Lockeretz, W., Shearer, G., Kohl, D.H. and Klepper, R.W., 1984. Comparison of organic and conventional farming in the corn belt. In: Organic Farming: Current Technology and Its Role in a Sustainable Agriculture. American Society of Agronomy, Madison, WI, pp. 37–48.

Lowrance, R., Hendrix, P.F. and Odum, E.P., 1986. A hierarchical approach to sustainable agriculture. Am. J. Alternative Agric., 1(4): 169–173.

Ribaudo, M.O., 1989. Water quality benefits from the conservation reserve program. US Department of Agriculture, Resources and Technology Division, Economic Research Service, Agricultural Economic Report No. 606, February 1989, p. 12.

Williams, J.R., Dyke, P.T., Fuchs, W.W., Benson, V.W., Rice, O.W. and Taylor, E.D., 1989. EPIC — Erosion/Productivity Impact Calculator: 2, User Manual. In: A.N. Sharpley and J.R. Williams (Editors), US Department of Agriculture Technical Bulletin No. 1768, US Dep. Agric., Washington, DC.

US Department of Agriculture, 1990. 1990 Farm Bill: Proposal of the Administration. Office of Publishing and Visual Communications, Washington, DC.

US House of Representatives, 1989. Sustainable Agricultural Adjustment Act of 1989, H.R. 3552, 101st Congress, First Session.

US House of Representatives, 1990. Food, Agriculture, Conservation, and Trade Act of 1990. Report 101-916, 101st Congress, Second Session.

World Commission on Environment and Development, 1987. Our Common Future. Oxford University Press, New York, p. 43.

Agriculture, Ecosystems and Environment, 46 (1993) 175–186
Elsevier Science Publishers B.V., Amsterdam

The sociology of agricultural sustainability: some observations on the future of sustainable agriculture

Frederick H. Buttel

Department of Rural Sociology, University of Wisconsin, Madison, WI 53706, USA

Abstract

It is argued that sociology and the other social sciences can contribute to agricultural sustainability in several ways, one of which is to help in assessing and understanding the social forces that affect agricultural research and agricultural policy. Many of the conditions that gave rise to expansion of sustainable agriculture in the 1980s are changing, and sustainable agriculture faces several fiscal and political vulnerabilities. A typology of sustainable agriculture research is developed to illustrate the point that scientists and proponents of sustainable agriculture can help to address these vulnerabilities by deversifying their approaches to developing environmentally sound agricultural technologies.

Introduction

The past 6 years have witnessed a significant reorientation of agricultural research institutions across the advanced industrial and developing countries. It was scarcely more than a decade ago that notions such as sustainable agriculture, alternative agriculture, low-input agriculture, and so on were dismissed within the world's agricultural scientific communities as being ideological, unscientific, utopian, too costly, or inconsistent with a productive agriculture. Over a very short span of time sustainable agriculture has been embraced by the agricultural research institutions in the developed and developing worlds. 'Sustainable development' has become commonly used in international development circles, and there is scarcely an international development agency or research center today that does not place a high priority on publicizing how its programs and strategies are consistent with sustainability. At the same time, the agricultural research establishments of the advanced countries are now adopting 'low-input' or 'sustainable systems' and establishing special centers, institutes and committees to coordinate and rationalize these research thrusts.

Sociologists and other social scientists have played a significant role in the emergence, institutionalization, and design of sustainable agriculture, both here and abroad. In the US, sociologists and other social scientists were amply represented among the small cadre of persons who pointed to the need for agricultural research institutions to take resource conservation and alternative approaches more seriously. Some of the major non-governmental organizations and public interest groups that have been pivotal in building the case for sustainable agriculture research (e.g. the Institute for Alternative Agriculture and the Center for Rural Affairs) have been staffed mainly by persons trained in the social sciences. Sociologists and other social scientists have done particularly significant research on the adoption of resource-conserving practices. They have also made major contributions through their research into identifying user needs and implementation strategies relating to sustainable agriculture technology (National Research Council (NRC), 1989; Office of Technology Assessment, 1990). As suggested by the title of this paper, these social science contributions are referred to as being sociology in sustainable agriculture.

The basic premise of this paper, however, is that sociology and the other social sciences such as anthropology and political science can play an equally important and constructive role in understanding and assessing agricultural sustainability. This kind of application of sociology may be referred to as the sociology of agricultural sustainability. Insights that can come from the sociology of agricultural sustainability are explored. The author's specific concern is to contribute to an understanding of the origins, bases, and future of sustainable agriculture. He discusses the nature of sustainable agriculture as a social movement, and how the 'movement character' of sustainable agriculture is reflected in how we conceive and research agricultural sustainability. The major social forces that will affect the long-term viability of sustainable agriculture are also stressed.

A good many people who do research on agricultural sustainability tend to see the growing interest in this work as being the result of growing scientific knowledge and appreciation of the need for agriculture to be more environmentally benign. This notion has a significant element of truth, but it is by no means clear that our knowledge on these matters was significantly greater in 1988 than it was in 1980, when the Council for Agricultural Science and Technology (CAST) (1980) did not hesitate to publish a report which belittled the notion that one could farm efficiently without use of agricultural chemicals.

The author's main argument is that while the current enthusiasm for sustainability is clearly promising, it is precarious. He considers the rise of sustainable agriculture to have occurred on account of some fortuitous or unanticipated political circumstances, such as farm crisis of the mid-1980s, a trend toward declining public trust in science, and the development of the international environmental movement. The farm crisis provided a compelling rationale for researching and adopting production systems based on minimizing the use of purchased inputs; in the process, the farm crisis atmosphere of this time also substantially shaped our definition, and thus our conceptualization, of what sustainable agriculture is, i.e. low (chemical)-input agriculture. Perhaps more fundamentally, sustainable agriculture was adopted for social reasons rather because of its technical merits. This is not to say that the technology of sustainable agriculture is without merit, but rather that the impetus to expansion of research on the topic was more by non-scientists, social values, and social momentum than by the scientific community or by major scientific breakthroughs. Specifically, the spectacular rise of sustainable agriculture as a symbol and as a concrete program of research and outreach had its origins in the growing environmental movement in American society, which itself was very strongly anchored in international environmental mobilization efforts focused on global warming, global environmental change, and loss of biodiversity.

This new environmental milieu affords many opportunities to accelerate resource conservation in agriculture. However, we should recognize that the circumstances of the rise of sustainable agriculture raise important questions about its long-term viability. Is the wave of sustainable agriculture based on temporary and non-replicable conditions of the 1980s that will inevitably be reversed sometime in the near future? Alternatively, does the agricultural sustainability thrust reflect a more enduring ideological redirection within scientific and policy-making circles? Has the rise of sustainable agriculture led to sufficient organizational restructuring of agricultural research institutions and the agricultural science communities so that it is substantially institutionalized and poised to achieve sustained advances? Is the research that has been done during this early phase of land-grant university sustainable agriculture development likely to lead to on-farm successes, justify further funding, and lead to knowledge that can help redesign agricultural ecosystems?

The social origins of sustainable agriculture

Prior to the late 1970s, the major environmental groups in the advanced countries paid relatively little attention to agriculture, being more fundamentally concerned with wildland preservation and with urban–industrial sources of pollution and environmental degradation. It is useful to recall that even Carson's widely circulated and provocative book, Silent Spring (1962), focused more on the threats of pesticides to fish, wildlife, and human health rather than on the impacts of pesticides on the ecological integrity of agricultural systems. The constituency for sustainable agriculture was confined mainly to a few groups and organizations such as the Rodale Institute and small organic farming movements in a few US states.

Increasingly, the environmental movement came to recognize that agriculture is a crucial environmental sector because it involves so much of the land base and is pivotal in affecting the quality and quantity of water resources.

Environmental attention to agriculture thus slowly but surely accelerated during the 1970s and early 1980s. This stage was typified by the fact that the Carter/Bergland Administration published the Report and Recommendations on Organic Farming (US Department of Agriculture (USDA), 1980).

Environmental efforts achieved sufficient momentum in the early to mid-1980s to move into a second phase. The early and mid-1980s were pivotal in several respects because this was the period of the height of the farm crisis, of rising real interest rates, heavy farm debt loads, collapse of the export market, and decapitalization of farmland assets. The growing incidence of farm financial stress was a propitious moment for proponents of low-input farming systems. The farm crisis gave credibility and lent urgency to the claim that these systems would help to minimize farmers' need to allocate scarce investment capital to purchased agrochemical inputs, and that sustainable agriculture could thus help debt-stressed farm operators to weather the crisis (Buttel et al., 1986, 1990a). Accordingly, there was a very significant growth in farmer interest in sustainable agriculture, which contributed to the growing pressure for more research. Sustainable agriculture research was essentially incubated in this period of farm crisis, was heavily promoted by farm crisis activists, and was given strong support by farmers heavily in debt, thereby placing a particular stamp on how we define sustainable agriculture. Sustainable agriculture has come to be largely synonymous with low chemical-input practices.

It was also significant that the farm crisis was not simply a crisis of high debt and low commodity prices. It was also a crisis of overproduction, the obvious solution to which — mandatory production control — was an anathema to the US administration. These conditions were thus also propitious for federal environmental legislation that could achieve some measure of production control. These factors opened the door for some major agricultural environment-related initiatives — the Conservation Reserve Program of the 1985 Farm Bill and the USDA Low-Input Sustainable Agriculture (LISA) initiative — that were strongly advocated by environmentalists, and which were not strongly opposed by the administration because they could serve essentially as disguised production control programs that would help increase farm commodity prices.

Agricultural sustainability was given particular momentum later in the decade as a result of several catalytic forces. In 1983 Gro Harlem Brundtland, then Prime Minister of Norway, was commissioned by the Secretary General of the United Nations to chair the Commission that prepared the Brundtland Report (World Commission on Environment and Development, 1987). The Commission, which included very strong representation of agricultural research policy reformers, including Robert Chambers, Calestous Juma, and Lester Brown among its advisory panels and authors of commissioned papers, had a significant impact on the environmental NGO communities – even before the report was published. The Brundtland Commission was particularly influential in conceptualizing environmental degradation as an impediment to agricultural and overall economic development. The Commission made the case that dealing with the growing threats posed by global environmental problems could be consistent with, rather than detract from, increases in economic development and productivity. The Commission was particularly influential in codifying the concepts of sustainability and sustainable development and their relevance to both socio–economic development and environmental conservation. In the US, these notions were adopted very rapidly within the environmental and farm policy reform groups that pushed for the LISA initiative and other pro-environmental provisions of the 1985 and 1990 Farm Bills.

A parallel factor in the rise of agricultural sustainability was the drought of the summer of 1988, and the role the drought played in precipitating the internationalization and expansion of environmental activism. Environmental groups had been aware for several years about the growing speculation within some quarters of the atmospheric science community that the world was in the early stage of a global warming trend — one concomitant of which would be rising temperatures and instability of rainfall in the temperate agricultural regions. The 1988 drought prompted environmental groups to publicize global warming and greenhouse gases, despite some misgivings such as being aware of the fact that reduction of greenhouse gases might compel greater reliance on nuclear power.

A global warming and global environmental change based program had several advantages from the standpoint of environmental activism. These themes could be very influential with the public and with policy-makers, since they could be buttressed with claims that calamities and disorder such as massive biospheric disruption, coastal inundation, destruction of the agricultural productivity of the temperate agricultural breadbaskets, and rising levels of skin cancer would occur if no action was taken to reduce pollution of the atmosphere. Global warming and associated themes, such as reducing global environmental change, stemming the depletion of stratospheric ozone, and promoting biodiversity conservation, could serve as a comprehensive, overarching justification for a lengthy agenda of environmental goals. The need to reduce industrial pollution, achieve greater energy conservation, conserve tropical rainforests, and so on could all be justified through one overarching imperative — to stem global warming — rather than each goal having to be justified on its own particular merits. This 'global packaging' was attractive, since it obviated the need for environmentalists to achieve multiple goals in many places simultaneously. Persuasive claims about global warming and global change could also provide a strong rationale to override the international, national, and subnational political dynamics that normally make environmental goals difficult to achieve.

The rise of international environmental activism, in which notions of 'sustainable development' and 'sustainability' were widely employed to engender confidence that environmental reform would not cause economic hardships, was of major benefit to the agricultural sustainability agenda. The sustainability notion was given particular legitimacy because of the urgency that arose concerning a response to the threats that global environmental change posed to human living standards and well-being.

Propelled by the unique combination of forces of the 1980s, sustainable agriculture had achieved consideration by the end of the decade. Sustainable agriculture programs have become common, though quite modest in scope, across land-grant and related institutions. Those scientists, administrators, and policy-makers who are ambivalent about or opposed to research strategies based on reducing chemical inputs have been partially neutralized. Even though some agricultural scientists privately express strong criticism about agricultural–environmental activism and the value of LISA-type research, they have had to tone down their opposition. For example, CAST's (1990) critique of the NRC Alternative Agriculture report included some favorable statements about the NRC report by several scientists. Interestingly enough, those opposed to the sustainability concept are often forced to couch their opinions, including criticisms of low-input agriculture, in sustainability terms, by stressing how the approaches and the technologies they favor are as, or more, 'sustainable' than the ones advanced by sustainability advocates. Moreover, the sustainable agriculture movement has no doubt been enhanced by the fiscal austerity in the land-grant system today; the availability of federal LISA funds has become particularly attractive now that formula funding for applied agricultural research is so constrained.

The future of sustainable agriculture

It seems fair to say that although gains have been achieved, the momentum behind sustainable agriculture has slowed, and there are growing concerns about the staying power of sustainable agriculture. This has occurred for several reasons.

First, the farm crisis has essentially passed from an acute to an almost invisible chronic condition as farm debt has been restructured, farmland assets have been devalued, real-estate related production costs have declined, and export revenues have increased. Thus, the input cost minimization rationale of sustainable agriculture may be declining, and the strength of the production control rationale for environmental protection is not as strong as it was a half dozen years ago. Hence, federal LISA funding has become vulnerable, and its termination would undercut sustainable agriculture efforts. It is not clear whether National Research Initiative funding will add significantly to, or be able to replace, LISA funding. In addition, the fiscal problems of state governments will probably preclude any major new special appropriation programs for sustainable agriculture over the foreseeable future.

Second, while global environmental activism has given agricultural sustainability increased visibility, this activism has been largely focused around atmosopheric pollution and on rainforests and rainforest biodiversity conservation. In some quarters of the movement, agriculture is actually seen more as a problem (i.e. agricultural destruction of rainforests) than an opportunity (Buttel, 1992). One of the potential concomitants of the growing fascination with rainforests — and of the declining interest in the ecological zones, such as most agricultural ones, where the majority of the Third World rural poor live — is that attention and resources may be diverted from agricultural research and development programs.

Third, we must recognize that the environmental movement has depended on its ability to sustain the stylized scientific facts of global warming, loss of biodiversity, and global environmental change in general. There are growing signs of hesitation within the scientific community about the empirical basis of many of global warming's received claims, especially about whether global circulation models are sufficiently robust to make confident predictions about global warming. The same can be said about claims of the biosphere-threatening impacts of the loss of biodiversity (Mann, 1991). It is no accident that some of this contradictory scientific evidence is being generated by persons and groups that are politically motivated to diminish the influence of environmental groups. However, much of this concern comes from persons who are not so motivated (Buttel et al., 1990b). For example, Bryson (1990), one of the pioneers of atmospheric modeling and early prophets of CO_2-induced global warming, maintains that the evidence from computer models of atmospheric pollution and global warming at this point is insufficient to justify claims of the inevitability of a major biosphere-threatening 'greenhouse effect'. Thus, it is unclear whether and how long the environmental movement will be able to sustain the major stylized facts of global environmental change and global warming. The questioning, and possible undermining, of the case for global warming could cause the rest of the global warming-linked agenda (such as sustainable development, biodiversity and rainforest conservation, and perhaps sustainable agriculture) to be diminished in tandem.

Fourth, the disappointing results of the June 1992 United Nations Conference on Environment and Development (UNCED), widely referred to as the 'Earth Summit', may have diminished the momentum that environmental awareness exhibited in the late 1980s. In addition to Bush Administration recalcitrance over the biodiversity convention, the story of the Earth Summit was one of acrimony between developed and developing countries. Developing country delegates to UNCED expressed resistance to proposals for forest conservation policies on the grounds that they are an unwarranted interference in their domestic policies, and hence with their national sovereignty. Most Third World delegates to UNCED also insisted strongly that their conservation efforts should be subsidized through expanded foreign aid programs by the advanced industrial countries. The advanced industrial countries, which see themselves entering a new era in which foreign aid no longer has either the domestic support or the geopolitical rationale it had during the height of the Cold War, were generally unwilling to make major commitments at UNCED for vast new infusions of 'green foreign aid'. While the long-term implications of UNCED are not yet clear, it seems likely that strong differences of interest and approach between developed and developing countries, along with the global recession of the early 1990s, will make environmental policies more difficult to implement.

The need to diversify the approach to sustainable agriculture

It was noted earlier that the predominant approach to agricultural sustainability research has been that which developed in the aftermath of the 1980's farm crisis. Given the new vulnerabilities faced by the sustainability movement, the time is ripe to consider whether new approaches to sustainability are needed. Table 1 shows that we can think of four major approaches to agricultural sustainability, along two dimensions: basic vs. applied research, and traditional land-grant productionist versus non-traditional research orientations. It identifies four major types of research that could and should be followed to pursue agricultural sustainability. These include (1) what we might call LISA-type agronomy, or in other words the dominant type of sustainable research, (2) what we might call chemical-substitutionist biotechnology, i.e. biotechnology research in which molecular biology tools are used, for example, to develop genetically engineered crop varieties and microorganisms (e.g. biopesticides, biofertilizers) that can substitute for purchased chemicals (Goodman, 1989), (3) agroecology (Carroll et al., 1990), and (4)

Table 1
A typology of approaches to sustainable agriculture

Basic versus applied research	Nature of the research	
	Traditional productionist research	Non-traditional research
Basic science	Input-substitution biotechnology	Agroecology
Applied science	LISA-type agronomy	Policy-led sustainable agriculture

'policy-led sustainable agriculture' (Faeth et al., 1991). Note that policy-led sustainable agriculture is only one example of an applied/non-traditional approach. By government policy-led sustainable agriculture, is meant an approach to sustainability that seeks to achieve a more environmentally sound agriculture through implementing public policies that give producers strong incentives to achieve sound environmental performance. The policy-led sustainable agriculture category is stressed here because it is often ignored in discussions of sustainable agriculture, and there is ample evidence to suggest that the American public policy environment is at least as important a barrier to agricultural sustainability as the lack of appropriate technology per se. Research on policy-led agricultural sustainability would thus consist of assessing comprehensively the social and ecological costs and benefits of various government policy approaches to encouraging sustainable practices (e.g. Faeth et al., 1991). Examples of such prospective policies might include applying a polluter-pays principle to agricultural sources of water pollution (similar to that employed now to industrial and municipal sources of pollution), including natural resource productivity criteria in assessments of agricultural commodity program alternatives, or imposing substantial flat-rate taxes (greater than 100%) on agricultural chemicals.

As noted earlier, sustainable agricultural research remains closely identified with applied LISA-type research of the sort that emerged during the US farm crisis atmosphere of the 1980s. There is much to be said for the typical LISA approach. It is generally based on relatively well-established lines of research and knowledge, often lends itself to on-farm trials, and accordingly has the potential to be delivered relatively quickly to farmers. Often this research is undertaken in very close interaction with extension and other outreach activities. Since organic and alternative farmer groups have long been among the more vocal clientele requesting this type of research and are understandably impatient for the land-grant system to deliver, this approach is best suited to satisfying those who want rapid results.

However, the fact that much of the land-grant sustainable agriculture program is focused on LISA-type agronomic research involves some weaknesses. There are two problems with the current approach. One is that the strengths of the LISA approach — that it is based heavily on employing existing knowledge about crop rotations and integration of crop and animal enterprises — may also be precisely its longer-term weakness. The gains that can be achieved may be exhausted fairly rapidly, and there will not be sufficient basic research and system-redesign knowledge in progress to serve as a basis of sustainable agriculture after the turn of the century. Despite the applied character of the research into sustainable systems based on LISA approaches, it still may be adequate if the public policy environment for sustainable agriculture remains adverse. LISA-type production systems may well have limited appeal if US commodity policies essentially continue to amount to subsidies of monoculture and continuous cropping (NRC, 1989; Faeth et al., 1991).

A second weakness is that each of the four types of agricultural sustainability work potentially has a significant role to play, meaning that some important opportunities are being missed. It seems that highly applied LISA-type agronomic research is also the type of research that is playing a declining role in the land-grant university and Agricultural Research Service systems as they continue to move toward basic research (Bush et al., 1991). There is the significant threat that this work would be among the first to be cut if financial support had to revert mainly to formula or other sources of institutional funding.

There are several other limitations to how we conceive of sustainable research today. It remains focused on reducing, or eliminating, the use of chemical inputs. Reduction of chemical inputs is an obvious and important place to start, but chemical use is not the only environmental problem of US agriculture. The emphasis on reduction of purchased inputs may cause us to ignore the fact that sustainable agriculture will, in all likelihood, have to be based on developing new inputs, such as new crop varieties and new biological control agents.

It is also surprising that cultivars tend to remain an exogenous component of most sustainable agriculture systems. It is well established from studies of environmentally benign, resilient traditional agroecosystems that the role of cultivars (and human-directed coevolution and indigenous knowledge) is integral to the overall structure of ecosystem processes. Sustainable agriculture research, however, is seldom built around cultivars, or even plant breeding. Instead, sustainable agriculture research more often than not tends to use commercially developed varieties, mostly developed for use under high-input conditions. This is not entirely the fault of sustainable agriculture scientists. Public breeding in general, and breeding of finished crop varieties in particular, has declined greatly in the land-grant system since the late 1970s. Public plant breeding is increasingly shifted toward basic research aimed at developing advanced breeding lines that are to be transferred to seed companies. Nonetheless, the next generation of sustainable agriculture research, both LISA and non-LISA in orientation, will need to treat more systematically the genetics and ecology of crops, biological inputs, and other biota at a variety of levels – from the macro (ecosystem) to micro (quantitative or molecular genetics, ecological genetics) levels. Both basic agroecology and biotechnology can have important roles to play here.

Sustainable agriculture in American agricultural research institutions has given insufficient attention to basic agroecological research (and to related areas such as evolutionary biology and conservation biology). This is by no means solely the fault of the land-grant and USDA systems. Agroecosystem research is not valued highly within the discipline of ecology and within non-USDA federal basic research funding programs. The sustainable agriculture community as a whole has given little stress to basic agroecology research, and the land-grant system has given little attention to building basic research in areas other than biotechnology. This has permitted the perpetuation of the misleading notion that basic research in agriculture can be considered coterminous with biotechnology. As much as biotechnology has the potential to contribute to sustainable agriculture, much of it — perhaps most — will not. Biotechnology can contribute substantially to agricultural sustainability only to the degree that it is quite deliberately harnessed in this direction.

Conclusions

The increased emphasis on agricultural sustainability is promising but has vulnerability. Just as sustainable agriculture needs to be built around biological diversity, to ensure a sound future for sustainable agriculture this field must be built on a diversity of rationales, appeals, and strategies. Proponents of sustainable agriculture should actively and vigorously debate the virtues of various approaches, but also recognize the the other approaches have the potential to contribute. Such diversity has not always been the case. There is a submerged, but pervasive, tendency for the sustainability community to be internally divided, mainly along the dimensions depicted in the typology of approaches in Table 1. Many agroecologists are skeptical of LISA-type research as being no less reductionist than conventional agronomy. LISA-based agronomists often tend to be skeptical of agroecology, and often say that agroecology is too esoteric, was a failure in its application to integrated pest management (IPM), and cannot yet be applied to solving the problems of intensively cultivated, high-productivity agroecosystems. There are all manner of criticisms of biotechnology (Hobbelink, 1991), on grounds of its reductionism, of concern that genetically modified organisms might involve ecological risks of their own, that it will exacerbate genetic uniformity, and that there is a close identification of biotechnology research with encouraging use of agrochemicals (e.g. herbicide-tolerant crop varieties, seed-chemical packages). However, different approaches to sustainable agriculture have both strengths and weaknesses, and it is not really possible at this early stage to predict the contributions each will be able to make over the next 20 to 30 years. Diversification of sustainable agriculture research is an insurance policy against possible failure or limitation of any one approach.

Finally, it is useful to note that the degree to which sustainability is achieved may depend, over the long-term, as much on public policy (and other social) factors as on the ability of the research community to deliver new technology. The most important of such factors will be the degree to which there is public regulation of agricultural–environmental performance of production agriculture and the role that agricultural (particularly commodity program) policies play in discouraging and encouraging resource-conserving and resource-degrading production systems. Sustainable, resource-conserving, environmentally sound technologies will have limited applicability and benefit as long as environmental and agricultural policies discourage their use.

Acknowledgments

This research was supported by the Cornell University Experiment Station Hatch Project No. 438, by USDA Regional Research Project NC-208, and by the Wisconsin Agricultural Experiment Station.

References

Bryson, R.A., 1990. Will there be a global 'greenhouse warming'? Environ. Conserv., 17: 97–99.

Busch, L., Lacy, W.B., Burkhardt, J. and Lacy, L.R., 1991. Plants, Power, and Profit. Basil Blackwell, Oxford, 275 pp.

Buttel, F.H., 1992. The 'environmentalization' of plant genetic resources: possible benefits, possible risks. Diversity, 8: 36–39.

Buttel, F.H., Gillespie, Jr., G.W., Janke, R., Caldwell, B. and Sarrantonio, M., 1986. Reduced-input agricultural systems: rationale and prospects. Am. J. Alternative Agric., 1: 58–64.

Buttel, F.H., Gillespie, Jr., G.W. and Power, A., 1990a. Sociological aspects of low-input agriculture in the United States: a New York case study. In: C.A. Edwards, R. Lal, P. Madden, R.H. Miller and G. House (Editors), Sustainable Agricultural Systems. Soil and Water Conservation Society, Ankeny, IA, pp. 515–532.

Buttel, F.H., Hawkins, A. and Power, A.G., 1990b. From limits to growth to global change: constraints and contradictions in the evolution of environmental science and ideology. Global Environ. Change, 1: 57–66.

Carroll, C.R., Vandermeer, J.H. and Rossett, P.M. (Editors), 1990. Agroecology. McGraw-Hill, New York, 641 pp.

Carson, R., 1962. Silent Spring. Fawcett. Greenwich, CT, 304 pp.

Council for Agricultural Science and Technology (CAST), 1980. Organic and Conventional Farming Compared. Report No. 84. CAST, Ames, IA, 32 pp.

Council for Agricultural Science and Technology (CAST), 1990. Alternative Agriculture: Scientists' Review. CAST, Ames, IA, 182 pp.

Faeth, P., Repetto, R., Kroll, K., Dai, Q. and Helmers, G., 1991. Paying the Farm Bill: US Agricultural Policy and the Transition to Sustainable Agriculture. World Resources Institute, Washington, DC, 70 pp.

Goodman, R.M., 1989. Biotechnology and sustainable agriculture: policy alternatives. In: J.F. MacDonald (Editor), Biotechnology and Sustainable Agriculture. National Agricultural Biotechnology Council, Ithaca, NY, pp. 48–57.

Hightower, J., 1973. Hard Tomatoes, Hard Times. Schenckman, Cambridge, MA, 268 pp.

Hobbelink, H., 1991. Biotechnology and the Future of World Agriculture. Zed Books, London, 159 pp.

Mann, C.C., 1991. Extinction: are ecologists crying wolf? Science, 253: 736–738.

National Research Council (NRC), 1989. Alternative Agriculture. National Academy Press, Washington, DC, 448 pp.

Office of Technology Assessment (OTA), 1990. Beneath the Bottom Line. OTA, Washington, DC, 337 pp.

US Department of Agriculture (USDA), 1980. Report and Recommendations on Organic Farming. USDA, Washington, DC, 94 pp.

World Commission on Environment and Development, 1987. Our Common Future. Oxford University Press, New York, 383 pp.

Agriculture, Ecosystems and Environment, 46 (1993) 187–196
Elsevier Science Publishers B.V., Amsterdam

Water quality and the environment

Managing pesticides for crop production and water quality protection: practical grower guides

Arthur G. Hornsby[a,*], Tasha M. Buttler[b], Randall B. Brown[a]

[a]*Soil Science Department, PO Box 110290, University of Florida, Gainesville, FL 32611-0290, USA*
[b]*MANTEQ (DowElanco), PO Box 1706, Midland, MI 48641-1706, USA*

Abstract

A decision aid has been developed for use by farmers and other pesticide users that permits the selection of pesticides on the basis of water quality impact in addition to efficacy and cost as is traditionally done. This 'kitchen table' decision aid uses information on environmental fate and toxicity of pesticides, soil leaching potential or run-off ratings in combination with pest management guidelines to make site-specific selections that minimize the potential impacts of pesticide use on water quality. Pesticide properties used include the sorption coefficient (K_{oc}), degradation half-life, lifetime health advisory level (HAL), and aquatic toxicity. The first two properties are used to derive a relative leaching potential index, RLPI, and a relative run-off potential index, RRPI, for each pesticide. Site-specific soil ratings for pesticide leaching and run-off, provided by the US Department of Agriculture Soil Conservation Service, are matched with these indices and toxicity values via a table of selection criteria. A pesticide selection worksheet is provided to organize the necessary information to make an informed decision. This decision aid assists the pesticide user in understanding the water quality impacts of alternative pesticide selections. These grower guides are customized to individual crops. They include a table of pesticides registered for use on the crop, organized by major use category (herbicides, insecticides, nematicides, fungicides, and soil fumigants). Within categories, trade names, common names, application type (soil vs. foliar), K_{oc}, RLPI, RRPI, HAL, and aquatic toxicity are listed. These grower guides can easily be adapted for use in other states or countries. This technology will not only help agricultural and urban pesticide users select pesticides with less water quality impact, but will also aid regulatory agencies in developing meaningful groundwater protection plans, and hopefully will result in more equitable public policy decisions.

Introduction

Until recently, environmental fate data have not been made available by either the US Environmental Protection Agency (USEPA) or product manufacturers that can be used to assist pesticide users in selecting appropriate products to minimize potential adverse impact on surface or groundwater quality. Wauchope et al. (1992) published an extensive database of pesticide environmental fate properties that can be used for making environmentally based decisions. Thus, only recently have comprehensive screening proce-

---

*Corresponding author.

dures based on the environmental fate data been feasible to aid pesticide users in a selection process.

Pesticide selection has historically been based on expected efficacy and cost of the product. The pesticide user is not likely to experiment with other products once a product has been found that controls the pests of concern. The possibility of selection criteria for products that are effective, cost efficient and environmentally benign is now an emerging reality. This emerging management capability will not only help the agricultural and urban pesticide users but will also help regulatory agencies understand the more subtle issues related to pesticide usage and, hopefully, result in more equitable public policy decisions.

Practical grower guides for pesticide selection

Both the US Department of Agriculture (USDA) Soil Conservation Service (USDA/SCS) and the Florida Cooperative Extension Service (FCES) have developed decision aids to assist their clients in selecting pesticides to avoid or reduce adverse water quality impacts. These decision aids differ slightly in their formulation but provide a much needed methodology that farmers and other pesticide users can easily use to include environmental considerations in their pesticide application decisions (Goss and Wauchope, 1990; Hornsby et al., 1991). Both approaches consider the environmental fate of pesticides and soil properties at the site of application.

The FCES methodology presented herein consists of a two-tier procedure that considers not only the potential for leaching and/or run-off but also toxicological impacts. Toxicological considerations are very important facets of this decision aid. The selection criteria are based on matching the soil properties at the application site to the environmental fate and toxicological parameters of the pesticides. These parameters include two derived indices (for leaching and run-off) and two toxicological parameters (USEPA lifetime health advisory levels and aquatic toxicity).

A 'pesticide selection worksheet' (Fig. 1) is used to organize the necessary information, to consider alternatives, and to make an informed decision. While this approach does not deal explicitly with efficacy, or with differences in application amounts and costs of alternative products, the user is reminded of these aspects in the FCES circulars that detail the procedure. Circulars have been developed for 55 crops grown in Florida (and elsewhere).

Pesticide parameters

Relative leaching potential index

The relative leaching potential index or RLPI defines the relative attenuation (reduction in mass as it moves through the soil) of each pesticide in soil,

PESTICIDE SELECTION WORKSHEET

Landowner/Operator Name: _____   Date: _____

Crop: _____   County: _____   Field ID _____

Farm ID: _____   Sheet ___ of ___

Target Pest (1)	IFAS Recommended Pesticides (2)	K_{oc} Value (3)	Realitive Losses Leaching RLPI (4)	Runoff RRPI (5)	Toxicity Lifetime MWLEQ* (6)	Aquatic Toxicity (7)	Soil Type (8)	Soil Leaching Rating (9)	Soil Runoff Level (10)	Selected Pesticide (11)	Comments (12)

If the K_{oc} value is 100 or less or if the RLPI value is 10 or less and the soil leach rating is high, then the pesticide has a high potential for leaching and should be used with extreme caution. Alternative pesticides and reduced rates should be considered if possible. Apply pesticide during periods with low potential for rainfall if possible.

Fig. 1. Example of the pesticide selection worksheet.

and therefore its potential to leach to ground water. The index is calculated by multiplying the ratio of the organic carbon sorption coefficient and the degradation 'half-life' by 10. The index is an integer. Values greater than 2000 are assigned a value '>2000'. Values between 1000 and 2000 are rounded to three significant digits. Values less than 1.0 are rounded up to 1. This ratio defines the relative attenuation of pesticides over a wide range of soils. Pesticides having high K_{oc} values (strongly adsorbed) and low degradation half-lives (low persistence) will have high RLPI values. Pesticides that are very mobile, for example, those having K_{oc} values less than 100 in sandy soils, or 50 or less in fine-textured soils, should be used with caution. There is some uncertainty in the data used to calculate this index. However, since the values are relative they can still be used. It is important to realize that the smaller the RLPI value of a pesticide the greater is its potential to leach.

Relative run-off potential index

The relative run-off potential index or RRPI defines the relative immobility of each pesticide in soil, and therefore its potential to remain near the soil surface and to be subject to loss in run-off either sorbed to eroded sediment or in the aqueous phase. This index represents the combined sediment and aqueous phase run-off potential and is calculated as follows.

(a) If K_{oc} is 1000 or greater, then the RRPI is 1 000 000 divided by the product of the sorption coefficient times the degradation half-life.
(b) If the K_{oc} is less than 1000, then the RRPI is the smaller of (i) the value calculated in (A), or (ii) the value calculated for the RLPI.

This index is an integer and values greater than 1000 are assigned the value '>1000'. Values less than 1.0 are rounded up to 1. As with the RLPI, there is some uncertainty in the data used to calculate this index. However, since the values are relative they can still be used. The smaller the RRPI value of a pesticide the greater is its potential to be lost in run-off.

Lifetime health advisory level or equivalent

The lifetime health advisory level or equivalent (HAL or HALEQ) provides a measure of pesticide toxicity to humans. The lifetime health advisory level as defined by USEPA (1989) is the concentration of a chemical in drinking water that is not expected to cause any adverse health effects over a lifetime of exposure (70 years), with a margin of safety. The values used are the USEPA lifetime health advisory level, HAL, or an equivalent value, HALEQ (denoted by a superscripted asterisk), calculated using the USEPA formula (HALEQ = RfD × 7000, where RfD is the reference dose determined by the USEPA). For non-carcinogenic pesticides the calculated HALEQ should not differ by more than a factor of 10 from the values forthcoming from the USEPA. HAL and HALEQ have units of micrograms per liter (ppb). The smaller the value the greater is the toxicity to humans.

Aquatic toxicity

The aquatic toxicity provides a measure of pesticide toxicity to aquatic species. The values used are the lethal concentration at which 50% of the test species die (LC_{50}). Unless otherwise noted by a lower-case letter following the value, the test species was rainbow trout. The smaller the value the greater is the toxicity to aquatic species.

Parameter convention

Data for K_{oc}, RLPI, RRPI, HALEQ, and aquatic toxicity are given for the active ingredient (common name) of a product as shown in an abbreviated form in Table 1 (Hornsby et al., 1991). Such data for the pesticides identified to control the pest of interest are then transferred to the pesticide selection worksheet (Fig. 1). When using a product that is a mixture of two or more active ingredients, one uses the RLPI, RRPI, HALEQ, and aquatic toxicity values for the most restrictive active ingredient in the mixture, i.e. one lists the smallest RLPI value, the smallest RRPI value, etc. among those for the active ingredients of the mixture.

Soil parameters

The following criteria were developed by the Florida USDA/SCS in collaboration with the FCES (Brown et al., 1991) to rate soils for leaching and run-off.

Leaching

Factors that determine the soil leaching rating are the soil permeability and the occurrence of mucky layers in the upper 2 m of the soil (Table 2). Exceptions are: (1) soils with a muck or peat layer are rated Low; (2) soils with a mucky layer are rated Medium unless the soil has a slowest permeability of less than 1.5 cm h^{-1}; then the soil is rated Low.

Run-off

The factors that determine the soil run-off rating are hydrologic group, permeability, and slope (Table 3). Exceptions are: (1) soils that are frequently flooded during the growing season are rated High; (2) soils rated Low

Table 1

Field corn — abbreviated pesticide parameter matrix for selecting pesticides to minimize water quality problems

Trade name[1]	Common name	Application type[2]		Sorption coefficient[3]	Relative losses		Toxicity	
		Soil	Foliar	K_{oc} (ml g^{-1})	Leaching	Run-off	HAL or HALEQ[4] (ppb)	Aquatic LC$_{50}$[5] (ppm)
					RLPI[6]	RRPI[7]		
Herbicide								
2,4-D Amine	2,4-D Amine		×	20	20	20	70	100
Aatrex	Atrazine	×	×	100	17	17	3	4.5
Atrazine	Atrazine		×	100	17	17	3	4.5
Banvel	Dicamba		×	2	1	1	200	28
Basagran	Bentazon		×	34	17	17	20	635
Bicep (M)	Metolachlor		×	200	22	22	100	2
Bicep (M)	Atrazine		×	100	17	17	3	4.5
Bladex	Cyanazine	×	×	190	136	136	10	9
Brominal	Bromoxynil		×	10000 E	>2000	14	100*	0.1
Bronco (M)	Glyphosate		×	24000 E	>2000	1	700	8.3
Bronco (M)	Alachlor		×	170	113	113	0.4	1.4
Buctril	Bromoxynil		×	10000 E	>2000	14	100*	0.1
Conquest (M)	Atrazine	×	×	100	17	17	3	4.5
Conquest (M)	Cyanazine	×	×	190	136	136	10	9
Dacthal	DCPA		×	5000	500	2	4000	100a
Dual	Metolachlor	INC	×	200	22	22	100	2
Eradicane	EPTC	INC		200	333	333	200*	17
Evik	Ametryn		×	300	50	50	60	3.2
Goal	Oxyfluorfen		×	100000 E	>2000	1	20*	0.2b
Gramoxone	Paraquat	×	×	100000 E	1000	1	30	15
Karmex	Diuron	×		480	53	23	10	4.9
Lasso	Alachlor	INC	×	170	113	113	0.4	1.4
Lexone	Metribuzin		×	60	15	15	200	76
Linex	Linuron	INC	×	400	67	42	10*	16
Lorox	Linuron		×	400	67	42	10*	16
Marksmen (M)	Atrazine		×	100	17	17	3	4.5
Marksmen (M)	Dicamba		×	2	1	1	200	28
Princep	Simazine	Inc		130	22	22	1	2.8
Prowl	Pendimethalin	×	×	5000	556	2	300*	0.199b
Ramrod	Propachlor	×		80	127	127	90	0.17
Roundup	Glyphosate		×	24000 E	>2000	1	700	8.3
Sencor	Metribuzin		×	60	15	15	200	76
Sutan+	Butylate	INC		400	308	192	350	4.2
Sutazine (M)	Atrazine	INC	×	100	17	17	3	4.5
Sutazine (M)	Butylate	INC		400	308	192	350	4.2
Tandem	Tridiphane		×	5600	2000	6	20*	0.53
Tomahawk (M)	Atrazine	×	×	100	17	17	3	4.5
Tomahawk (M)	Butylate	INC		400	308	192	350	4.2
Treflan	Trifluralin	INC		8000	1330	2	5	0.041

Trade name[1]	Common name	Application type[2]		Sorption coefficient[3]	Relative losses		Toxicity	
		Soil	Foliar	K_{oc} (ml g^{-1})	Leaching	Run-off	HAL or HALEQ[4] (ppb)	Aquatic LC_{50}[5] (ppm)
					RLPI[6]	RRPI[7]		
Insecticide/miticide								
Ambush	Permethrin		×	100000	>2000	1	400*	0.0041
Asana	Esfenvalerate	×	×	5300	1510	5	nd	0.00069j
Broot	Trimethacarb	INC		400 E	200	125	nd	nd
Comite	Propargite		×	4000 E	714	4	100*	0.12
Counter	Terbufos	INC		500	1000	400	0.9	0.01
Cygon	Dimethoate		×	20	29	29	1*	6.2
Cythion	Malathion		×	1800	>2000	556	200	0.2
Defend	Dimethoate		×	20	29	29	1*	6.2

[1]Trade name, (M) indicates that the product is a mixture of two or more active ingredients.
[2]Application type: INC, incorporated; ×, applied to soil surface or foliage.
[3]Sorption coefficient: E, estimated.
[4]HAL or HALEQ, lifetime health advisory level or lifetime health advisory level equivalent; *lifetime health advisory equivalent.
[5]Aquatic toxicity LC_{50}, value is for rainbow trout 48 or 96 h exposure time, unless otherwise specified; a, channel catfish; b, bluegill; j, fat head minnow.
[6]Relative leaching potential index (RLPI), smaller number indicates greater leaching hazard.
[7]Relative run-off potential index (RRPI), smaller number indicates greater run-off hazard.
nd, no data available.

Table 2
Criteria for rating soils for pesticide leaching

Rating	Criteria
High	Slowest permeability is 15.2 cm h^{-1} or more.
Medium	Slowest permeability is between 1.5 and 15.2 cm h^{-1}.
Low	Slowest permeability is 1.5 cm h^{-1} or less.

Table 3
Criteria for rating soils for pesticide runoff

Rating	Criteria
High	Soils in Hydrologic Group D in their natural, undrained state.
Medium	Soils in Hydrologic Group C, and any soils in Hydrologic Group B (in their natural, undrained state) that have a permeability of less than 15.2 cm h^{-1} within 51 cm of the soil surface.
Low	Soils in Hydrologic Group A, and any soils in Hydrologic Group B (in their natural, undrained state) that have a permeability of 15.2 cm h^{-1} or greater in all of the upper 51 cm of the soil profile.

Table 4
Abbreviated example of soil ratings for Manatee County

MUID[1]	SEQ NUM[2]	MUSYM[3]	Soil name[4]	Soil leach[5]	Soil run-off[6]
81001	1	1	Adamsville variant	High	Medium
81002	1	2	Beaches	High	High
81003	1	3	Braden	Medium	Medium
81004	1	4	Bradenton	Low	High
81005	1	5	Bradenton	Low	High
81006	1	6	Broward variant	High	High
81007	1	7	Canova	Low	High
81007	2	7	Anclote	Medium	High
81007	3	7	Okeelanta	Low	High
81008	1	8	Canaveral	High	Medium
81009	1	9	Canaveral	High	Medium
81010	1	10	Canaveral	High	Medium
81011	1	11	Cassia	High	Low
81012	1	12	Cassia	High	Low
81013	1	13	Chobee	Low	High
81014	1	14	Chobee variant	Low	High
81015	1	15	Delray	Low	High
81016	1	16	Delray	Low	High
81017	1	17	Delray	Low	High
81017	2	17	Eaugallie	Low	High

[1]MUID, Soil Conservation Service's map unit identifier.
[2]SEQ NUM, Sequence Number, indicating a particular soil name among one or more names constituting a map unit name.
[3]MUSYM, Map Unit Symbol from the soil map and legend in the Soil Survey of Manatee County, Florida. Note that if a MUSYM appears more than once in this list it signifies that two or more soils are co-dominant in that map unit, and each such soil is rated separately here.
[4]Soil name, Name of soil or other landscape component (e.g. urban land, pits, dumps, water).
[5]Soil leach, The rating of the soil for leaching of pesticides through the soil profile.
[6]Soil run-off, The rating of the soil for run-off of pesticides from the soil surface.

are changed to a rating of Medium where the slope is greater than 12%; (3) soils rated Medium are changed to a rating of High where the slope is more than 8%.

These criteria were used by the Florida SCS and FCES to rate soils in each county that has a published or interim soil survey. Table 4 is an abbreviated example of soil ratings for Manatee County, Florida (Hurt et al., 1991). The pesticide user need only locate the pesticide application site on the county soil survey map to identify the map unit(s) (MUSYM) shown on the field or area being treated and then find the corresponding MUSYM(s) in Table 4 to obtain the soil ratings (High, Medium, or Low) for leaching and run-off for the application site. These ratings are then transferred to the pesticide selection worksheet. These criteria may need to be modified for use in other states. The USDA/SCS National Water Quality Staff has developed ratings for all states using slightly different criteria. These soil ratings are available from local or state USDA/SCS offices.

Pesticide selection criteria

After the chemical data and soil ratings have been transferred to the pesticide selection worksheet (Fig. 1), criteria presented in Table 5 can then be used to select pesticides for the application site that will have the least impact on water quality. Table 5 provides decision criteria for the different soil ratings. By first looking at the soil ratings, selection is made by matching the soil ratings with appropriate selection criteria.

The 'Selection Criteria' encourage the user to 'move away' from the 'worst case', as defined by the smallest RLPI/RRPI and HAL/aquatic toxicity values, rather than defining the 'best choice'. The philosophy of the FCES is to provide information to enable the user to make an informed decision, but not to make the decision for the user. For some combinations of crops and pests there are few alternative selections. In this case, the user may not be able to select a product using these criteria. Nevertheless, the user is apprised of the likely consequences of their use and can proceed with prudent use of these products. One should note that use of this procedure in no way preempts requirements set forth on the product label.

The methodology set forth in this section has been published in a series of Extension Fact Sheets and Circulars by the FCES for joint use by the SCS in conservation plans and by the FCES with agricultural and urban pesticide users.

Table 5
Pesticide selection criteria

If soil ratings are		Then select pesticide with
Leach	Run-off	
High	Low	Larger RLPI value and larger HALEQ value
Medium	Low	Larger RLPI value and larger HALEQ value
Low	Low	Larger RLPI and RRPI values and larger HALEQ and aquatic toxicity values
High	Medium	Larger RLPI and RRPI values and larger HALEQ and aquatic toxicity values
Medium	Medium	Larger RLPI and RRPI values and larger HALEQ and aquatic toxicity values
Low	Medium	Larger RRPI value and larger aquatic toxicity value
High	High	Larger RLPI and RRPI values and larger HALEQ and aquatic toxicity values
Medium	High	Larger RRPI and RLPI values and larger aquatic toxicity and HALEQ values
Low	High	Larger RRPI value and larger aquatic toxicity value

References

Brown, R.B., Hornsby, A.G. and Hurt, G.W., 1991. Soil ratings for selecting pesticides for water quality goals. Circular 959. Florida Cooperative Extension Service, Institute of Food and Agricultural Sciences, University of Florida, Gainesville, FL, 4 pp.

Goss, D.W. and Wauchope, R.D., 1990. The SCS/ARS/CES pesticide properties database. II. Using it with soils data in a screening procedure. In: D.L. Weigmann (Editor), Pesticides in the Next Decade: The Challenges Ahead. Virginia Water Resources Research Center, VPI, Blacksburg, VA, pp. 471–493.

Hornsby, A.G., Buttler, T.M., Colvin, D.L., Sprenkel, R.K., Dunn, R.A. and Kucharek, T.A., 1991. Field corn: managing pesticides for crop production and water quality protection. Circular 982. Florida Cooperative Extension Service, Institute of Food and Agricultural Sciences, University of Florida, Gainesville, FL, 12 pp.

Hurt, G.W., Hornsby, A.G. and Brown, R.B., 1991. Manatee County: soil ratings for selecting pesticides. Soil Science Fact Sheet SL-86. Florida Cooperative Extension Service, Institute of Food and Agricultural Sciences, University of Florida, Gainesville, FL, 4 pp.

US Environmental Protection Agency (USEPA), 1989. Drinking water health advisory: pesticides. United States Environmental Protection Agency, Office of Drinking Water Health Advisories, Lewis Publishers, 121 South Main Street, Chelsea, MI 48118, 819 pp.

Wauchope, R.D., Buttler, T.M., Hornsby, A.G., Augustijn-Beckers, P.W.M. and Burt, J.P., 1992. The SCS/ARS/CES pesticide properties database: select values for environmental decision making. Rev. Environ. Contam. and Toxicol., 123: 1–155.

Agriculture, Ecosystems and Environment, 46 (1993) 197–215
Elsevier Science Publishers B.V., Amsterdam

The Lake Erie Agroecosystem Program: water quality assessments

David B. Baker

Water Quality Laboratory, Heidelberg College, Tiffin, OH 44883, USA

Abstract

In contrast with the watersheds draining into Lakes Superior, Michigan, Huron, and Ontario, where forestry is the dominant land use, the dominant land use in Lake Erie's watershed is row crop agriculture. Consequently, the tributaries draining into Lake Erie carry, on average, much larger loads of sediments, nutrients, and pesticides than do the tributaries entering the other Great Lakes. To support the development, operation, and assessment of agricultural non-point pollution control programs in the Lake Erie Basin, the major tributary watersheds are analyzed as large-scale agroecosystems, using mass balance approaches. Material export from the watersheds is based on detailed tributary loading studies which were initiated in the mid-1970s. The monitoring programs have now been in operation for up to 18 years, producing data that serve: (1) to assess the effectiveness of non-point pollution control efforts, (2) to guide future non-point pollution control programs, and (3) to illustrate many of the regional water quality impacts of agricultural land use. The water quality data illustrate the large day-to-day, season-to-season, and year-to-year variability in both pollutant concentrations and loads, which is characteristic of non-point pollution. The data also illustrate systematic shifts in pollutant concentration and loading patterns that occur in relation to watershed size. Although gross erosion rates in northwestern Ohio tributaries are relatively low, the phosphorus and nitrate export rates are high in comparison with other US streams and rivers. Analysis of the water quality data reveal significant downward trends in time for total and soluble phosphorus and significant upward trends in nitrate. The reductions in phosphorus export apparently reflect the effectiveness of agricultural pollution abatement programs that combine more careful fertilizer management with increasing use of conservation tillage. The increasing nitrate concentrations may reflect a trade-off associated with the adoption of conservation tillage.

Introduction

Row crop agriculture is the dominant land use in the watersheds draining directly into Lake Erie (International Joint Commission, 1978). This contrasts with the drainage basins of the other four Great Lakes, where forestry is the dominant land use (Fig. 1). The Lake Erie Basin also has more urban and industrial land use than the other Great Lakes. Because of these more intensive land uses in the Lake Erie Basin, coupled with the soil types and geology of this region, Lake Erie receives far larger amounts of sediment and both point and non-point phosphorus from its watershed than the other lakes receive from their watersheds (Fig. 1) (International Joint Commission,

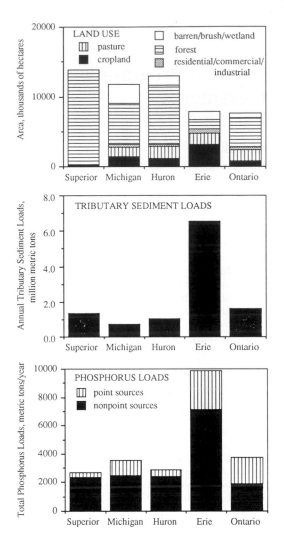

Fig. 1. A comparison of land use, tributary sediment loads, and point and non-point source phosphorus loads for the five Great Lakes.

1978, 1987). This large phosphorus loading, coupled with the shallowness of Lake Erie, has made this lake far more subject to the problems of eutrophication than the other Great Lakes.

As part of the water quality agreement between the US and Canada, a phosphorus target load of 11 000 metric tonnes per year has been set for Lake Erie (International Joint Commission, 1988). Large investments in phosphorus removal at municipal treatment plants have reduced point source loading from 15 260 t in 1972 to 2449 t in 1985 (International Joint Commission, 1987).

However, to meet the target load for Lake Erie, it has long been recognized that reductions in phosphorus loads from agricultural non-point sources would be necessary (US Army Corps of Engineers, 1982). Consequently, agricultural phosphorus reduction programs have been mounted throughout Lake Erie's crop production areas. To reduce particulate phosphorus loading associated with sediment transport, programs have been initiated to reduce cropland erosion by accelerating farmer adoption of conservation tillage. These programs also advocate more careful management of phosphorus fertilizers.

While the initial focus of agricultural pollution abatement programs was phosphorus load reduction for Lake Erie, recognition of the magnitude of water resource impacts from sediments, nitrates and pesticides in Lake Erie tributaries, as well as in the lake itself, soon followed (Baker, 1985). In addition, impacts on aquatic biota became apparent through the stream biomonitoring programs of the Ohio Environmental Protection Agency (US Environmental Protection Agency, 1990). Consequently, agricultural pollution abatement programs have shifted toward more comprehensive management programs that attempt to address the full range of water resource degradation associated with agricultural production in this region.

As a tool to support programs aimed at reducing the water resource impacts of agriculture in the Lake Erie Basin, we are organizing relevant data within the framework of large-scale agroecosystem analyses (Fig. 2). In this framework, agricultural management practices and weather conditions are viewed as the major inputs into the soil and land resources of Lake Erie watersheds. These interact to yield the agricultural products of this region and to pollute water resources. The ongoing tracking of these input and output variables provides a mechanism to assess progress in efforts to minimize the adverse impact of food production on water resources and to guide ongoing agricultural pollution abatement programs.

A unique aspect of the Lake Erie Agroecosystem Program is that it can draw upon unusually detailed water quality data for Lake Erie tributaries. Our laboratory initiated detailed tributary monitoring programs in the mid-1970s

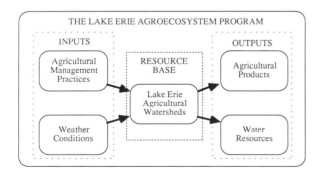

Fig. 2. Major components of the Lake Erie Agroecosystem Program.

and these programs have continued up to the present. In this paper, the impacts of agriculture on the water resources of this region, as revealed by these monitoring programs will be reviewed.

Study area and methods

The locations of our study watersheds are shown in Fig. 3. Each of the watersheds is subtended by a US Geological Survey (USGS) stream gaging station. The USGS station number, drainage area, period of record, land use, and numbers of nutrient and pesticide samples analyzed at each station are listed in Table 1. The agricultural watersheds of northwestern and north–central Ohio are representative of the Eastern Corn Belt Plain and Huron/Erie Lake Plain Ecoregions (Omernik and Gallant, 1988). Soybeans and corn are the dominant crops. The relatively high clay content of soils and the extensive use of subsurface tile drainage have important impacts on the quality of water in the tributaries draining these regions. The more urbanized and forested watersheds of the Cuyahoga and Grand Rivers are located within the Erie/ Ontario Lake Plain Ecoregion. Additional details of the sampling programs have been described by Baker (1988).

Fig. 3. Locations of Lake Erie tributary monitoring stations.

Table 1
Characteristics of the Lake Erie Basin tributary loading stations, including drainage area, period of record, land use and database

Station	Name	USGS station number	Drainage area (km²)	Years of record	Period of record	Land use (by percent)[1]					Nutrient samples analyzed	Pesticide samples analyzed
						C	P	F	W	O		
1	Maumee River	04193500	16395	13	1976–1978 1982–1991	76	3	8	4	9	6312	599
2	Sandusky River	04198000	3240	17	1975–1991	80	2	9	2	7	7600	649
3	Huron River	04199000	961	9	1975–1979 1988–1991	73	4	12	2	8	3478	257
4	Cuyahoga River	04208000	831	9	1983–1991	4	43	29	3	21	4029	171
5	Grand River	04212100	774	4	1988–1991	26	9	44	4	17	1227	22
6	River Raisin (MI)	04176500	2699	10	1982–1991	67	7	9	3	14	2337	159
7	Honey Creek	04197100	386	16	1976–1991	83	1	10	1	6	7648	848
8	Rock Creek	04197170	88.0	9	1983–1991	81	2	12	1	4	4558	765
9	Lost Creek	04185440	11.3	10	1982–1991	83	0	11	1	5	5106	576
10	Bayou Ditch	04195830	7.3	4	1988–1991	97	–	–	–	3	1197	220
11	LaCarpe Creek	04195825	7.6	4	1988–1991	96	–	–	–	4	1110	230
12	Old Woman Creek at Berlin Rd.	04199155	57.2	3	1988–1990	51	16	21	0	13	772	181
13	Old Woman Creek at Route 6	04199165	68.6	3	1988–1990	49	16	21	2	13	500	188

[1]Land use categories indicate percent of basin in: C, cropland; P, pasture; F, forest; W, water/wetland; O, other. Data from US Army Corps of Engineers (1978).

At each station, automatic samplers are used to collect three to four discrete samples per day for nutrient and sediment analyses. The pumps for the automatic samplers subsample from small reservoirs in the gaging stations that are fed by continuously operating submersible pumps located in the streams. All stations operate on a year-round basis. At weekly intervals, the samples are returned to our laboratory for chemical analyses. During periods of high flow and/or high turbidity, all of the samples from each automatic sampler are analyzed, while at most other times, one sample per day is analyzed. During the pesticide run-off season of mid-April through August, second automatic samplers, equipped with glass sample containers, are placed in the stream gaging stations for the agricultural watersheds. During the September through March period for the agricultural watersheds, and year-round at the non-agricultural watersheds, two grab samples per month are collected for pesticide analyses.

Each sample is analyzed for suspended sediment, total phosphorus, soluble reactive phosphorus, nitrate plus nitrite nitrogen, nitrite nitrogen, ammonia nitrogen, total Kjeldahl nitrogen, chloride, sulfate, silica, and conductivity. Automated colorimetric procedures are used for the nutrient analyses (US Environmental Protection Agency, 1979). Dual-column capillary gas chromatography with nitrogen–phosphorus detectors is used for the pesticide analyses (Kramer and Baker, 1985).

Results and discussion

Pollutant concentrations

A basic characteristic of river systems is that pollutant concentrations are highly variable depending on stream flow conditions and season. This variability is illustrated in Fig. 4 where stream discharge and the concentrations of suspended sediments, total phosphorus, nitrate nitrogen and atrazine are shown for the Sandusky River for 1985. The frequency of occurrence of various pollutant concentrations in the Sandusky River for the period of record is shown in the concentration exceedency curves of Fig. 5.

The flow-weighted mean concentrations (FWMC) and time-weighted mean concentrations (TWMC) of sediment and nutrients at the major transport stations are shown in Table 2. These averages are based on the entire period of record for each station (Table 1). Where pollutant concentrations increase with increasing stream flow, FWMCs are higher than TWMCs. This is the case for suspended solids, total phosphorus, and total Kjeldahl nitrogen at all stations, and for soluble reactive phosphorus and nitrate in the agricultural watersheds. For the Cuyahoga River, with its more urbanized watershed, soluble reactive phosphorus and nitrate have higher TWMCs than FWMCs, suggesting that point sources are more important in this watershed. In all of the

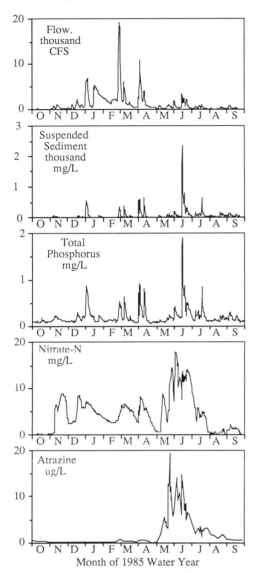

Fig. 4. Annual hydrograph, sediment graph, and chemographs for the Sandusky River monitoring station during the 1985 water year.

watersheds the TWMC of chloride is higher than the FWMC, reflecting higher chloride concentrations in groundwater-dominated baseflow conditions than in surface run-off water during high flow periods. The nitrate concentrations in northwestern Ohio tributaries are considerably higher than those reported for major rivers in the US (Hem et al., 1990) as well as for most of the rivers

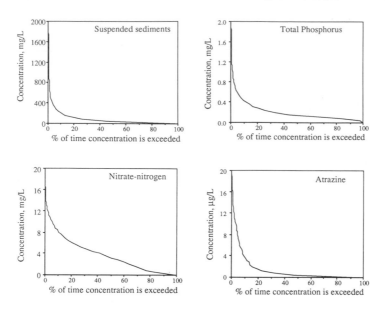

Fig. 5. Concentration exceedency curves of suspended sediment, total phosphorus, nitrate nitrogen, and atrazine for the Sandusky River over the entire period of record.

included in the Global Environment Monitoring System (Meybeck et al., 1989).

Some of the most striking differences among our study watersheds are for nitrate concentrations and for the ratios of particulate phosphorus to suspended sediment. The nitrate concentrations in northwestern Ohio rivers, such as the Sandusky River, are particularly high because of the extensive use of tile drainage systems in their watersheds. In Fig. 6, the nitrate concentrations of the Sandusky River are compared with nitrates in the Cuyahoga and Grand Rivers, which have more urbanized and forested watersheds. At all of our transport stations, the ratios of particulate phosphorus to suspended sediment in individual water samples decrease as sediment concentrations increase (Baker, 1984). However, the ratios are much lower in the Grand River than in the Maumee River (Fig. 7). The sediment in the agricultural watersheds is enriched with phosphorus, due, in part, to historical applications of phosphorus fertilizer aimed at building up phosphorus soil test values.

The TWMCs of various herbicides for the 1983–1991 period at these river stations are shown in Table 3. In all cases, the long-term TWMCs are below the maximum contaminant levels (MCLs) and the lifetime health advisory levels set for drinking water by the US Environmental Protection Agency (US Environmental Protection Agency, 1989). Although the monitoring programs include analyses for various insecticides, they are seldom detected be-

Table 2
Summary of concentrations, annual average exports, and unit area exports of sediments and nutrients from selected Lake Erie Basin transport stations for their periods of record

Parameter	River	TWMC $(mg\,l^{-1})$	FWMC $(mg\,l^{-1})$	Average annual export (t)	Unit area export $(kg\,ha^{-1})$
Suspended solids	Maumee	89	211	1010010	610
	Sandusky	85	221	246000	760
	Cuyahoga	90	206	206000	1000
	Honey Creek	57	193	21800	565
	Rock Creek	45	243	5760	643
Total phosphorus	Maumee	0.27	0.44	2130	1.30
	Sandusky	0.21	0.43	459	1.41
	Cuyahoga	0.33	0.39	346	1.89
	Honey Creek	0.20	0.41	47.4	1.23
	Rock Creek	0.14	0.41	10.2	1.14
Soluble reactive phosphorus	Maumee	0.063	0.070	348	0.21
	Sandusky	0.046	0.068	74.5	0.23
	Cuyahoga	0.090	0.057	58.3	0.32
	Honey Creek	0.054	0.065	7.73	0.20
	Rock Creek	0.005	0.020	0.92	0.13
Nitrate plus nitrite nitrogen	Maumee	4.1	5.4	26300	16.0
	Sandusky	3.7	5.0	5130	15.8
	Cuyahoga	2.4	1.8	1587	8.7
	Honey Creek	5.1	6.3	594	15.4
	Rock Creek	2.3	3.5	103	11.5
Total Kjeldahl nitrogen	Maumee	1.5	1.9	9800	6.0
	Sandusky	1.7	1.7	2040	6.3
	Cuyahoga	1.2	1.4	1000	5.5
	Honey Creek	1.2	1.8	215	5.6
	Rock Creek	0.8	1.9	47.8	5.3
Chloride	Maumee	42.1	28.7	128792	78.6
	Sandusky	41.6	27.9	25198	77.8
	Cuyahoga	106.7	94.8	85283	466.0
	Honey Creek	28.8	20.8	2415	62.6
	Rock Creek	33.7	20.6	551	61.4

cause of their lower application rates and more rapid degradation rates than those of the herbicides.

While seasonal concentration patterns are evident for suspended solids, phosphorus and nitrate (Baker, 1988), they are most obvious for herbicides (Baker and Richards, 1989). In Fig. 8, the monthly and annual average TWMCs of atrazine in the Sandusky River are plotted from 1983 to 1991. During May, June, and July, the monthly TWMCs frequently exceed the MCL

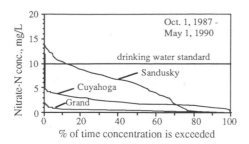

Fig. 6. A comparison of nitrate nitrogen concentrations for the Sandusky, Cuyahoga, and Grand Rivers using concentration exceedency curves.

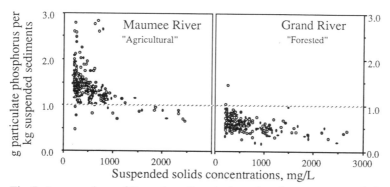

Fig. 7. A comparison of the ratios of particulate phosphorus to suspended solids concentrations in relation to the suspended solids concentration for the Maumee and Grand Rivers.

for atrazine ($3.0 \, \mu g \, l^{-1}$). The relatively low concentrations of atrazine during other months bring the annual average atrazine concentrations below the drinking water standard. Since compliance with pesticide MCLs is based on annual average concentrations, atrazine concentrations comply with the standard even though the average monthly concentrations often exceed the MCL during the months immediately following atrazine application.

Pollutant export

Both the average annual export and the average unit area export of sediments and nutrients at major transport stations are shown in Table 2. The frequency of sample collections for both sediments and nutrients allows loads to be calculated using the mid-interval technique as employed by the US Geological Survey at their daily sediment stations (Porterfield, 1972). Details of our load calculation procedures have been described elsewhere (Baker, 1988).

The unit area export rates of phosphorus and nitrates from northwestern Ohio watersheds appear to be above the mean/median values reported for agricultural watersheds in the US (Omernik, 1977; Frink, 1991). There are

Table 3
Concentrations of major herbicides at the monitoring stations, April 1983 to present. Listed in each block, from top to bottom, are the maximum observed concentration, the 95th percentile and the median concentration (time-weighted), TWMC and FWMC. Concentrations have not been adjusted for recovery. All concentrations are reported in micrograms per liter

River Basin area (km^2)	Atrazine	Alachlor	Metolachlor	Metribuzin	Cyanazine	Linuron
Maumee 16395	21.45	18.35	26.20	5.77	9.96	7.29
	7.47	3.00	5.32	1.83	1.97	0.00
	0.58	0.00	0.28	0.013	0.031	0.00
	1.61	0.54	1.16	0.29	0.38	0.047
	1.77	0.84	1.14	0.39	0.46	0.020
Sandusky 3240	24.61	36.13	36.76	9.26	19.87	6.86
	8.84	3.76	8.59	1.68	1.73	0.29
	0.53	0.001	0.35	0.00	0.00	0.00
	1.78	0.66	1.65	0.28	0.35	0.046
	1.69	0.65	1.49	0.23	0.21	0.032
Honey Creek 386	54.04	54.87	95.75	10.52	17.47	15.5
	10.85	4.44	9.08	1.28	2.07	0.68
	0.66	0.11	0.35	0.00	0.028	0.00
	2.33	0.89	1.80	0.24	0.40	0.17
	2.47	1.13	1.57	0.25	0.38	0.20
Rock Creek 88	48.63	23.40	96.92	15.95	24.77	12.01
	6.61	2.16	8.15	1.20	0.71	0.68
	0.21	0.00	0.17	0.00	0.00	0.00
	1.34	0.39	1.62	0.23	0.18	0.15
	1.69	0.48	1.47	0.19	0.25	0.16
Lost Creek 11.3	68.40	64.94	63.64	25.15	22.62	13.44
	5.67	1.07	3.08	0.80	1.64	0.00
	0.27	0.00	0.00	0.00	0.00	0.00
	1.30	0.48	0.62	0.20	0.50	0.05
	2.44	1.26	1.17	0.29	0.90	0.08
Cuyahoga 1831	6.80	1.16	5.39	1.49	1.36	5.04
	0.99	0.24	0.63	0.28	0.27	0.056
	0.090	0.00	0.00	0.00	0.00	0.00
	0.31	0.042	0.15	0.067	0.053	0.085
	0.23	0.026	0.028	0.088	0.010	0.00
Raisin 2699	12.46	7.52	5.91	2.46	3.75	1.92
	3.91	2.02	1.50	0.37	1.11	0.18
	0.30	0.00	0.00	0.00	0.00	0.00
	0.76	0.37	0.32	0.11	0.21	0.044
	1.30	0.75	0.44	0.20	0.33	0.082

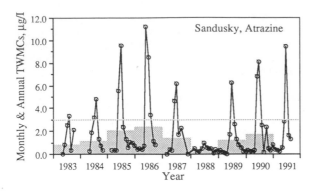

Fig. 8. A comparison of monthly and annual average atrazine concentrations for the Sandusky River from 1983 and 1991.

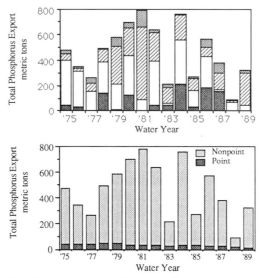

Fig. 9. Annual variability in total phosphorus export from the Sandusky River, (A) showing seasonal variability and (B) contributions of point and non-point sources. In (A) for each bar, the seasonal sequence from the base of the bar upward is: fall (O,N,D), winter (J,F,M), spring (A,M,J), and summer (J,A,S).

few large watershed studies of the detail and duration of these, that permit direct comparisons of export coefficients. Most export coefficients are based on field and plot level studies rather than on long-term watershed export studies.

The extent of annual variability in pollutant export from northwestern Ohio tributaries is illustrated by phosphorus export data for the Sandusky River (Fig. 9). In Fig. 9a, the seasonal contributions to the yearly phosphorus export are shown. Most phosphorus export occurs during winter and spring storm events. In Fig. 9b, the annual point source phosphorus inputs into the Sandusky River, as derived from sewage treatment plant data, have been subtracted from the total phosphorus export to estimate the contributions of non-point sources to the annual export. This graph illustrates the low and decreasing contributions of point sources and the large and highly variable annual contributions of non-point sources.

Long-term studies are necessary to quantify pollutant loadings from non-point sources because of the extensive annual variability associated with weather conditions. This same annual variability greatly complicates the task of assessing the effectiveness of non-point control programs in reducing pollutant loading.

Watershed scale effects

With study watersheds ranging in size from 7.3 km^2 to 16 395 km^2, the data sets reveal systematic shifts in characteristics of both pollutant concentrations and export in relation to watershed size. In part, these scale effects reflect the consequences of the mixing of waters in differing stages of their hydrographs and chemographs as storm run-off waters move through the drainage network. Such scale effects are evident even within relatively uniform landscapes. Other aspects of scale effects reflect shifting average composition of watersheds, in terms of chemical use, soils, slopes, crops and land use, as watersheds increase in size from edge-of-field conditions to small watersheds and to large river basins.

Scale effects relative to pollutant concentrations can be summarized in terms of changing chemograph patterns (Fig. 10). In small watersheds, the chemographs have higher peak concentrations but shorter durations. As watersheds

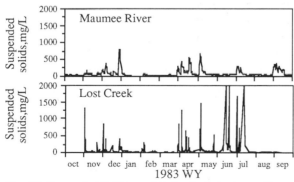

Fig. 10. Example of effects of watershed size on patterns of sediment concentrations. Data for Lost Creek and the Maumee River for the 1983 water year.

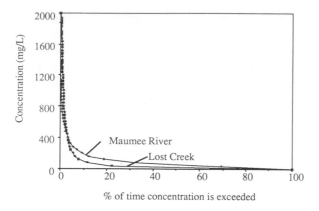

Fig. 11. Comparison of suspended solids concentrations for Lost Creek and the Maumee River over the period of record, using concentration exceedency curves.

increase in size, the chemographs flatten out with lower peak concentrations and extended durations. These shifts are also evident in concentration exceedency curves (Fig. 11). In this figure, the concentration exceedency curves of suspended sediments are compared for the Maumee River and Lost Creek.

Scale effects relative to pollutant export are reflected in the shifting importance of short time intervals in accounting for total pollutant export (Baker, 1988). For example, the 1% of time with the highest sediment transport rates accounts for 60% of the total sediment export from the 88 km^2 Rock Creek watershed but it accounts for only 27% of the total sediment export from the 16 395 km^2 Maumee River watershed. The magnitude of these scale effects shifts for various pollutants, depending on the processes that move them from land surfaces into waterways. They are much more evident for suspended sediments and particulate-associated pollutants (e.g. phosphorus) than they are for soluble chemicals (e.g. nitrate).

In designing demonstration projects that involve efforts to document the water quality benefits of agricultural pollution abatement projects, the importance of scale effects must be taken into account. Relatively small watersheds are often chosen because the potential for achieving a high proportion of adoption of best management practices is considered more feasible. However, small watersheds require more intensive sampling efforts to document material export and they are inherently more variable from year to year than are large watersheds.

Trends in nutrient export

Analysis of data from the long-term transport stations suggests that significant changes in nutrient export have occurred. The data have been evaluated using both regression and Seasonal Kendall trend tests (Richards, 1993). Regressions were calculated for monthly average TWMCs and FWMCs. Upon adjustment for both seasonal and flow-associated relationships, both total phosphorus and soluble reactive phosphorus showed highly significant declines ($P \leq 0.001$) for the Maumee, Sandusky and Cuyahoga Rivers. Nitrate concentrations increased significantly ($P < 0.05$) in the Maumee and Sandusky Rivers while they decreased in the Cuyahoga River ($P \leq 0.001$). The Seasonal Kendall test was in all cases in agreement with the direction of trends indicated by the regression analyses, but was more conservative in its assessment of significance.

The changes in nutrient export are not only statistically significant, but also of such magnitude as to be ecologically important. The reductions in total phosphorus export indicated by the regression studies are shown in Table 4. The reductions in phosphorus export are much larger than can be accounted for in terms of point source controls that were implemented during the study period. The data suggest that agricultural controls have been very effective in reducing total phosphorus export and even more effective in reducing soluble phosphorus export. However, nitrate concentrations in the agricultural tributaries have been increasing on average at rates of about 0.06 to 0.10 mg l^{-1} year^{-1} for the past 16 years.

The likely causes of the decreased non-point source phosphorus export include reductions in phosphorus fertilizer application rates, changes in fertilizer application methods, and increases in conservation tillage. The increases in nitrate concentrations and export may reflect a 'trade-off' related to phosphorus control programs. With conservation tillage, more water percolates into the soil. This increases the proportion of nitrate-laden tile effluent to low-nitrate surface run-off water in the streams. Nitrogen fertilizer application rates peaked in the early 1980s and have declined since that time. The amounts of cropland in northwestern Ohio, as well as the proportions of soybeans, corn, and wheat have shown little change since 1970.

Since the environmental problems posed to rivers in this region by phosphorus and sediment are more severe than those posed by nitrate, the net benefits of trade-offs associated with the use of conservation tillage are posi-

Table 4
Reductions in total phosphorus export from major Lake Erie tributaries during the period of record

River	Initial phosphorus load (t)	Estimated total load reduction (t)	Point source reduction (t)	Estimated non-point source reduction (t)
Maumee	2600	936	222	714
Sandusky	600	604	29	175
Cuyahoga	450	274	147	127

tive. Most of the municipal water supplies have developed ways of meeting the nitrate N drinking water standards of 10 mg l^{-1} during periods of high nitrate concentrations in rivers by diluting river water with ground water or reservoir water having low nitrate content. Conservation tillage not only reduces particulate phosphorus export but also reduces the environmental costs associated with erosion.

Approaches to agricultural non-point pollution control

In the Lake Erie Basin, as elsewhere in the US, the major efforts to reduce agricultural non-point source pollution are focusing on accelerating voluntary farmer adoption of best management practices (BMPs) with respect to tillage, fertilizer and pesticide management, and livestock waste handling. These efforts essentially seek to modify the high-input production systems that have come to characterize much of US agriculture in such ways as to make it less polluting and more sustainable. In recent years, there has also been increasing attention to more fundamental changes in agricultural production systems that have come to be called 'low input, sustainable agriculture'. This approach would generally rely much less on fertilizers and pesticides and focus more attention on rotations, mixed farming (combining livestock and crop production), and alternative crops.

Assuming that grain production will continue to be the major role of agriculture in the Lake Erie Basin, the alternatives for meeting a given production goal are summarized in Fig. 12. These alternative production systems might be termed (1) a relatively high input, sustainable agricultural system with widespread use of BMPs, and (2) a low input, sustainable agricultural system. Under the higher input, sustainable system, the production goals could be met with fewer acres devoted to the production system. In theory, such a system could allow more area for re-establishment of riparian corridors and

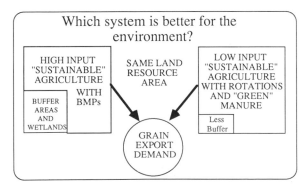

Fig. 12. Alternative approaches for meeting grain production goals and addressing agricultural pollution problems in Lake Erie watersheds.

wetlands than would a low input sustainable system. Degradation of the riparian corridors and wetland destruction are viewed as major factors contributing to impaired aquatic communities of this region (Ohio Environmental Protection Agency, 1990). In Ohio, current generation pesticides are not considered to be a major factor leading to degraded aquatic communities (Ohio Environmental Protection Agency, 1990).

Conclusions

Tributary monitoring programs in the Lake Erie Basin clearly document impacts of agriculture on the water resources of this region. While there have been a variety of agricultural non-point control demonstration projects in the Lake Erie Basin for many years, it has only been in recent years that basin-wide efforts to reduce agricultural pollution have been mounted. Tributary monitoring programs in the Lake Erie Basin indicate that agricultural best management practices can result in significant reductions in phosphorus loading from agricultural systems. While nitrate concentrations have increased in area rivers, municipal water supplies using area rivers as raw water sources are developing alternative sources of water that can be used to dilute river water during periods when nitrate concentrations in rivers exceed drinking water standards. The continued monitoring of these watersheds will provide a way to assess the effectiveness of comprehensive agricultural pollution abatement programs in reducing the adverse impacts of agriculture on the water resources of this region.

Acknowledgments

The Lake Erie Agroecosystem Program is supported by grants to the Heidelberg College Water Quality Laboratory from the George Gund Foundation of Cleveland, Ohio and the Joyce Foundation of Chicago, Illinois. The monitoring programs on Lake Erie tributaries have been supported over the years by many organizations, including the US Army Corps of Engineers, the US Environmental Protection Agency, the US Department of Agriculture (Soil Conservation Service), the Ohio Department of Natural Resources, and various industries, including pesticide and detergent manufacturers. The author gratefully acknowledges the help of WQL staff in the data analyses and preparations (R.P. Richards, N.L. Creamer, L.K. Wallrabenstein), in the operation of the sampling and analytical programs (J.W. Kramer, D.E. Ewing, B.J. Merryfield), and in the review of this manuscript (K.A. Krieger).

References

Baker, D.B., 1984. Fluvial transport and processing of sediments and nutrients in large agricultural river basins. EPA-600/3-88-054. US Environmental Protection Agency, Environmental Research Laboratory, Athens, GA, 168 pp.

Baker, D.B., 1985. Regional water quality impacts of intensive row-crop agriculture: a Lake Erie basin case study. J. Soil Water Conserv., 40: 125–132.

Baker, D.B., 1988. Sediment, nutrient and pesticide transport in selected lower Great Lakes tributaries. EPA-905/4-88-001. US Environmental Protection Agency, Region 5, Great Lakes National Program Office, Chicago, IL 60604, 225 pp.

Baker, D.B. and Richards, R.P., 1989. Herbicide concentration patterns in rivers draining intensively cultivated farmlands of northwestern Ohio. In: D. Weigmann (Editor), Pesticides in Terrestrial and Aquatic Environments. Proceedings of a National Research Conference, 11–12 May 1989, Virginia Polytechnic Institute and State University, Richmond, Virginia, pp. 103–120.

Frink, C.R., 1991. Estimating nutrient exports to estuaries. J. Environ. Qual., 20: 717–724.

Hem, J.D., Demayo, A. and Smith, R.A., 1990. Hydrogeochemistry of rivers and lakes. In: M.G. Wolman and H.C. Riggs (Editors), Surface Water Hydrology. The Geological Society of America, Boulder, CO, pp. 189–231.

International Joint Commission (IJC), 1978. Environmental Management Strategy for the Great Lakes System. Final Report to the IJC from the International Reference Group on Great Lakes Pollution from Land Use Activities. Windsor, Ont., 115 pp.

International Joint Commission (IJC), 1987. 1987 Report on Great Lakes Water Quality. Great Lakes Water Quality Board report to the IJC, Windsor, Ont., 130 pp.

International Joint Commission (IJC), 1988. Great Lakes Water Quality Agreement of 1978 as amended by Protocol signed 18 November 1987. Consolidated by the IJC, Windsor, Ont., 130 pp.

Kramer, J.W. and Baker, D.B., 1985. An analytical method and quality control program for studies of currently used pesticides in surface waters. In: J.K. Taylor and T.W. Stanley (Editors), Quality Assurance for Environmental Measurements. American Society for Testing and Materials, Philadelphia, pp. 116–132.

Meybeck, M., Chapman, D.V. and Helmer, R. (Editors), 1989. Global Freshwater Ecology — A First Assessment. Basil Blackwell, Cambridge, MA, 306 pp.

Ohio Environmental Protection Agency, 1990. 1990 Ohio Water Resource Inventory. Ohio Environmental Protection Agency, Columbus, OH, 136 pp.

Omernik, J.M., 1977. Nonpoint source-stream nutrient level relationships: a nationwide study. EPA-600/3-77-105, National Technical Information Service, Springfield, VA, 151 pp.

Omernik, J.M. and Gallant, A.L., 1988. Ecoregions of the Upper Midwest States. EPA/600/3-88/037. Environmental Research Laboratory, US Environmental Protection Agency, Corvallis, OR, 56 pp.

Porterfield, G., 1972. Computation of fluvial-sediment discharge. Chapter C3 in Book 3 Applications of Hydraulics. Techniques of Water-Resources Investigations of the United States Geological Survey. United States Government Printing Office, Washington DC, 66 pp.

Richards, R.P., 1993. Trends in nutrient and sediment concentrations in Lake Erie tributaries, 1975–1991. J. Great Lakes Res., 19(2): (in press).

US Army Corps of Engineers, 1978. Application of the Universal Soil Loss Equation in the Lake Erie Drainage Basin, Appendix 1. Lake Erie Wastewater Management Study. US Army Corps of Engineers, Buffalo, NY, 519 pp.

US Army Corps of Engineers, 1982. Lake Erie Wastewater Management Study. Final Report. US Army Corps of Engineers, Buffalo, NY, 225 pp.

US Environmental Protection Agency, 1979. Methods for chemical analysis of water and wastes. EPA-600/4-79-020, US Environmental Protection Agency, Cincinnati, OH, 350 pp.

US Environmental Protection Agency, 1989. Drinking Water Health Advisory: Pesticides. Lewis Publishers, Chelsea, MI, 819 pp.

US Environmental Protection Agency, 1990. Water Quality Program Highlights — Ohio EPA's Use of Biological Survey Information. Assessment and Watershed Protection Division, Office of Water, Washington, DC, 4 pp.

Agriculture, Ecosystems and Environment, 46 (1993) 217–222
Elsevier Science Publishers B.V., Amsterdam

Non-point source programs and progress in the Chesapeake Bay

Lynn R. Shuyler

Chesapeake Bay Program Office, US Environmental Protection Agency, 140 Severn Ave., Annapolis, MD 21403, USA

Abstract

Early research studies resulted in the first Chesapeake Bay Agreement, signed in 1983 by the States of Pennsylvania, Maryland, Virginia, the District of Columbia and the Environmental Protection Agency (EPA). These early studies confirmed that a significant source of nutrients delivered to the bay are from non-point sources (NPS). Early bay program strategies at both the State and Federal level increased point source controls and created a new level of non-point source programming and funding for NPS programs which are detailed in the paper. In 1987 a second Chesapeake Bay Agreement was signed by the agreement States, DC and the EPA. This agreement established several goals for the restoration of the bay, including a nutrient reduction goal of a 40% reduction of the controllable nitrogen and phosphorus entering the mainstem of the Chesapeake Bay. The re-evaluation of the nutrient reduction strategy and the modeling to support it are discussed along with the NPS abatement and control progress made through 1990. Also presented are challenges for the NPS programs to address the findings that are coming from the re-evaluation process along with recommendations for change that were contained in a 1990 report on the effectiveness of the NPS programs in the bay.

Introduction

Chesapeake Bay is the largest and most productive estuary in the nation. It has been the subject of intensive study and implementation of programs to control pollution into the bay for the past 16 years. The decline of the living resources within the bay was first noticed in the 1960s; the cry of alarm went largely unanswered, however, until the 1970s, when Congress directed the US Environmental Protection Agency (EPA) to study the bay and to determine the cause of the decline. These early studies resulted in the first Chesapeake Bay Agreement, signed in 1983 by the States of Pennsylvania, Maryland, Virginia, the District of Columbia and the EPA. This first agreement signaled the beginning of a cooperative effort to restore the living resources of the bay. The early investigative efforts did not pinpoint any one source of pollution as the major reason for the declining living resources in the bay; rather the studies indicated that all sources would have to be reduced. Both nitrogen and phosphorus were implicated as nutrients to be controlled. Sediment and a few toxic hot spots in Baltimore (MD) and Norfolk (VA) harbors were also identified as needing control programs.

In 1987, a second Chesapeake Bay Agreement was signed by the three States, the District of Columbia and the EPA representing the US government. This agreement went much farther than just pledging to restore the living resources of the bay; it set 29 different goals for the jurisdictions to reach in the following 12–24 months. One of these goals was the development of a nutrient reduction strategy for the basin. This strategy called for a 40% reduction of the 1985 controllable nitrogen and phosphorus loads to the bay by the Year 2000. The non-point source (NPS) goal is stated as the controllable average year load, based on 1985 land use. It recognizes that not all of the non-point source loads are man induced; some portion of the load from a land use that cannot be changed. As an example, the loads from mature forest lands cannot be managed, nor are they man induced. This strategy realized the need to update and revise the goal, and calls for a re-evaluation during 1991. The re-evaluation process uses a three-dimensional model of Chesapeake Bay to predict the water quality in the bay and a watershed model to predict loads to the bay from various management scenarios. Also, bay water quality goals are being linked more closely with the habitat requirement of the living resources in the bay. This linkage will help direct the targeting of NPS programs to better respond to needs within the bay.

NPS loads in the Chesapeake Basin

To understand the complexity of the NPS control programs in the bay, one has to realize the size of the drainage basin and the mix of land uses. The Chesapeake Bay drainage Basin contains about 64 000 square miles and drains parts of six States and the District of Columbia. The bay covers about 5000 square miles, bringing the total area to 69 000 square miles. Table 1, provides a distribution of land uses for the Chesapeake Bay Drainage Basin that are used in the watershed model. This information, when linked to nutrient loading values, indicates the relative contribution of nutrients to individual land

Table 1
Distribution of land use in the Chesapeake Basin

Land use	Total acreage	% of total basin
Crop land	8237391	20.00
Pasture	3739158	8.96
Forest	24457144	60.00
Urban	4160082	10.00
Water	526115	1.00
Animal waste	14473	0.04

Source: Chesapeake Bay Program, 1991a.

uses regardless of the aerial extent. For example, even though forests comprise 60% of the land in the basin, the nutrient contributions from forest lands are low compared with total agriculture (crop land, pasture and animal wastes), which comprises a much smaller land area (29%).

Table 2, provides a summary of model results for total phosphorus and total nitrogen loadings from non-point and point sources for the entire watershed model. The summary data show that non-point sources (including animal wastes, crop land, pasture, urban, forest and atmospheric deposition) contribute 66% of the phosphorus and 77% of the nitrogen to the bay, while point sources (primarily municipal waste-water treatment plants) contribute the difference (34% and 23%, respectively). In the watershed model, animal waste loads are only the run-off loads from the land area representing animal production facilities and do not include manure, which is spread on crop land and is included in the load from crop land. Therefore, the loads from animal waste seem low when the total animal numbers are considered.

Table 3 compares model results of total agricultural NPS loads with the animal waste portion of that total. Total agricultural NPS load is defined as the sum of the loads from all crop land, all pasture land and animal waste. Animal waste makes up 18% and 5%, respectively, of the phosphorus and nitrogen loads that come from all agricultural non-point sources combined.

The controllable fraction of the non-point source load is always in question; however, in the bay program, the man-induced loads are defined as 'benchmark loads'. These are derived by converting all land uses to forest and subtracting the all-forest loads from the 1985 baseline loads for all land uses. This provides the benchmark loads from which all progress is measured and is the basis for the reduction goals. The 1985 point sources are also considered to be benchmark loads. Therefore, in the Chesapeake Bay program these bench-

Table 2

A comparison of phosphorus and nitrogen loadings from non-point and point sources (million pounds per year based on 1985 land use)

	Phosphorus		Nitrogen	
	NPS	PS	NPS	PS
Farmland	10.83		127.93	
Animal waste	2.99		19.36	
Forest	0.84		70.66	
Urban	2.22		32.91	
Atmo. to water	1.71		40.73	
Point sources		8.63		84.72
Total	18.59	8.63	291.59	84.72

Source: Chesapeake Bay Program, 1991a.

Table 3

A comparison of phosphorus and nitrogen loadings from total agricultural non-point and animal waste sources (million pounds per year based on 1985 land use)

	Phosphorus		Nitrogen	
	Agriculture	Animal	Agriculture	Animal
Farmland	10.83		127.93	
Forest	0.84		70.66	
Urban	2.22		32.91	
Animal waste		2.99		19.36
Total	13.89	2.99	231.50	19.36

Source: Chesapeake Bay Program, 1991b.

Table 4

Phosphorus and nitrogen loadings to Chesapeake Bay (million pounds per year based on 1985 land use)

	Phosphorus	Nitrogen
'1985 Baseline'	27.22	376.31
Total benchmark	20.27	180.66
NPS benchmark	11.64	95.94
PS benchmark	8.63	84.72

Source: Chesapeake Bay Program, 1991b.

mark loads can be used to set reduction goals and as the basis for measuring progress. Progress is discussed later in this paper.

Table 4 contains 1985 baseline loads from all states in the basin and benchmark loads from only the bay agreement states (Pennsylvania, Maryland, Virginia, and the District of Columbia). Loads from the other non-agreement states (New York, Delaware and West Virginia) are excluded from the benchmark analysis because they have not signed the Chesapeake Bay Agreements and do not participate in the funding from the bay program.

Control programs

From the early days of the bay NPS programs, pollution from animal wastes has been a high priority. Early emphasis in bay NPS programs was on the management of animal manure. Pennsylvania developed a manure management manual and required nutrient management plans for lands receiving manure. Virginia and Maryland quickly followed suit, developing nutrient management programs of their own. All programs began by targeting land that received manure as the first priority, with nutrient management plans becoming a mandatory component of a manure management system in all three States. In Virginia, Maryland and Pennsylvania, a manure management system must contain a nutrient management plan before State cost share funds can be used for the system. Also, in Pennsylvania nutrient management plans are required before any management practice is eligible for assistance from the State NPS program.

Managing the manure from animal production is an important part of the control of pollution from animal production, but it is not the total answer. Controlling the run-off from the animal production facilities is a very important component of the total equation to control pollution from animal production. Within the Chesapeake Bay Basin, there are over 3.25 million animal units housed in over 6000 production units.

In a recent report on the effectiveness of the NPS programs in the Chesapeake Bay Basin (Non-point Source Evaluation Panel, 1990), an independent panel made several recommendations for improvement to the NPS programs in the basin. One recommendation was to reduce the threshold number for animal production facility permits at both the state and federal level. Another was mandatory nutrient management plans for watersheds exporting excessive nutrient loads. These, and the other recommendations will require better methods of targeting program dollars and technical assistance.

The watershed model can produce permissible loads for segments within each river basin for each land use, thereby further refining the location of a load source. The model does not contain small enough cells to pinpoint small problem areas, but along with GIS information about land use, it can be very useful in developing basin and subbasin targeting strategies. Some of the scenarios that are being run with the watershed model will evaluate different cut-off limits for NPDES permits for animal production and several nutrient management and erosion control scenarios for both urban and agriculture. These results will provide program managers several options to evaluate based on an analysis of costs for each option. All of these factors help shape the NPS programs within the Chesapeake Bay Basin and may have a strong impact on animal production in the region. One additional factor that has not yet come into play in the bay basin is the mandatory management measures for Coastal Zone Management areas (US Environmental Protection Agency, 1991). This program could make management of non-point sources mandatory for that portion of the Chesapeake Bay Basin that lies within the coastal zone management boundary.

Progress

Annually the three States and the District receive approximately $7 million in Federal grant funds from the EPA for the abatement and control of NPS pollution. These funds are matched dollar for dollar by the States, bringing the total annual grant funding to about $14 million. The total NPS program spending by these States far exceeds the Federal grant funding, plus, both ASCS and SCS have conservation and special water quality programs that address NPS problems in the basin. Since the beginning of the agricultural cost share programs by the States in 1984–1985, installation of agricultural best management practices have been tracked by the States and reported to the EPA Chesapeake Program Office. EPA obtains similar tracking information from the US Department of Agriculture for the Agricultural Conservation Program which operates in all the States and counties in the basin. Combined data from these sources indicate that about 12–14% of the crop land acres that need treatment to reduce erosion have been controlled and that about 10–12% of the animal wastes have been controlled. A watershed model progress run using these tracking data for agriculture indicates a 12% and 8% reduction in total nitrogen and total phosphorus, when compared with the NPS benchmark loads shown in Table 4.

It is clear from the independent panel report and progress implementation data that we are not reaching enough landowners at the current pace to reach our goal by the Year 2000. It is also clear that money is not the total answer. It has been suggested that outreach and education are the answer. One fact is clear however, if voluntary cooperation does not increase many times in the next few years, the programs will have no option but go regulatory. This is what Congress has told us with the Coastal Zone Management Act and the proposed Section 319 language. I believe it is time we heard the message and moved to control our own destiny before someone else does it for us, if it is not too late now. These actions from Congress just may be the push that moves the Chesapeake Bay Basin into a more regulatory approach in the very near future.

References

Chesapeake Bay Program, 1991a. 1990 Progress Report for the Baywide Nutrient Reduction Strategy. Chesapeake Bay Program Office, Annapolis, MD, pp. 5–31.
Chesapeake Bay Program, 1991b. Material for Draft Reevaluation Report for the Baywide Nutrient Reduction Strategy. Chesapeake Bay Program Office, Annapolis, MD.
Non-point Source Evaluation Panel, 1990. Report and Recommendation of the Non-point Source Evaluation Panel. Chesapeake Bay Program Office, Annapolis, MD, pp. 1–28.
US Environmental Protection Agency, 1991. Proposed Guidance Specifying Management Measures for Sources of Nonpoint Pollution in Coastal Waters. Proposed under the authority of Section 6217(g) of the Coastal Zone Act reauthorization amendments of 1990. Office of Water, Washington, DC, pp. 2-1–2-121.

Agriculture, Ecosystems and Environment, 46 (1993) 223–231
Elsevier Science Publishers B.V., Amsterdam

Agricultural best management practices for water pollution control: current issues

Terry J. Logan

Department of Agronomy, The Ohio State University, 2021 Coffey Road, Columbus, OH 43210, USA

Abstract

From the first awareness of agricultural sources of water pollution in the US in the 1960s, we finally see in the 1990s a commitment at the national level for agricultural non-point source (NPS) pollution control. This has been occasioned by a growing awareness that, with point sources of some pollutants largely controlled by waste-water treatment, greater attention must be paid to NPS pollution control, a large percentage of which is agricultural. The 1985 Farm Security Act mandated several national erosion control programs that will have some impact on water quality, and there is opportunity to supplement these programs with best management practices (BMPs) specifically designed to address agricultural water pollutants, primarily nitrate, phosphorus and modern pesticides. This paper discusses fundamental processes affecting transport of agricultural pollutants in surface and ground water and suggests how knowledge of these processes can be used to evaluate existing agricultural NPS BMPs and to develop supplemental practices.

Introduction

National awareness of agriculture as a significant source of environmental contamination dates back at least to 1962 with the publication of Rachel Carson's 'Silent Spring', if not earlier. The role of agriculture in water pollution emerges in the same period with the finding that non-point source (NPS) phosphorus from agricultural run-off may be as significant a contributor to eutrophication of lakes and streams as point sources of untreated domestic waste water. National studies of lake eutrophication, and more regional studies of important water bodies such as Lake Tahoe, California and the lower Great Lakes in the early 1970s, implicate NPS phosphorus from soil run-off and erosion, and from livestock waste run-off. In the same period, run-off and leaching from fertilizers and livestock waste are shown to contribute to high nitrate levels in some rivers and water wells in agricultural areas. Concern for pesticide contamination of water in this period focused on sediment contamination by persistent chlorinated insecticides such as DDT, aldrin, dieldrin, heptachlor, endrin, chlordane, and toxaphene. By the late 1970s, extensive regional studies had clearly identified the causes and extent of agricultural NPS surface water pollution by sediment and nutrients. Planners at the State and national level had, by this time, developed remedial plans (e.g. Pollution from Land Use Activities Reference Group, 1978; Lake Erie Wastewater Management Study, 1982 for the lower Great Lakes) to deal with agricultural NPS phosphorus loads, based on voluntary adoption by farmers of conservation tillage and livestock waste management practices, but these were largely unfunded. Modest efforts in States such as Wisconsin achieved some on-farm treatment, but it was for the most part business as usual. Demonstration projects such as the Rural Clean Water Projects, Model Implementation Projects and others (e.g. the Army Corps of Engineers Honey Creek Project in Ohio) fine-tuned field methodologies for quantifying loadings, assessing land use changes and achieving voluntary participation in remedial programs. These efforts were not strongly institutionalized in the US Department of Agriculture (USDA) or State management agencies and had minimal impact.

By the early 1980s several forces combined to change national priorities for agricultural NPS water pollution control. First, the public was environmentally sensitized by incidents such as Love Canal in New York and Times Beach in Missouri to the probable contamination of ground water by chemicals. The US Environmental Protection Agency (USEPA) issued several reports on groundwater contamination including its 'Ground-water protection strategy' (USEPA, 1984). For reasons not clearly understood, the public was much more concerned by the prospect of groundwater contamination in the 1980s than they were with surface water contamination in the 1960s and 1970s. Drinking water wells were surveyed for nitrate and pesticide contamination, and wells in agricultural areas in 23 States were found to have detectable levels of at least 17 pesticides (Cohen et al., 1986). Public and congressional pressures resulted in strong policy statements for groundwater protection by USDA agencies. Section 319 of the Clean Water Act finally provided limited funding to States for implementation of agricultural best management practices (BMPs) for water pollution control. In addition, the 1985 Farm Security Act mandated sweeping changes in the control of farmland erosion that will have impact on agricultural NPS pollution (see below).

Where do we stand in 1992? With respect to phosphorus-induced eutrophication, advanced waste-water treatment in the last 15 years has resulted in dramatic reductions in point source P loadings to lakes and streams with corresponding improvements in water quality. Yet, eutrophication persists in many water bodies and further reductions will have to be made in agricultural NPS loads. Most plans call for this to be achieved by a combination of conservation tillage and other soil erosion control practices, livestock waste management, and P fertility management. Survey data on conservation tillage use in the US (McCain, 1990) suggest that adoption of these practices has slowed in recent years with about 73 million acres under some form of conservation tillage practice. The impact that the 1985 Farm Security Act provisions will have on this figure is discussed later. Most efforts in livestock waste management have focused on cost sharing for waste storage facilities. However, there is a persistent problem in many livestock areas with inadequate land base for agronomic utilization of livestock waste nutrients. With respect to P fertility, management, significant acreages of agricultural soils in the US have available P levels in the excessive range. It is heartening to note that P fertilizer use in the US has decreased in recent years (Wallingford, 1991), and Baker (1993) has shown from trend analysis of long-term tributary loading data for northern Ohio that both dissolved and total P loads attributable to non-point sources may have declined in the last few years; greatest reductions were seen for dissolved P. This suggests that watershed loadings of the most bioavailable P form, dissolved inorganic P, may decline more rapidly than previously thought.

Unlike P loads, Baker has found that nitrate levels have increased in tributaries in agricultural watersheds in northern Ohio, and the recent USEPA national water well survey (USEPA, 1990a) shows that 2.4% of rural private and 1.2% of community wells have nitrate N levels above the maximum contaminant level (MCL) of 10 mg l^{-1}. Other indications nationwide suggest that high nitrate in surface water is a persistent seasonal problem (from late winter to early summer), far more so than groundwater contamination which is more localized, being found primarily in areas of high rainfall or irrigation use, intensive agriculture and on highly permeable soils and bedrock geology. Unlike P, there is no indication at the national level that N use rates on major crops has declined (Wallingford, 1991), and there is some evidence to suggest that farmers will be utilizing more of the agricultural and domestic wastes produced in the nation as communities move towards complete recycling of wastes rather than disposal.

Pesticides found in water today are not the persistent chlorinated insecticides found in sediments in the 1960s and 1970s. They are the high-volume use herbicides, such as atrazine, alachlor, metolachlor, metribuzin, and cyanazine, and widely used fungicides and nematicides such as aldicarb. These compounds are sufficiently persistent, water soluble and low in soil attenuation that they can move readily in surface run-off and leach to tile drains (or to deeper groundwater aquifers if they are sufficiently conducting). Baker (1993) has shown that tributaries draining agricultural watersheds in northern Ohio can have seasonal concentrations above the MCL for atrazine and alachlor. He also found that less than 1% of private wells in Ohio had atrazine concentrations above the MCL; atrazine was the most commonly detected pesticide. The national water well survey found that the most commonly detected pesticide residues were the acid metabolites of DCPA, a turfgrass herbicide (common name is Dacthal) (USEPA, 1990a). Atrazine was the next most commonly detected compound. Several of the compounds found in earlier well screening (Cohen et al., 1986) have been banned or removed from the market by the manufacturer, including aldicarb which was found in ground water throughout the US wherever it was used to any significant degree. Developing factors which will determine which of the leachable compounds will retain their registration in the future are: (1) the overall process of retesting and registration under Federal Insecticide, Fungicide and Rodenticide Act (FIFRA); (2) development by EPA of MCLs for a wider range of pesticides than presently exists; (3) implementation of USDA–Soil Conservation Service (SCS) technical criteria which specify that farmers with approved conservation plans under the Conservation Compliance (CC) provisions of the 1985 Farm Security Act may have to select alternative pesticides if the compounds they are presently using are shown to have potential to move to ground water in their local environment (Hornsby et al., 1993).

Livestock waste disposal has evolved in the last 20 years from being primarily a problem of lack of or improper storage of manure to one of inadequate land base for efficient reutilization of manure nutrients. This is evidenced in areas of large beef, dairy and poultry operations where the livestock enterprise controls only a fraction of the land base required. Even where there are State-level controls on manure utilization at agronomic rates, rates are usually based on crop N needs. The result is that available P levels can increase rapidly to excessive levels. A few States, including Ohio, are attempting to impose a P limit based on an upper bound soil test (300 kg ha^{-1} Bray P1 in Ohio) but in many areas soil P levels already exceed these values. Another emerging factor is the national movement towards beneficial re-use of a wide range of domestic organic wastes including municipal sewage sludge, sludge compost, yard waste compost, municipal solid waste (MSW) compost, and food processing waste. This trend has occurred for a number of reasons: (1) the quality of municipal sewage sludge with respect to pathogen levels and to the content of trace elements and trace organics (USEPA, 1990b) has improved in recent years, thereby increasing the percentage of sludge that can safely be recycled; (2) new technologies such as sludge composting, alkaline stabilization and pelleting can transform sludge into products that are acceptable to communities and potentially marketable; (3) new national sludge regulations to be released in 1992 will greatly increase the beneficial re-use of sludge (USEPA, 1989); (4) ocean dumping of sludge has been banned as of 1992, and there is great community pressure to exclude sludges, yard waste and MSW from declining landfill space; (5) community acceptance of sludge and MSW incineration has declined. Farmers with access to these materials are using them now and will continue to do so as long as they are economically attractive. Lack of data and understanding on nutrient supply from these materials can result in overfertilization if farmers do not adjust fertilizer rates to compensate for organic waste nutrients.

Evaluation of agricultural best management practices for water quality

Agricultural BMPs for NPS pollution control have focussed primarily on soil erosion control. Logan (1990) reviewed the SCS technical guide practices and classified them as structural, cultural or management. Structural practices include such things as terraces and grassed waterways and their impact is primarily to reduce run-off through increased infiltration and to reduce soil erosion. Cultural practices include conservation tillage, contour cropping and cover cropping. They protect the soil surface and reduce erosion; however, they may or may not increase infiltration and decrease run-off, depending on the hydraulic conductivity of the soil (Logan, 1990). Management practices for fertilizer, pesticide and livestock waste application, and, more generally, integrated pest management and integrated fertility management primarily affect the source of a potential contaminant by increasing use efficiency.

These practices can be evaluated as to their potential to decrease (or even increase) contaminant losses by run-off, erosion (contaminant attached to sediment) or leaching. Two factors must be considered: (1) Is there a reduction in the amount of the potential contaminant in the soil as a result of the practice (e.g. rotation of corn with a small grain can reduce average N use versus continuous corn)? (2) How will the practice affect mass distribution of the contaminant between eroded sediment, run-off water and percolating water? The latter information can be obtained by combining simple (universal soil loss equation and SCS run-off curve number) or more complex erosion/hydrology models (AGNPS, GLEAMS) with knowledge of the fate of the contaminant, particularly the partitioning of the contaminant between soil/sediment and water. Figure 1 illustrates how partitioning can be used to evaluate mass transfer to run-off, sediment or leachate. Use of this approach suggests that structural practices that reduce run-off by increasing infiltration can reduce sediment P losses and pesticide run-off losses, but could increase

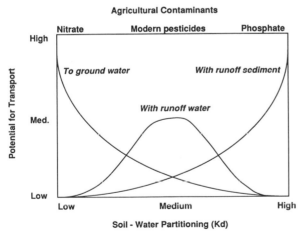

Fig. 1. Potential of agricultural contaminants to be transported in run-off and leaching as a function of their soil/water partitioning.

Table 1

Classification of conservation practices and agricultural BMPs by environmental objective, pollutant type, and medium impacted (Logan, 1990)

BMP	Primary environmental objective[1]	Pollutant type[2]	Medium impacted[3]			
			Surface water	Ground water	Air	Soil
Structural						
Terraces, hillside ditches	E	E, P	P	N/A	N	P
Grass waterways	E	E, P	P	N	N	P
Subsurface (tile) drains, water table management	S	E, P, N, S	P/A	P	N	P
Irrigation systems	S, E	S, N	P	P	P	P
Chemigation backsiphon devices	Q	C, N	N	P	N	N
Sediment and water retention basins	L, Q	E, P, N	P	A	A	N
Surface drains	N, Q	N	A	P	N	P
Manure storage, run-off control, filter strips	W, L, Q	N, P, B, O, M	P	P	P	P
Irrigation tailwater recovery systems	E	E, C, S	P	N	N	P
Cultural						
Conservation tillage	E, L, Q	E, P	P	N/A	N/A	P
Contour cropping	E, L	E	P	N/A	N	P
Stripcropping	E, L	E	P	N/A	N	P
Contour stripcropping	E, L	E	P	N/A	N	P
Cover cropping	E, L, Q	E	P	P	N	P
Crop rotation	E	E	P	N/P	N	P
Subsoiling	S	S, E	P	N/A	N	P
Land grading	S, E	S, E	A	N/P	N/P	P
Critical area planting	E, L, Q	E, P, N	P	P	P	P
Stream bank protection	E	E	P	N/P	N/P	P
Low input farming	E, L, Q	E,C	P	N/P	P	P
Management						
Integrated pest management	Q	C, M	P	P	P	P
Animal waste management	L, W, Q	N, P, M	P	P	P	P
Fertilizer management	L, Q	N, P	P	P	P	P
Pesticide management	Q	C, M	P	P	P	P
Irrigation management	S, L, Q	S. N. P	P	P	P	P

[1]E, erosion control; L, eutrophication; W, animal waste management; Q, water quality; S, salinity; N, none, for example, surface drains primarily eliminate wetness problems.

[2]E, sediment; P, phosphorus; N, nitrogen; C, pesticide; B, biological oxygen demand; S, salt; M, heavy metals; O, pathogenic organisms.

[3]P, positive impact; A, adverse impact; N, no impact.

nitrate leaching losses. Conservation tillage on a permeable soil would reduce sediment P losses, reduce run-off pesticide losses (provided pesticide application rates did not increase substantially), and increase nitrate leaching. On

Table 2
Summary of proposed practices to modify or supplement conservation compliance practices for control of agricultural water contamination (Logan, 1991)

Practice (Pollutant)	Objective	Description
(1) Yield goals (N, P)	Restrict nutrient application to actual crop utilization	Establish long-term potential of climate and soil to produce economic crop yields
(2) Soil and manure tests (N, P)	Restrict nutrient application to actual crop needs	Establish actual nutrient supply capacity of soil, and nutrient content of manure through chemical analysis
(3) Fertilizer management (N, P)	Optimize nutrient utilization by the crop, reduce run-off and leaching losses	Lower application rates to meet realistic yield goals; time fertilizer application to optimize crop utilization; inject or incorporate fertilizer to reduce run-off losses
(4) Residue nutrient credits (N, P)	Prevent excess fertilizer application by better utilizing nutrients in residue, green manure crops	Determine nutrient carry-over from crop residues and green manure crops and adjust fertilizer rates accordingly
(5) Management of green manure crops (N, P)	Optimize nutrient utilization from green manure crops by subsequent crops	Method and timing of incorporation of green manure crops to optimize nutrient utilization by the subsequent crop and to avoid losses
(6) Manure management (N, P)	Optimize manure nutrient utilization by the crop and minimize losses	Rate, timing and method of application of manure to optimize nutrient utilization by the crop, minimize nutrient build-up, and reduce losses
(7) Pesticide management (pesticide)	Reduce the potential for pesticide movement by run-off or leaching	Pesticide selection for mobility and half-life; formulation; and timing and method of application to reduce potential for run-off and leaching losses
(8) IPM (pesticide)	Reduce pesticide use	Field scouting and other IPM practices to reduce pesticide use
(9) Management of permanent vegetation (N, P, pesticide)	Reduce or eliminate fertilizer, manure or pesticide use on permanent plantings	Selection of appropriate species, mowing, and other practices which reduce or eliminate the need for nutrient or pesticide additions to permanent plantings
(10) Irrigation management (N, P, Pesticide)	Manage irrigation to reduce excessive run-off and leaching	Rate and timing of irrigation to reduce excessive run-off or leaching; coordinating irrigation with fertilization and spraying to minimize chemical loss; installation of chemigation back-siphon devices

a slowly permeable soil, run-off volume and pesticide losses in run-off would be little affected as would nitrate leaching. P losses with sediment would still be decreased because of the erosion control provided by residue cover. Logan (1990) used this approach to evaluate a wide range of potential BMPs for contaminant control. Table 1 summarizes those findings. They suggest that, whereas traditional soil erosion control practices will be effective in reducing sediment P losses, pest and fertility management approaches will also be required to achieve significant reductions in pesticide and nitrate contamination of surface and ground water. In a study conducted for the USEPA, Logan (1991) evaluated the potential of CC practices, mandated by the 1985 Farm Security Act to reduce phosphate, nitrate and pesticide contamination of surface and ground water. Specifically, proposed alternative conservation systems (ACSs) were reviewed for 29 states. The ACSs are the sets of minimum practices that farmers can adopt to be in compliance with the CC program. Few of the states included structural measures in the ACSs except for ephemeral gully erosion control. Most states relied on cultural practices, including conservation cropping sequence (close-grown crops in rotation), residue management, contour cropping and cover crops. Permanent plantings were also specified by a few states. The conclusion from evaluation of the ACSs (Logan, 1991) is that they would achieve significant reductions in sediment P at the national level, but would have little impact on pesticide or nitrate levels in surface or ground water.

Logan (1991) also identified practices that could modify or supplement State ACSs to reduce surface and groundwater contamination by agricultural contaminants. These are summarized in Table 2. They primarily involve nutrient and pest management with emphasis on establishing accurate and realistic yield goals and taking nutrient credits for organic wastes, residues and cover crops. It is important that they be integrated with selected ACSs and not used in isolation.

Conclusions

Agricultural NPS pollution of surface and ground water by nutrients and pesticides has been identified as a major problem in the US since the 1960s, but, in spite of major research efforts to quantify the problems and develop solutions in the 1970s, significant programs and federal funds to reduce these contaminants have not been forthcoming until now.

Existing agricultural BMPs are designed primarily to control soil erosion and will have little impact on NPS pollutants except for sediment-bound P which will be significantly reduced. Significant reductions in surface and groundwater contamination by nitrate, phosphate and pesticides will only be achieved by use of fertility and pest management practices integrated with other BMPs. Integration of agricultural BMPs to achieve water pollution control must be made at the level of technical assistance. Too often, agencies present technical information piecemeal, leaving it to the farmer to integrate this data into workable systems.

References

Baker, D.B., 1993. The Lake Erie Agroecosystem Program: water quality assessments. Agric. Ecosystems Environ., 46: 197–215.

Carson, R., 1962. Silent Spring. Houghton Mifflin, Boston, MA.

Cohen, S.Z., Eiden, C. and Lorber, M.N., 1986. Monitoring ground water for pesticides. In: W.Y. Garner et al. (Editors), Evaluation of Pesticides in Ground Water. Am. Chem. Soc., Symp. Ser. 315, pp. 170–196.

Hornsby, A.G., Buttler, T.M. and Brown, R.B., 1993. Managing pesticides for crop production and water quality protection: practical grower guides. Agric. Ecosystems Environ., 46: 187–196.

Lake Erie Wastewater Management Study, 1982. Final Report. US Army Corps of Engineers, Buffalo District, Buffalo, NY, 22 pp.

Logan, T.J., 1990. Agricultural best management practices and groundwater quality. J. Soil Water Conserv., 45: 201–206.

Logan, T.J., 1991. Water quality evaluation of conservation compliance program. Report to the U.S. Environmental Protection Agency, Office of Water, 36 pp. (unpublished).

McCain, D., 1990. U.S. farmers increase conservation tillage acreage. Conserv. Impact, 8(10): 1.

Pollution From Land Use Activities Reference Group, 1978. Environmental management strategy for the Great Lakes system. Final Report to the International Joint Commission. IJC, Windsor, Ontario, 115 pp.

US Environmental Protection Agency, 1984. Groundwater protection strategy. Office of Groundwater Protection, Washington, DC.

US Environmental Protection Agency, 1989. 40 CFR Parts 257 and 503. Standards for the disposal of sewage sludge: proposed rule. Fed. Regist., 54: 5746–5902.

US Environmental Protection Agency, 1990a. National pesticide Phase I report. Executive summary. Office of Ground-Water Protection, WH-550G, Washington, DC, 15 pp.

US Environmental Protection Agency, 1990b. National sewage sludge survey. Fed. Regist., 55: 47210–47213.

Wallingford, G.W., 1991. The U.S. nutrient budget is in the red. Better Crops Plant Food, 75(4): 16–18.

Agriculture, Ecosystems and Environment, 46 (1993) 233–243
Elsevier Science Publishers B.V., Amsterdam

Environmental aspects of integrated pest management

Economic injury level concepts and their use in sustaining environmental quality

Leon G. Higley[a,*], Larry P. Pedigo[b]

[a]*Department of Entomology, 202 Plant Industry Bldg., University of Nebraska–Lincoln, Lincoln, NE 68583-0816, USA*
[b]*Department of Entomology and Leopold Center for Sustainable Agriculture, Iowa State University, Ames, IA 50011, USA*

Abstract

The economic injury level (EIL) concept is the basis for decisions in most integrated pest management (IPM) programs. These IPM programs are fundamentally different from control approaches to handling pest problems by focusing on tolerating pest effects. The EIL is essential for IPM programs because it indicates which levels of pests can be tolerated and which cannot. By increasing our ability to tolerate pests, it is possible to eliminate or reduce the need for management tactics. Moreover, we can improve and maintain environmental quality through better decisions on the use of those tactics. EILs help maintain environmental quality by reducing unnecessary use of management tactics, especially of pesticides. However, including environmental considerations explicitly in the decision-making process could greatly improve the ability of IPM to sustain environmental quality. To understand the current and potential roles of EILs in maintaining environmental quality, it is first necessary to examine the components of the EIL. These include economic damage, economic thresholds, and the EIL itself. Of these elements, the economic threshold has been the most problematic because it depends on predictions of pest population growth rates. Most approaches to economic threshold development can be grouped into subjective and objective categories, with the objective category being based on calculated EILs. Increased availability of calculated EILs and their related economic thresholds would reduce unnecessary use of management tactics. Additional attention to the EIL itself could further improve the responsiveness of IPM programs to environmental concerns. In particular, by including direct considerations of environmental factors, as well as economic and biological information, into EILs, it would be possible to develop environmental EILs. An environmental EIL evaluates a management tactic based not only on its direct costs and benefits to the user but also on its effects on the environment. Activities to support greater environmental responsiveness in the EIL include accounting for environmental costs in the C variable, reducing damage per injury by increasing plant tolerance in the D variable, and developing an effective, yet environmentally responsible, K variable by reducing pesticide application rates. Focusing research efforts on these aspects of the EIL concept offers the prospect of improved responsiveness to environmental concerns in IPM, without a reliance on new tactics.

---

*Corresponding author.

Introduction

Modern, progressive approaches for managing pests depend upon integrated pest management (IPM) programs. A key feature of these programs is the use of formal criteria for making management decisions about pests, and the economic injury level (EIL) concept is the foundation of these criteria. Most often, the tactics used in IPM programs are emphasized over procedures developed to employ those tactics. But the strength of IPM, particularly in maintaining environmental quality, comes from an objective consideration of economic and biological factors in managing pests. The EIL concept is central to such considerations. This paper explores the role of EILs and IPM programs, examines how they function to maintain environmental quality, and considers new directions for improving environmental quality through the uses and modifications of existing EILs.

At the outset, it is necessary to understand the nature of IPM. Although aspects of integrated control and pest management have been discussed by many authors, there is no single paper that fully defines the basis of IPM. Consequently, there has been considerable debate and confusion over what IPM entails. IPM has been viewed as a philosophy, a theoretical background, a set of rules, a scheme for judicious pesticide use, or even an excuse for the status quo. Although IPM may encompass some of these features, it is best identified as a formal system for managing pests based on economics, ecology, and environmental considerations. The formal basis for IPM depends on its having specific goals and procedures. These goals are: (1) to minimize the economic impacts of pests, (2) to produce sustainable solutions to pest problems, and (3) to maintain environmental quality. Among the procedures used to achieve these goals are: (1) identifying and evaluating pest problems, (2) deciding on the need for and choice of management action, and (3) taking appropriate management action. Unlike other approaches, IPM allows for a consideration of direct and indirect consequences of both pest activities and management actions.

Prior to the development of IPM, approaches to pest problems focused on control rather than management. Control refers to having influence over something versus management which refers to the judicious use of means to accomplish a desired end. The objective of pest control is to reduce pest numbers to whatever degree desired (usually to zero). In contrast, pest management focuses on reducing or modifying the impact of pests to tolerable levels. The crucial role of EILs in pest management is to indicate which levels of pests are tolerable and which are not. Typically, control focuses only on the presence or absence of pests, whereas IPM is based on evaluations of numbers and the effects of pests. Moreover, control is not founded on considerations of economic, ecological, or environmental factors. Therefore, control is a far simpler approach to pest problems than IPM. Control is successful only as long as it is possible to dramatically alter pest numbers or their effects. As a philosophy for addressing pest problems, pest control has failed because of the failure of the tactics employed to impose control.

The advantage of pest control over management is that it requires far less information and knowledge than IPM. However, pest control is a dangerously naive philosophy that ignores much biological reality. Furthermore, using tactics in an effort to control pests increases the likelihood of deleterious side-effects from those tactics and frequently reduces the useful life of the tactics used. For these reasons, trying to control pests may often cause more harm than good. Nevertheless, concepts associated with control continue to be influential, even within IPM programs.

The most important research activity in a control philosophy is to identify the most powerful tactics for killing pests. For much of this century, work on pests very much focused on identifying effective tactics. The most striking success in this search was the discovery of the insecticidal properties of DDT. For a time, DDT and other synthetic pesticides were considered to be 'silver bullets', panaceas for pest problems. Ultimately, the fallacies in this approach became obvious. Shortly after the introduction of chemical pesticides, many insect populations developed resistance to those pesticides. Additionally, harmful side-effects from pesticide use, such as environmental contamination and impacts on non-target species, became evident.

Although chemical pesticides like DDT have fallen into disfavor, the idea that another 'silver bullet' control tactic lies over the horizon has not. The development of IPM did not eliminate the pest control philosophy, and the continued influence of pest control is illustrated by ongoing efforts to find magic solutions to pest problems. Some of these efforts may be couched within the framework of IPM but trying to address pest problems with single solutions remains as ill-founded today as the chemical approach was decades ago. Biological control, pheromones, and, most recently, biotechnology all have been promoted as solutions to all future pest problems. Such claims ignore the biological realities associated with handling pest problems. In particular, they disregard the ability of pest populations to respond to strong selection pressure and circumvent control tactics. Moreover, searching for 'silver bullets' draws emphasis and resources away from more appropriate activities that would allow better management of pests.

Certainly, new management tactics are needed to have an important role in IPM. Although it is unrealistic to expect such new tactics to solve all pest problems, new approaches improve our ability to deal with those pest problems. In particular, alternatives to pesticides are needed to minimize the deleterious effect of the use of those chemicals.

However, new tactics are not the only means of improving environmental quality. Within IPM, EILs provide an objective formal basis for making decisions on managing pests. These decisions are at the heart of IPM programs, because they indicate the degree to which it is possible to tolerate pest activities. By increasing our ability to tolerate pests, it is possible to eliminate or reduce the need for management tactics. Moreover, we can improve and maintain environmental quality through better decisions on the use of those tactics. To date, IPM decision-making has directly addressed only economic criteria. In this context, EILs help to maintain environmental quality by reducing unnecessary use of management tactics, especially of pesticides. However, including environmental considerations explicitly in the decision-making process could greatly improve the ability of IPM to sustain environmental quality. Consequently, focusing on the EIL concept offers the prospect of improved responsiveness to environmental concerns in IPM, without a reliance on new tactics.

The EIL concept

To understand the current and potential roles of EILs in maintaining environmental quality, it is first necessary to examine the components of the EIL. The history of the EIL concept is interwoven with forward thinking in biological control and the eventual formulation of ideas in integrated control, a precursor to IPM. Although producer profits and efficiency were important motives for development of sound decision-making, these incentives were probably not the reason for inception of the EIL concept. Rather, the idea developed from the concerns of perceptive scientists and practitioners about excessive or inappropriate uses of pesticides. Thus, the idea was developed largely as a means of applying pesticides in a rational and judicious manner, one that would help alleviate ecological problems within agroecosystems and associated habitats. Three factors are fundamental in the concept: economic damage, the economic injury level itself, and the economic threshold.

Economic damage

Economic damage is the most basic component of the EIL concept. It was defined by Stern et al. (1959) as "the amount of injury which will justify the cost of artificial control measures". A significant improvement of this purely subjective definition was made by Southwood and Norton (1973), who related cost and benefit as

$$C(a) = Y[s(a)] \times Ps(a)] - Y(s) \times P(s)$$

where Y is yield, P is price per unit of yield, s is level of pest injury, a is control action $[s(a)]$ is level of injury as modified by a suppressive tactic, and C is cost of the control action. Considering this expression, economic damage begins when cost of the control tactic equals the yield times the price when the tactic is applied minus the yield times the price without the tactic.

EIL

The EIL is the guideline for determining a final decision rule. Stern et al. defined it as "the lowest population density that will cause economic damage". The EIL is a theoretical value that, if realized by the pest population, will result in economic damage. As such, this value is a measure against which the destructive status and potential of a pest population is determined. Although first expressed as a pest density, the EIL is actually a level of injury indexed by pest numbers, i.e. pest numbers are used for practicality.

A refinement to the EIL concept is to consider injury in standard units termed injury equivalents (Pedigo et al., 1986). An injury equivalent is the total injury produced by a single pest over an average lifetime. This is a potential value; therefore, a pest dying prematurely would achieve only a partial equivalent. One use of injury equivalents is when making management decisions for a pest guild, such as for a complex of defoliators as was done by Hutchins et al. (1988). These authors considered consumption of five soybean-defoliating insects and established injury equivalents to establish a multispecies EIL.

Whether expressed as numbers or injury equivalents, the EIL is governed by five primary variables: cost of the management tactic per production unit (C), market value per production unit (V), injury units per pest (I), damage per injury unit (D), and the proportionate reduction of injury averted by the application of a tactic (K). The general statement of the EIL based on these variables is:

$$\text{EIL} = C/VIDK.$$

Economic threshold

Unlike the EIL, the economic threshold (ET) is a practical or so-called operational rule, rather than a theoretical or potential value (Mumford and Norton, 1984). It was defined by Stern et al. (1959) as "the population density at which control action should be determined to prevent an increasing pest population from reaching the economic injury level". The ET is an action level, used mostly when monitoring a pest population in a previously undamaged crop. If called for, a tactic is applied to prevent injury from reaching the EIL. Although measured in insect density, the ET is actually a time to take action, i.e. numbers are simply an index of that time.

Although the calculation of the EIL is simple and rather straightforward, that for the ET is not. Developing ETs is problematic because of the complex variables involved. Included among others are: (1) the EIL variables (this is because the ET is based on the EIL), (2) pest and host phenology, (3) population growth and injury rates, and (4) time delays associated with the IPM tactics utilized. Because of the uncertainties involved, particularly in pest population growth rates, most ETs are relatively crude, not carrying the same quantitative resolution as EILs.

In developing ETs, several methods have been used to give reasonable accuracy and yet deal with the uncertainty in predicting natural phenomena. These approaches can be grouped as subjective determinations or objective determinations.

Subjective determinations represent our least sophisticated approach to developing ETs. Subjective ETs have been termed nominal thresholds (Poston et al., 1983), and they are not based on objective criteria, such as EILs. Instead, they are simply chosen based on expert opinion. Unfortunately, because we lack crop loss functions, these are the most common form of ETs currently found in IPM recommendations. Although static and possibly inaccurate, these still are more progressive than using none at all. Therefore, their use can often result in reduced pesticide applications.

Conversely, objective ETs are based on calculated EILs, and they can be updated easily with changes in EIL variables. For objective ETs, a current EIL is calculated, and projections are made on the potential of pest injury to exceed the EIL. Then, decisions on actions are taken based on expected injury rate and logistical delays in applying tactics, as well as activity rate of tactics. Different forms of objective ETs are found in recommendations, including: (1) fixed ETs, (2) descriptive ETs, and (3) dichotomous ETs (Pedigo et al., 1989).

Fixed ETs are used when estimates of injury rates or dynamics of pest population are unknown. Therefore, the ET is set at a fixed level, often 25–50% below the EIL. Use of the term 'fixed' does not mean that these are unchanging; it means only that the percentage of the EIL is a constant. Although fixed ETs are crude, they may be our most advanced form when predictions of pest population dynamics are inaccurate. There are many examples of fixed ETs, including those for pests of grapes, beans, soybeans, sorghum, rice, and apples (Pedigo et al., 1986).

Descriptive ETs are much less common than fixed ETs, usually because predictive information on pest population ecology is lacking. Descriptive ETs are based on predicting pest population growth and/or injury rates to decide on the need for action and, if needed, proper timing. These ETs have been derived both by stochastic and deterministic procedures. An example of the stochastic type can be seen in ETs for spider mites, *Tetranychus* spp., on cotton in the western US. Here, a computer model was developed for mite ETs that considers early rates of population growth and physiological time to estimate damage trajectories and pest potentials (Wilson, 1985). When establishing descriptive ETs with deterministic methods, estimates of future pest population growth and injury rates are determined from life-table information or mechanistic models of a population process, e.g. predator/prey model.

An example of the deterministic procedure was observed in work with the green cloverworm, *Plathypena scabra*, in soybeans in the midwestern US (Ostlie and Pedigo, 1985).

Finally, dichotomous ETs have been developed for the green cloverworm on soybean using a classifying procedure called time-sequential sampling (Pedigo and Van Schaik, 1984). With this approach, class limits (m_0, m_1) were established for typical economic and a subeconomic populations using weekly larval counts through time (Pedigo et al., 1989). This program was both accurate and efficient, giving a 25% time saving, compared with a conventional sampling program for making decisions.

Environmental EILs

Decreasing dependence on pesticides and improving responses to environmental concerns can and should be met by developing environmentally based EILs and their concomitant ETs. An environmental EIL is an EIL that evaluates a management tactic based on not only its direct costs and benefits to the user but also its effects on the environment. The EIL equation integrates many management elements, each of which may have a role in making pest management more environmentally sustainable. Elsewhere (Pedigo and Higley, 1992), we have discussed the practicality of manipulating variables in the EIL equation to make pest management decisions more responsive to environmental concerns and to transform the conventional EIL into an environmental EIL, and this issue will be reviewed in the remainder of the paper.

Assigning realistic management costs (C)

Although C accounts for the direct economic costs of management actions, neither C nor other components of the EIL address the indirect environmental costs associated with environmental risks from management actions. If we can put a monetary value on these environmental risks, however, it is possible to include these costs in variable C of the EIL. Including environmental costs guarantees that pest management decisions based on EILs reflect environmental considerations better. Such additions in cost increase EILs and, therefore, result in less frequent pesticide applications. Moreover, by assigning more realistic costs to pesticide use, alternative methods with lower or negligible environmental costs may become more economically feasible.

For EILs to reflect environmental risks, it is necessary to identify risks from specific pesticides, to determine the relative importance of these risks, and to make monetary estimates of the benefits of avoiding different risk levels. The central problem is that determining both the relative importance of environmental risks and the monetary value of avoiding these risks are highly subjective tasks. Thus, we cannot objectively determine environmental costs. It is possible, however, to use formal procedures in these subjective determinations.

Higley and Wintersteen (1992) have proposed one approach for estimating the environmental costs of pesticides through the economic technique of contingent valuation. Contingent valuation is an established method used by economists for estimating, through opinion surveys, the value of non-market goods, such as environmental quality. Briefly, Higley and Wintersteen estimated the level of risk posed by 32 field crop insecticides to different environmental elements (surface water, ground water, aquatic organisms, birds, mammals, beneficial insects) and to human health (acute and chronic toxicity). They also estimated from survey data the relative importance of avoiding risk to each of these elements. Additionally, survey respondents (producers) indicated how much they would be willing to pay, in either higher pesticide costs (for safer pesticides) or yield losses, to avoid different levels of risk (high, moderate, or low) from a single application of a pesticide. With these data, individual environmental costs were calculated for each pesticide and environmental EILs were calculated.

Including environmental costs in EILs to produce environmental EILs incorporates a level of environmental protection into IPM decisions. Additionally, the data on environmental costs provide information that producers can use to choose the least environmentally hazardous pesticide, when a pesticide application is appropriate. The procedure of Higley and Wintersteen (1992) represents a first attempt to incorporate environmental costs into EILs; other approaches also may be possible. A strong incentive for additional work in this area is that the development and widespread use of environmental EILs would undoubtedly reduce pesticide inputs and environmental risks associated with pesticide use. Consequently, environmental EILs hold much promise as a method for sustaining environmental quality.

Manipulating crop market value (V)

The variable V is extremely important in setting limits to IPM actions; crops with lower market values can tolerate more pests and are most suitable for EILs, and crops with higher market values can tolerate few pests and benefit less from EILs. However, V is probably least amenable of the EIL components to purposeful change. In a free market system, crop market values depend on supply and demand and, therefore, cannot be manipulated easily. Thus, although producers might be able to use safer, but more expensive, management practices, these increased costs cannot be reflected in higher market prices. Consequently, potential manipulations of V do not seem to offer much promise for improving environmental quality.

Reducing injury per pest (I)

Reducing injury per pest is an approach that is being considered for some medical pests (for example, by identifying and manipulating genotypes in disease vector populations that cannot transmit pathogens). In most instances, however, reducing injury per pest is difficult, and the ecological effects of such may be unknown or may carry unacceptable environmental risks. Because entire pest populations must be changed, this approach seems unlikely to have broad application.

Reducing damage per pest (D)

Understanding how pest injury alters yield is central to the EIL concept. Indeed, the fundamental biological basis for the EIL is in component D — that is, how much damage (yield loss) is produced from a given amount of injury. Unfortunately, D is very difficult to establish and is poorly defined for most insect species. Greater emphasis on insect and plant relationships will improve and expand the use of EILs, thus promoting environmentally sound IPM.

Another important prospect is that we may be able to reduce D itself by reducing plant susceptibility to injury. Reducing D implies that less yield loss occurs for a given amount of injury. Such results may be possible as we identify those mechanisms plants use to tolerate injury and compensate for it. The striking theoretical advantage to IPM programs that rely on a reduced value of D is that pests cannot develop resistance. Plants that can tolerate or compensate for injury do not place selection pressures on pest populations. Therefore, the benefits of tolerance and compensation in plants are sustainable and permanent.

Developing an environmentally responsible K value

The variable K is the proportion of total pest injury averted by the timely application of a management tactic (Pedigo and Higley, 1992). Achieving a K value of 1.00 is usually the goal of IPM practitioners. Lower values are often not a matter of choice; rather, they may reflect the less-than-desired effectiveness of the tactic employed — for example, achieving only 60% mortality when 95% or greater is desired. Although achieving a K value of 1.00 may be a worthy goal, at least from a profit perspective, such attempts may foster the overuse of pesticides — particularly when the goal of mortality is substituted for that of avoiding damage.

Pest overkill can occur even when label rates of pesticides are followed by producers. Label rates form the basis for most recommendations in IPM and usually are established for an array of environmental conditions and application methods. Unfortunately, because of product liability fears and other factors, even the minimum rate recommended may be too high. Consequently, an argument can be made for more testing to determine real minimal effective rates. In such testing, the goal should be to determine the lowest pesticide rate achieving a K value virtually equal to 1.00. Achieving this value may mean reducing rates to attain 60–70% mortality and still producing crop yields that are not significantly different from those with 100% pest mortality. In addition to reducing pesticide use, rate reductions could make pesticide applications more compatible with natural and biological controls.

Conclusions

One of the chief merits of the IPM approach to handling pest problems is its objective of sustaining environmental quality. Unlike alternatives, such as simple control philosophies, IPM is based on the concept of tolerating pests and considering the direct and indirect impact of management actions. Because the EIL concept is a unifying principle in IPM, broadening the environmental scope of EILs could dramatically improve the sensitivity of IPM programs to environmental concerns. An environmental scope for our EILs involves focused efforts on manipulating EIL variables relative to environmental issues — namely, developing environmental EILs. Such environmental EILs depend on our ability to quantify and include environmental costs in calculations, develop tolerant plant varieties, and develop environmentally responsible management tactics. Expanding the use of EILs provides additional opportunities for improving environmental quality. In contrast to efforts to find new magic solutions to pest problems, which have failed in the past and are virtually certain to fail in the future, continued work within and upon the EIL concept is a powerful new direction for conserving environmental quality and sustaining agricultural production.

Acknowledgments

We thank R. Peterson, Dept. of Entomology, University of Nebraska–Lincoln for his review of this manuscript. This is Journal Paper No. J-14750 of the Iowa Agriculture and Home Economics Experiment Station, Ames, Project Nos. 2580 and 2903 and paper 9785 of the journal series of the Nebraska Agricultural Research Division, Contribution No. 785 of the Department of Entomology, University of Nebraska–Lincoln. This work was supported in part by University of Nebraska Agricultural Experiment Station Projects 17-053 and 17-055.

References

Higley, L.G. and Wintersteen, W.K., 1992. A new approach to environmental risk assessment of pesticides as a basis for incorporating environmental costs into economic injury levels. Am. Entomol., 38: 34–39.

Hutchins, S.H., Higley, L.G. and Pedigo, L.P., 1988. Injury equivalency as a basis for developing multiple-species economic injury levels. J. Econ. Entomol., 81: 1–8.

Mumford, J.D. and Norton, G.A., 1984. Economics of decision making in pest management. Annu. Rev. Entomol., 29: 157–174.

Ostlie, K.R. and Pedigo, L.P., 1985. Incorporating pest survivorship into economic thresholds. Bull. Entomol. Soc. Am., 33: 98–101.

Pedigo, L.P. and Higley, L.G., 1992. A new perspective on the economic injury level concept and environmental quality. Am. Entomol., 38: 12–21.

Pedigo, L.P. and van Schaik, J.W., 1984. Time-sequential sampling: a new use of the sequential probability ratio test for pest management decisions. Bull. Entomol. Soc. Am., 30: 32–36.

Pedigo, L.P., Hutchins, S.H. and Higley, L.G., 1986. Economic injury levels in theory and practice. Annu. Rev. Entomol., 31: 341–368.

Pedigo, L.P., Higley, L.G. and Davis, P.M., 1989. Concepts and advances in economic thresholds for soybean entomology. In: A.J. Pascale (Editor), Proceedings of the World Soybean Research Conference IV, 5–9 March 1989, Buenos Aires, Argentina, Vol. 3, Realization Orientación, Graficá Editora, Buenos Aires, pp. 1487–1493.

Poston, F.L., Pedigo, L.P. and Welch, S.M., 1983. Economic injury levels: reality and practicality. Bull. Entomol. Soc. Am., 29: 49–53.

Southwood, T.R.E. and Norton, G.A., 1973. Economic aspects of pest management strategies and decisions. Ecol. Soc. Aust., Mem., 1: 168–184.

Stern, V.M., Smith, R.F., van den Bosch, R. and Hagen, K.S., 1959. The integrated control concept. Hilgardia, 29: 81–101.

Wilson, L.T., 1985. Developing economic thresholds in cotton. In: R.E. Frisbie and P.L. Adkisson (Editors), Integrated Pest Management on Major Agricultural Systems. Texas Agric. Exp. Stn. Publ. MP-1616, College Station, TX, pp. 308–344.

Agriculture, Ecosystems and Environment, 46 (1993) 245–256
Elsevier Science Publishers B.V., Amsterdam

Reorganizing to facilitate the development and use of integrated pest management

Frank G. Zalom

Statewide IPM Project, University of California, Davis, CA 95616, USA

Abstract

Integrated pest management (IPM) is the systems approach to reducing pest damage to tolerable levels using biological controls, cultural controls, genetically resistant hosts, and, when appropriate, chemical controls, especially those which are selective and do not contribute to environmental contamination and human health problems. Basic elements of an IPM program include research and extension to produce a body of information on farming practices and pest biology, a program for monitoring pests, natural enemies and crop status throughout the season, control action thresholds, IPM tactics for pest control, and crop consultants to apply the program. IPM programs are far from being universally adopted, even when available and validated. Important constraints to IPM use have been categorized as technical, financial, educational, institutional and social issues. Many of these factors are not mutually exclusive. In many instances, the basic biology of pests, beneficial organisms, and their interaction in agricultural ecosystems is not understood. IPM researchers should be aware of the ultimate use of their work. Risk is probably the most important financial obstacle to IPM adoption by growers. Availability of funding is also a limitation of IPM research. User education is a key element in successful implementation of complex technologies such as IPM. When comprehensive and well-organized programs have been attempted, the level of adoption has often been high and the amount of pesticide use demonstrably reduced. Such programs have helped to accelerate the development of the private pest management industry which provides information services to growers on crop production and pesticide use. Institutional constraints are extremely important, and their structures may actually hinder IPM development and use. The development and use of IPM could be enhanced through institutional reorganization. Better analysis of the impact of regulations and farm policy on pesticide use should be instituted so that such government efforts do not actually increase pesticide use. Private sector incentives for developing and marketing biological controls and biological pesticides should be made. Development of the private pest management industry should be encouraged, possibly linking the issue to stronger controls on the purchase and use of pesticides. Better marketing of IPM should be both studied and encouraged.

Introduction

Pests, including insects, mites, weeds, disease-causing organisms and vertebrates, lower the quality and yield of agricultural products. Before World War II, farmers relied heavily on non-chemical pest control methods such as crop rotation, tillage, bait trapping and hand-removal of pests. The pesticidal materials available contained metals such as copper, lead, antimony and arsenic, and botanical compounds such as nicotine and pyrethrum. These materials were toxic and expensive to produce in quantity, therefore their availability was limited. Equipment for their application was relatively unsophisticated or lacking. Overall pesticide use was low relative to current levels.

The development of synthetic pest control chemicals during World War II, in combination with improvements in application technology, increased the potential for pest control dramatically. Pest control researchers overwhelmingly embraced and perfected the technology. Governmental policies including tax incentives made the production and use of pesticides attractive to businesses and farmers. Extension agents and pesticide salesmen promoted pesticide use. Pesticide development, production and use became institutionalized, and farmers became increasingly dependent upon them.

Problems associated with pesticide use

The impact of pesticide use on the environment is a complex issue, and several factors have raised scientific and public concern regarding pesticide use. Genetic resistance to chemicals has been documented for more than 450 pest species (Georghiou, 1986). Outbreaks of secondary pests, those which have been released from natural control, has been well documented. For example, it is widely believed that spider mites have emerged as a serious agricultural and forestry pest since 1946 because their predators have been reduced by chemical sprays applied for primary pests. Non-target species are often affected, because pesticides generally kill a broad spectrum of organisms, only a fraction of which are the target pests. Off-target effects on native plants, mammals, birds, fish and other wildlife are of concern. Pesticide-induced losses of honey-bees, natural enemies, alternative food sources for biological control agents, and those organisms which function in decomposition and mineral recycling in the soil, and phytotoxic effects on crop plants can result in reduced crop yields or quality.

Human exposure to pesticides is an important health and social issue. Agricultural workers have the greatest risk of exposure, and may come into contact with pesticides during the application process or when entering recently treated areas. The non-agricultural public may also be exposed to small doses of pesticides if they live near treated areas (e.g. through air-drift and contaminated drinking water), eat contaminated food, or touch recently treated livestock, foliage or stored food products.

Environmental and human health issues have typically been addressed through label restrictions imposed during the pesticide registration process, and this has provided a good degree of protection. However, it can be argued that source reduction would significantly increase the predictability of protection. One approach to source reduction would be the widespread adoption of integrated pest management (IPM) strategies and tactics.

IPM

IPM is the systems approach to reducing pest damage to tolerable levels using biological controls, cultural controls, genetically resistant hosts, and, when appropriate, chemical controls, especially those which are selective and do not contribute to environmental contamination and human health problems. A comprehensive IPM program for an agricultural system is comprised of the following basic elements: (1) dedicated research and extension personnel to produce a body of timely information concerning farming practices, pest biology and pest management tactics to formulate the program, (2) a program to monitor pests and natural enemy population levels and the state of the crop throughout the season, (3) control action thresholds — levels of pest abundance at which some control action must be taken to protect the crop from unacceptable economic loss, (4) IPM tactics which involve a spectrum of agents and materials used to suppress pest populations, (5) crop consultants to apply the program in the field, and (6) willing and cooperative growers. The application of IPM is especially relevant today as agricultural producers, consultants and scientists try to move away from reliance upon pesticides. It is ironic that although IPM theory is fairly well understood and promoted it has not been widely translated into practice.

As early as 1964, concerns were voiced by IPM researchers (Van den Bosch, 1964) about the slow rate of adoption of integrated control, and particularly biological control, by farmers. Some authors (e.g. Lincoln and Parencia, 1977; Norton, 1982; Pimentel, 1982) have acknowledged the problem and have recognized that the technical complexity of IPM would delay its adoption. Others have undertaken more detailed studies focussing on external factors impacting on IPM use.

Constraints to IPM adoption

A large number of constraints to IPM use have been identified in various studies (e.g. Corbet, 1981; Grieshop et al., 1988). Wearing (1988) categorized these obstacles as technical, financial, educational, organizational (institutional), and social. In his study of IPM in the US, Europe and Australia/New Zealand, all of these general obstacles were identified as important, but their relative rank varied.

Technical constraints

In many instances, the basic biology of pests, beneficial organisms, and their interaction in agricultural ecosystems is not understood. Similarly, the application of this knowledge to the management of pests in cropping systems through tactics such as monitoring guidelines, control action thresholds, biological controls, cultural controls, and host plant resistance is lacking. Research in these areas is essential for the continued development of IPM. In spite of our general lack of knowledge, technical obstacles are often regarded as less important than other general obstacles to IPM adoption (Gruys, 1982; Perkins, 1982).

The need to simplify IPM methodology is especially important, particularly with respect to developing monitoring and sampling guidelines and control action thresholds (Corbet and Smith, 1976; Hussey, 1980; Goodell, 1984; Arends and Robertson, 1986). Wearing (1988) found that private pest management consultants (50%) and extension agents and specialists (52%) perceived these to be more important technical constraints than researchers (27%), probably reflecting closer relationship to practical application. Similarly, Flint and Klonsky's (1989) study of California pest control advisors rated research on monitoring and thresholds to be their highest rated technical need, followed in order by cultural control methods, biological control methods, economic analysis of controls, pesticide resistance studies, crop and pest interaction studies, and computer simulation models. Lack of selective pesticides and selecting IPM compatible pesticides are also mentioned as major technical obstacles in some studies (see Newsom, 1975; Croft, 1978; Gruys, 1982).

Financial constraints

IPM often increases net profits for growers who adopt it. However, there is still a perception by growers that IPM does not offer short-term economic advantages compared with conventional control, particularly because of additional labor costs from sampling and monitoring (Poe, 1981). Private pest management consultants and others who provide IPM services must charge for providing what many growers consider to be pest control advice, a service which has traditionally been provided, with no direct cost, by representatives of farm supply dealers who also sell farm chemicals. Growers expect to budget for tangible items such as machinery, pesticides, and fertilizers, but the concept of purchasing advice is less acceptable. Further, paying for advice in advance of pest problems or if no pest problem occurs is difficult for growers to accept. Growers must be convinced that IPM advice has value. Private pest management consultants are constrained from adopting more time-consuming IPM technologies in order to provide their services at the lowest possible cost. In California, where the crop-consulting industry is as well developed as in most places in the US, consultants receive $3.00–8.00 per acre per season for field crops and $12.00–25.00 per acre per season for perennial crops depending on the location, crop value, and competition. While some consultants have developed successful businesses, long hours, lack of liability insurance, and low to moderate pay in relation to the level of education needed has kept many qualified individuals from entering the field (Whalon and Weddle, 1986). Cooperative IPM programs in which farmers share costs and losses have been tried successfully for citrus in California (Graebner et al., 1984) and for cotton in Texas (Frisbie et al., 1983).

Risk is probably the most important financial obstacle to IPM adoption. Growers value pesticides for reducing production risk as well as contributing to profit. It is very important that IPM can be shown to decrease this risk (Way, 1977; Gruys, 1982; Antle and Park, 1986). In reality, IPM strategies, such as monitoring, are tools for managing risk. The more growers learn about pests in their fields and the likelihood of resulting damage, the less is the uncertainty in their minds about the state of their crop and the more likely it becomes that they will not choose to make a preventative pesticide application. Antle (1988) believes that the value of IPM in terms of risk reduction should actually increase in relation to the grower's level of risk aversion.

A grower's perception of risk can be a function of loan and contractual commitments of growers, which may require growers to follow what is considered prudent practices to meet yield or quality objectives (Van den Bosch, 1978). Even when contracts do not specify prudent grower practices (implying among other things pesticide use), the degree to which a grower is leveraged by debt is an important factor.

Lack of funds for university research and extension programs has been cited as a constraint to IPM adoption in many studies (see Whalon and Croft, 1984; Frisbie and Adkisson, 1985; Tauber et al., 1985), and there has in fact been a consistent erosion of the base budget for agricultural research and extension programs in the US over the past 20 years. Indeed, enhanced IPM development and adoption can be documented (Zalom et al., 1991) as having been achieved as a result of such targeted IPM programs as the Huffaker and Adkisson Projects, the federal extension pilot projects, and state-supported IPM projects including those of California, New York, and Texas. However, federal funds for IPM have not increased during the 1980s, and only a few states have provided budget augmentations for IPM programs.

Other financial obstacles shown to affect IPM development and use include the value of a commodity, which can affect the production and marketing objectives of a grower. Higher value often equates to higher chemical inputs. In addition, lack of funds for commercial development of biological control agents (Hussey, 1985) and biological pesticides including bacteria, fungi, and viruses, has limited their availability and increased their price.

Educational constraints

IPM is a complex set of behaviors, decision-making procedures, methods, technologies, and values organized to provide efficient alternative methods of pest management (Apple and Smith, 1976). Implementation of any such complex innovation requires intensive education of users. In the case of IPM this includes students, growers, pest management consultants, and extension staff. Traditional programs and educational structures may not be adequate to address the needs of IPM education. Because there is no formula for applying IPM to a given situation, IPM education probably requires more intensive visual or interactive methodology. Because IPM addresses pest management within a cropping system, an interdisciplinary structure is necessary.

It has been suggested that a lack of education of IPM developers about the perceptions and needs of growers could be a greater obstacle to adoption than the reverse (see Goodell, 1984). Such a lack of understanding can lead to development of inappropriate technology which will never be adopted.

Institutional constraints

The structure and edicts of regulatory, educational, and corporate or industrials institutions influence the development and use of IPM programs. Lack of coordination can be problematic (Kuhr, 1981; Perkins, 1982), especially among organizations, personnel, and disciplines.

Efforts to mandate or regulate IPM specifically have not heretofore been very successful. Grieshop et al. (1990) studied a mandated IPM program for lessees on state-owned land in California in which adoption quickly declined when no enforcement action was taken for non-compliance. It was found that the principal state agency involved, the Department of Parks and Recreation, lacked experience in dealing with agricultural production, and failed to adequately address grower concern for risk. Growers lacked confidence in research and extension efforts supported by the regulatory agency. California regulations mandating the licensing of anyone making pest control recommendations were intended in part to foster the use of IPM strategies and tactics. However, because distinctions were not made between private pest management consultants and the majority of licensees who work for farm supply companies, these regulations may not have had the desired effect (Willey, 1978).

Some government programs can indirectly affect IPM adoption. Examples include commodity price supports which reward growers for maximizing production, often at the expense of increased pesticide use which would not be profitable in the absence of support, and marketing orders which may use low pest damage standards as tools which regulate supply. Farm subsidy programs encourage growers to plant the same crop each year to qualify for benefits. This discourages crop rotation, an effective cultural control for many species of insects, diseases, nematodes and weeds.

Even geopolitical issues such as pest quarantines can affect pesticide use. Many pesticides are applied to control pests both in the field and through the postharvest period to meet phytosanitary requirements for specific foreign markets. Similarly, eradication programs for exotic pests are often conducted to preserve foreign markets.

The lack of interdisciplinary collaboration in IPM research, extension, and teaching has been a major constraint voiced by many individuals (see Brader, 1980; Barnett, 1981; Miller, 1983). There is a tendency for research and educational activities at the Land Grant Universities to be conducted within strong disciplinary-oriented departmental units (e.g. agronomy, plant pathology and entomology) which have evolved in response to institutional pressures for specialization. Similarly, degree programs which would lead to interdisciplinary professional rather than research degrees in plant health and pest management are few, and not well supported (Kendrick, 1988).

Individual achievements rather than team accomplishments are typically rewarded (Pimentel, 1982), leading to the predominance of such efforts at the expense of true cropping system studies involving biological, social and economic components. The rewards structure also leads to an emphasis on basic biological studies, often on a biochemical or molecular level, rather than on an organismal level. This research emphasis is driven to a large extent by Federal funding programs which place emphasis on fundamental research. Such funding, while critical from the standpoint of scientific discovery, leverages great amounts of resources in the Land Grant Universities.

Other organizational obstacles also exist, most notably cosmetic standards imposed by government agencies such as the Federal Food and Drug Administration, the US Department of Agriculture (USDA), and state departments of agriculture, corporations including processors, packers, and retailers, and commodity associations such as cooperatives and marketing orders (Corbet, 1981; Fenemore and Norton, 1985). These quality standards have largely been imposed in response to consumer concerns, but have also been used to regulate markets.

Social constraints

Wearing (1988) found that social and marketing obstacles rated above all others for their impact on IPM use. In agriculture, the rate at which the IPM adoption process occurs and the ultimate level of adoption may be affected by many factors including demographic attributes of potential users (Rajotte et al., 1987; Zalom et al., 1987), land ownership (Grieshop et al., 1988), communication channels used by growers or managers, and growers' perceptions of the technology.

Growers receive pest management information from a variety of sources. In this regard, chemical controls have a competitive advantage over IPM. There is a well-established infrastructure for pesticide supply and use, with a high ratio of chemical sales personnel, technical representatives, and farm supply dealers, to private pest management consultants or extension IPM staff (Van den Bosch, 1978; Smith and Pimentel, 1978). Further, corporations expend large amounts of resources in marketing agricultural chemicals, and use market analysis effectively to identify and improve market fit (Tait, 1982). Wearing (1988) cited lack of marketing expertise by IPM extension staff and consultants as a major obstacle to implementation.

Pesticides give growers immediate reinforcement in terms of pest control, and growers have developed confidence in their use (Way, 1977; Corbet, 1981; Hussey, 1985). IPM often takes longer to realize benefits. In addition, the concept of economic thresholds, which implies tolerating a number of pests in the crop, is perceived as risky by many growers (Pimentel, 1982). Experience with IPM can change this perception. Grieshop et al. (1988) found that growers they studied who used IPM had significantly more confidence and satisfaction in IPM than did growers who did not use IPM. IPM users also perceived IPM as less risky, less time consuming, and easier to use than did non-users.

Reorganizing to address constraints

If we hope to hasten the development and use of IPM, it is necessary for us to address the constraints present in our agricultural system from research and extension through production and marketing.

Agricultural research and extension

Although many research and extension professionals are involved in IPM activities in Land Grant Colleges, the USDA, and private companies, the level of activity is insufficient to rapidly develop and implement IPM programs universally. The obvious response would be to direct more funding to the present system. While this would undoubtedly be beneficial, perhaps a more careful examination of institutional structure and incentives to researchers is needed as well. Most aspects of IPM research require an extensive understanding of the crop ecosystem, and strong interdisciplinary programs. While some of this research is considered quite technical, especially to those growers or consultants trying to apply the research, development of biological controls, cultural controls, and decision rules in IPM are not considered highly technical by many university colleagues reviewing promotion or tenure packages, or by large funding programs such as the USDA Competitive Grants, National Science Foundation and National Institutes for Health, which tend to sponsor research in more fundamental areas such as molecular biology or biochemistry. Therefore, some researchers and many new positions at the Land Grant Universities are directed away from applied field ecology because of a lack of institutional support. Even for those researchers who have an interest in pursuing such studies, there is little incentive to do the adaptive research necessary to lead to commercial use. Adaptive research in many cases is conducted by individuals in Cooperative Extension or the private sector. Unfortunately, unless these individuals were part of the original research, or unless the original researcher is willing to participate in the adaptation of their research, the probability for failure is great.

Greater investment in research and extension, while necessary, should be accompanied by mechanisms which would address both short- and long-term problems associated with conversion to an agricultural production system based on lower pesticide use. Research and extension funding programs should be more goal oriented, no matter what the time frame of the proposals or the fundamental scientific question being addressed. Individuals designing and administering these programs should clearly identify the goals of the programs, and research and extension should be closely linked in both the development and implementation of the programs. Perhaps position descriptions should be required of individuals with Agricultural Experiment Station or Cooperative Extension appointments, and the merit and promotion process linked to those documents.

More positions in Cooperative Extension should be directed specifically towards IPM including biological and cultural pest controls, and pesticide safety. In most states, there are far more staff addressing production than pest management issues and, of those addressing pest management, even fewer specifically work toward the implementation of IPM. The agricultural chemical industry has an established infrastructure for marketing products to growers which gives it an advantage over public and private promoters of IPM.

Private sector incentives such as grants or tax incentives for developing and marketing biological controls and microbial pesticides should be available in order to interest more companies in their development and production, and to hasten their use. This is especially important because the selectivity and short persistence of these materials, which are considered environmentally beneficial attributes, also limit their potential market when compared with more traditional pesticides.

Pest management consultants

Implementing IPM requires individuals well trained in pest management and the plant sciences. The success of private pest management consultants in reducing pesticide use through implementing IPM strategies and techniques is well documented (see Hall, 1977). However, the number of private pest management consultants is limited. Increasing the utilization of private pest management consultants could be accomplished by increasing grower incentives, or by licensing these individuals and requiring their direct input in the pesticide use process, by prescribing the use of restricted agricultural chemicals in response to careful monitoring of the crop health and in accordance with label requirements. Degree programs which would lead to professional accreditation in plant health and pest management should be supported at Land Grant universities, and they could be structured as either graduate groups or professional schools at those institutions.

Marketing

Quality standards imposed by government agencies, marketing orders and private packers or processors are warranted when health standards are at issue. They are also necessary to some extent when consumer preference would make producers less competitive. The imposition of quality standards should be carefully considered and should be based solely on health benefits and cost/benefit analysis.

Commodity price supports reward growers for maximizing production and encourage growers to plant the same crop each year to qualify for benefits. An emphasis on production often results in increased pesticide use which would not be economical at lower commodity prices. These programs also discourage good pest management practices such as crop rotation. Impact upon pesticide use should be a consideration in support or incentive programs.

Certification might be an incentive for consumers to purchase products grown using IPM practices. This would require establishing a set of crop-specific IPM standards which a grower would need to follow in order to meet requirements, and might be similar to those regulations defining and certifying organically grown products. Compliance could be achieved through examination boards or by requiring use of private pest management consultants.

The development and use of IPM is affected by a number of technical, financial, educational, institutional and social issues. Compelling biological issues such as pest resistance and economic issues such as increased cost of pesticides and consumer concerns about pesticides will continue to make IPM an increasingly popular approach for producers. However, the rapid development and use of IPM may not occur unless many of the factors restricting IPM are addressed. Science and education alone may not adequately address these factors. They may require intervention in the form of new public policies, regulations and institutional structures.

Acknowledgment

This manuscript was formulated in part from the literature reviewed for a commissioned paper submitted to the Office of Technology Assessment, United States Congress. The support of that agency is gratefully acknowledged.

References

Antle, J.M., 1988. Integrated pest management: it needs to recognize risks, too. Choices, 3(3): 8–11.

Antle, J.M. and Park, S.K., 1986. The economics of IPM in processing tomatoes. Calif. Agric., 40 (3 & 4): 31–32.

Apple, J.L. and Smith, R.F., 1976. Integrated Pest Management. Plenum Press, New York.

Arends, J.J. and Robertson, S.H., 1986. Integrated pest management for poultry production. Implementation through integrated poultry companies. Poult. Sci., 65: 675–682.

Barnett, W.W., 1981. Practical aspects of pest management. HortScience, 16: 515–516.

Brader, L., 1980. Advances in applied entomology. Ann. Appl. Biol., 94: 349–365.

Corbet, P.S., 1981. Non-entomological impediments to the adoption of integrated pest management. Prot. Ecol., 3: 183–202.

Corbet, P.S. and Smith, R.F., 1976. Integrated control: a realistic alternative to misuse of pesticides? In: C.B. Huffaker and P.S. Messenger (Editors), Theory and Practice of Biological Control. Academic Press, New York, pp. 661–682.

Croft, B.A., 1978. Potentials for research and implementation of integrated pest management on deciduous tree fruits. In: R.F. Smith and D. Pimentel (Editors), Pest Control Strategies. Academic Press, New York, pp. 101–115.

Fenemore, P.G. and Norton, G.A., 1985. Problems of implementing improvements in pest control: a case study of apples in the U.K. Crop Prot., 4: 51–70.

Flint, M.L. and Klonsky, K., 1989. IPM information delivery to pest control advisers. Calif. Agric., 43(2): 18–20.

Frisbie, R.E. and Adkisson, P.L., 1985. IPM: definitions and current status in U.S. agriculture. In: M.A. Hoy and D.C. Herzog (Editors), Biological Control in Agricultural IPM Systems. Academic Press, New York, pp. 41–50.

Frisbie, R.E., Phillips, J.R., Lambert, W.R. and Jackson, H.B., 1983. Opportunities for improving cotton insect management programs and some constraints on beltwide implementation. US Dep. Agric., Agric. Handb., 589: 521–557.

Georghiou, G.P., 1986. The magnitude of the resistance problem. In: Pesticide Resistance: Strategies and Tactics for Management, National Academy Press, Washington, DC.

Goodell, G.E., 1984. Challenges to international pest management research and extension in the third world: do we really want IPM to work? Bull. Entomol. Soc. Am., 30: 18–26.

Graebner, L., Moreno, D.S. and Barttelle, J.L., 1984. The Fillmore Citrus Protective District: a case story in integrated pest management. Bull. Entomol. Soc. Am., 30: 27–33.

Grieshop, J.I., Zalom, F.G. and Miyao, G., 1988. Adoption and diffusion of integrated pest management innovations in agriculture. Bull. Entomol. Soc. Am., 34(2): 72–78.

Grieshop, J.I., MacMullen, E., Brush, S., Pickel, C. and Zalom, F.G., 1990. Extending integrated pest management by public mandate: a case study from California. Soc. Nat. Resour., 3: 33–51.

Gruys, P., 1982. Hits and misses. The ecological approach to pest control in orchards. Entomol. Exp. Appl., 31: 70–87.

Hall, D.C., 1977. An economic and institutional evaluation of IPM. US Environ. Prot. Agency Publ. 68-01-2982, Washington, DC.

Hussey, N.W., 1980. Crop protection: a challenge in applied biology. Ann. Appl. Biol., 96: 261–274.

Hussey, N.W., 1985. Biological control — a commercial evaluation. Biocontrol News Inf., 6: 93–99.

Kendrick, J.B., 1988. A viewpoint on integrated pest management. Plant Dis., 82: 647.

Kuhr, R.J., 1981. Regional planning and coordination of integrated pest management programs. HortScience, 16: 514–515.

Lincoln, C. and Parencia, C.R., 1977. Insect pest management in perspective. Bull. Entomol. Soc. Am., 23: 9–14.

Miller, A., 1983. Integrated pest management: psychosocial constraints. Prot. Ecol., 5: 253–268.
Newsom, L.D., 1975. Pest management: concept to practice. In: D. Pimentel (Editor), Insects, Science and Society. Academic Press, New York, pp. 257–277.
Norton, G.A., 1982. A decision-analysis approach to integrated pest management. Crop Prot., 1: 147–164.
Perkins, J.H., 1982. Insects, Experts and the Insecticide Crisis. Plenum Press, New York.
Pimentel, D., 1982. Perspectives of integrated pest management. Crop Prot., 1: 5–26.
Poe, S.L., 1981. An overview of integrated pest management. HortScience, 16: 501–506.
Rajotte, E.G., Kazmierczak, R.F., Lambur, M.T., Norton, G.W. and Allen, W.A., 1987. The national evaluation of extension's integrated pest management (IPM) programs. Va. Coop. Ext. Serv. Publ. 491-010 through 491-024, Blacksburg, VA.
Smith, E.H. and Pimentel, D., 1978. Pest Control Strategies. Academic Press, New York.
Tait, E.J., 1982. Farmers' attitudes and crop protection decision making, In: R.B. Austin (Editor), Decision Making in the Practice of Crop Protection. Monograph No. 25. British Crop Protection Council, London, pp. 55–60.
Tauber, M.J., Hoy, M.A. and Herzog, D.C., 1985. Biological control in agricultural IPM systems: a brief overview of the current status and future prospects. In: M.A. Hoy and D.C. Herzog (Editors), Biological Control in Agricultural IPM Systems. Academic Press, New York, pp. 41–50.
Van den Bosch, R., 1964. Practical application of the integrated control concept in California. Proc. 12th Int. Congr. Entomol., London, Vol. 12, pp. 595–597.
Van den Bosch, R., 1978. The Pesticide Conspiracy. Doubleday, Garden City, NY.
Way, M.J., 1977. Integrated control-practical realities. Outlook Agric., 9: 127–135.
Wearing, C.H., 1988. Evaluating the IPM implementation process. Annu. Rev. Entomol., 33: 17–38.
Whalon, M.E. and Croft, B.A., 1984. Apple IPM implementation in North America. Annu. Rev. Entomol., 29: 435–470.
Whalon, M.E. and Weddle, P., 1986. Implementing IPM strategies and tactics in apple: an evaluation of the impact of CIPM on apple IPM. In: R.E. Frisbie and P.L. Adkisson (Editors), CIPM Integrated Pest Management on Major Agricultural Systems. Texas Agricultural Experiment Station, College Station, TX, pp. 619–637.
Willey, W.R.Z., 1978. Barriers to the diffusion of IPM programs in commercial agriculture. In: E.H. Smith and D. Pimentel (Editors), Pest Control Strategies. Academic Press, New York, pp. 285–308.
Zalom, F.G., Klonsky, K. and Barnett, W.W., 1987. Evaluation of California's almond IPM program. Univ. of Calif. Integr. Pest Manage. Publ. 6, Davis, CA.
Zalom, F.G., Stimmann, M.W. and Smilanick, J.M., 1991. Integrated pest management and ground water contamination. In: Beneath the Bottom Line: Agricultural Approaches to Reduce Agrichemical Contamination of Groundwater. Volume II, Part D: Pest and Pesticide Management. Office of Technology Assessment, Congress of the United States, Washington, DC, NTIS PB91-168 278.

Agriculture, Ecosystems and Environment, 46 (1993) 257–272
Elsevier Science Publishers B.V., Amsterdam

Ethnoscience and biodiversity: key elements in the design of sustainable pest management systems for small farmers in developing countries

Miguel A. Altieri

Division of Biological Control, University of California, Berkeley, 1050 San Pablo Avenue, Albany, CA 94706, USA

Abstract

Biodiversity is a salient feature of traditional farming systems in developing countries and performs a variety of renewal processes and ecological services in agroecosystems. It is of fundamental importance to understand the role biodiversity can play in reducing pest problems, if vegetation management is to be used effectively as a primary IPM tactic in small-scale sustainable agriculture. The maintenance of biodiversity in traditional agroecosystems is not random, but depends on a complex set of indigenous technical knowledge systems (ethnoscience). Thus, the ensemble of traditional crop protection practices used by indigenous farmers represents a rich resource for modern workers seeking to create IPM systems that are well adapted to the agroecological, cultural and socio–economic circumstances facing small farmers throughout the developing world.

Introduction

Traditional farming systems in developing countries exhibit two salient features: a high degree of vegetational diversity (biodiversity) and a complex system of indigenous technical knowledge (ethnoscience). Both elements are obviously highly interrelated since the maintenance of biodiversity is dependent upon local farmers' knowledge about the environment, plants, soils and ecological processes (Toledo et al., 1985). Despite acknowledging the importance of both elements, most agriculturalists have yet to take full advantage of the benefits of biodiversity and ethnoscience in the implementation of rural development projects.

Nowhere is the potential of biodiversity and ethnoscience more applicable than in the realm of pest management. On the one side, the regulating effects of vegetational diversity on pest populations are well known. On the other, there are clear indications that certain ethnic groups of farmers, some more than others, have a thorough knowledge of the history, biology and bionomics of a variety of insect pests (Altieri, 1990). This paper offers a synthesis of current knowledge about the relationships between biodiversity, ethnoscience and pest management, along with an agroecological basis for designing pest control methods tailored to the socio–economic and cultural circumstances of small farmers throughout the developing world.

Biodiversity and pest management

Biodiversity refers to all species of plants, animals and microorganisms existing and interacting within an ecosystem. Polycultural and agroforestry systems typical of most traditional farming systems exhibit a high degree of biodiversity. This biodiversity performs a variety of renewal processes and ecological services in these agroecosystems. The diversity of crops and wild plants provides vegetative cover which prevents soil erosion, regulates the water balance and nutrient cycling and aids in the control of the abundance of undesirable organisms (Altieri and Letourneau, 1982). When these natural services are lost, due to biological simplification through adoption of monocultures or use of high-input technologies, the social, economic and environmental costs can be quite significant.

Ecological theory suggests that diversified cropping systems contain natural elements of pest control. The majority of agroecological studies show that structural (i.e. spatial and temporal crop arrangement) and management (e.g. crop diversity and input levels) attributes of agroecosystems influence herbivore dynamics. Most experiments that mixed other plant species with the primary host of a specialized herbivore showed that, in comparison with diverse crop communities, simple crop communities have greater population densities of specialist herbivores (Altieri and Letourneau, 1984). In these less diverse systems, herbivores exhibit greater colonization rates, greater reproduction, less tenure time, less disruption of host finding and lower mortality by natural enemies (Andow, 1991).

So far, two hypotheses have been proposed to explain the commonly observed lower herbivore abundance in polyculture (Altieri and Letourneau, 1982): the resource concentration hypothesis and the natural enemies hypothesis.

The resource concentration hypothesis

This hypothesis states that crop monocultures represent a concentrated resource for specialized herbivores, which increases the attraction and accumulation of these species, the time they spend in the system and their reproductive success. Visual and chemical stimuli, from host and non-host plants in a polyculture, affect the rate at which herbivores colonize a polyculture and their behavior in these habitats. In a polyculture, non-host species may mask the chemical attractants of the host, reduce the contrast between the host and its background or simply hide the plants from view (i.e. make them less apparent).

The natural enemies hypothesis

This hypothesis predicts an increased abundance of arthropod predators and parasitoids in polycultures due to the increased availability of alternate prey, nectar sources, and suitable microhabitats. The net effect of natural enemies on pest abundance in polycultures will depend upon whether natural enemies are governed more in their behavior by prey density or by background plant density, although at times both factors operate simultaneously.

There is general agreement that these hypotheses are not mutually exclusive since a particular herbivore population may be affected simultaneously by both the concentration of resources and by natural enemies. In fact, herbivore regulation in polycultures may involve other mechanisms not considered by the hypotheses, such as microclimate, differences in levels of secondary compounds or in plant quality, and so on. The point is that it is important to identify key mechanisms, differences in mechanisms between cropping systems, and which plant assemblages enhance regulatory effects and which do not, and under what management and agroecological circumstances.

Ecological theory and pest management in polycultures

The most vegetationally diverse traditional agroecosystems are those under extensive shifting cultivation and/or those under intensive subsistence farming in the tropics. These systems are usually characterized by complex cropping systems (e.g. intercropping, agroforestry and rotations) with crop sequences and associations managed in a variety of ways in time and space.

Theoretically these systems contain built-in elements of natural pest control. The question is then: how can emergent ecological theories help in designing polycultures that offer even better or more effective herbivore protection features? In other words, is there anything that ecologists can do to help traditional farmers improve the polycultural systems they already have?

From a practical standpoint it is easier to design insect manipulation strategies in polycultures using the elements of the natural enemies hypothesis than those of the resource concentration hypothesis, mainly because we cannot yet identify the ecological situations or life history traits that make some pests sensitive (i.e. their movement is affected by crop patterning) and others insensitive to cropping patterns (Andow, 1991). The recognition that crop monocultures are difficult environments in which to induce the efficient operation of beneficial insects, because these systems lack adequate resources for effective performance of natural enemies and because of the disturbing cultural practices often utilized in such systems, can offer useful insights to biological control practitioners. Polycultures already contain specific resources provided by plant diversity, and are usually not disturbed with pesticides (especially when managed by resource-poor farmers who cannot afford high-input technology), and are thus more amenable to manipulation. By replacing or adding diversity to existing systems, it may be possible to exert changes in habitat diversity that enhance natural enemy abundance and effectiveness (Altieri and Letourneau, 1982; Powell, 1986) by (1) providing alternative host-prey at times of pest host scarcity; (2) providing food (pollen and nectar) for adult parasitoids and predators; (3) providing refuges for overwintering, nesting, and so on; (4) maintaining acceptable populations of the pest over extended periods to ensure continued survival of beneficial insects.

The specific effect or the strategy to use will depend on the species of herbivores and associated natural enemies, as well as on properties of the vegetation, the physiological condition of the crop, or the nature of the direct effects of a particular plant species. In addition, the success of enhancement measures can be influenced by the scale upon which they are implemented (i.e. field scale, farming unit or region), since field size, within-field and surrounding vegetation composition, and level of isolation (i.e. distance from the source of colonizers) will all affect immigration rates, emigration rates, and the effective tenure time of a particular natural enemy in a crop field.

Perhaps one of the best strategies for increasing the effectiveness of predators and parasitoids is the manipulation of non-target food resources (i.e. alternate hosts, prey and pollen-nectar) (Altieri and Liebman, 1986). Here it is not only important that the density of the non-target resource be high, to influence enemy populations, but that the spatial distribution and temporal dispersion of the resource also be adequate. Proper manipulation of the non-target resource should result in the enemies colonizing the habitat earlier in the season than the pest, and frequently encountering an evenly distributed resource in the field, thus increasing the probability of the enemy remaining in the habitat and reproducing (Andow, 1991). Certain polycultural arrangements increase, and others reduce, the spatial heterogeneity of specific food resources; thus particular species of natural enemies may be more or less abundant in a specific polyculture. These effects and responses can only be determined experimentally across a whole range of traditional agroecosystems. The task is indeed overwhelming since enhancement techniques must necessarily be site specific (Altieri, 1991a).

Although polycultures seem to exhibit pest-suppression potential, most of the data from field tests include polyculture treatments that mimic farmers' practices. Thus, at this point, it is not clear whether these patterns hold across cropping systems managed under farmers' conditions. In their surveys of traditional maize cropping systems in Tlaxcala, Trujillo-Arriaga and Altieri (1990) found that certain crop associations would reduce populations of the scarab beetle *Macrodactylus* sp., while others would increase them. In a survey of insect communities associated with maize grown in association with other annual crops and with trees or shrubs in Indonesia, it was found that pest damage and abundance of natural enemies varied considerably between fields. It was not clear whether these differences were related to the vegetational structure of the systems, or were merely a consequence of differential management, location, or chance (Altieri and Liebman, 1986). It is clear that pest dynamics will vary significantly between systems depending on insect species, location and size of the field, vegetational composition, and cultural management.

Ethnoscience and pest management

Classification of animals, especially insects and birds, is widespread among traditional farmers and indigenous groups. In their survey of pest control practices used by local farmers in the Philippines, Litsinger et al. (1980) found that farmers had local names in separate dialects in each location for most pests attacking rice, corn and grain legumes. Farmers were not aware of some pests considered as problems by entomologists, and consequently did not attempt control measures.

Insects and related arthropods have major roles as crop pests, causes of disease, food and medicinals, and are important in myth and folklore. In many regions, agricultural pests are tolerated because they also constitute agricultural products; that is, traditional agriculturalists may consume plants and animals that would otherwise be considered pests. In Indonesia, a grasshopper pest in rice is trapped at night and eaten (with salt, sugar, and onions) or sold as bird food in the market. In northeast Thailand, rural inhabitants commonly eat termites and a crab that damages rice stalks. Ants, some of which may be major crop pests, are one of the most popular insect foods gathered in tropical regions (Brown and Marten, 1986).

In his studies of Kabba farmers in Nigeria, Atteh (1984) not only found that farmers could identify the pests affecting their crops, but that they could also rank the pests according to the degree of damage they caused to crops. In addition, further research revealed that, for each pest, farmers had knowledge of (1) the history of the pest, including dates when the pest was noticed, when it became a menace, peak periods of occurrence in the past, and type of damage done; (2) the biology of the pest, including the life cycle of the pest, its breeding behavior, and ecological and climatic conditions facilitating or discouraging increase in numbers; (3) the bionomics of the pest — the feeding preferences and the severity of damage done to plants attacked.

A good example of farmer knowledge of the biology and bionomics of pests is the case of the variegated grasshopper, *Zonocerus variegatus*, in southern Nigeria. Richards (1985) found that local farmer knowledge was equivalent to that of his scientific team concerning the grasshoppers' food habits, life cycle, mortality factors, degree of damage to cassava, and the egg-laying behavior and egg-laying sites of the females. Farmers were aware that, numerous as these insects are, they congregate under only a few shaded areas on the farm or in an area to lay eggs at a particular period, and that these eggs are kept in pods and inserted an inch or so below the soft ground surface. Farmers had discovered on their own that the egg-laying sites can be marked and the egg pods dug up. Once exposed to the hot sun the eggs die. They had in fact tried this as a control measure. Farmers had also established a close relationship between the presence of a weed (*Eupatorium odoratum*) and the advance and severity of the pests (Richards, 1985). In this particular case, farmers' knowledge added facts to that of the researchers in regard to the dates, severity and geographical extent of some of the outbreaks, plus the fact that the grasshopper was eaten and sold and was of special importance to women, children, and poor people. Thus, the final control recommendation by scientists, of clearing the egg-laying sites from a block of farms, did not require most farmers to learn new concepts, and for some the practice was nothing new.

Indigenous pest control methods

Traditional farmers rely on a variety of management practices to deal with agricultural pest problems. Two main strategies can be distinguished. One is the use of direct, non-chemical pest control methods (i.e. cultural, mechanical, physical and biological practices) (Table 1). The second is reliance on built-in pest control mechanisms, inherent to the biotic and structural diversity of complex farming systems, commonly used by traditional farmers (Brown and Marten, 1986). This ensemble of cultural practices can be grouped into three main strategies, depending on which element of the agroecosystem is manipulated.

(1) Manipulation of crops in time

Farmers often manipulate the timing of planting and harvest carefully and use crop rotations to avoid pests. These techniques obviously require considerable ecological knowledge of pest phenology. Although these techniques often have other agronomic benefits (e.g. improved soil fertility), the farmers sometimes explicitly mention that they are done to avoid pest damage. For example, in Uganda, farmers utilize time of planting to avoid stem borers and aphids in cereals and peas, respectively (Richards, 1985). Many farmers are aware that planting out of synchrony with neighboring fields can result in heavy pest pressure and therefore use a kind of 'pest satiation" to avoid extensive damage. In the central Andes, a potato fallow rotation is carefully

Table 1
Pest management strategies and specific practices used by traditional farmers throughout the developing world

Strategy	Practices
Mechanical and physical control	Scarecrows, sound devices Wrapping of fruits, pods Painting stems, trunks with lime or other materials Destroying ant nests Digging out eggs/larvae Hand picking Removal of infested plants Selective pruning Application of materials (ash, smoke, salt, etc.) Burning vegetation
Cultural practices	Intercropping Overplanting or varying seeding rates Changing planting dates Crop rotation Timing of harvest Mixing crop varieties Selective weeding Use of resistant varieties Fertilizer management Water management Plowing and cultivation techniques
Biological control	Use of geese and ducks Transfer of ant colonies Collecting and/or rearing predators and parasites for field release Manipulation of crop diversity
Insecticidal control	Use of botanical insecticides Use of plants or plant parts as repellents and/or attractants Use of chemical pesticides
Religious/ritual practices	Addressing spirits or gods Placement of crosses or other objects in the field Prohibition of planting dates

observed, apparently to avoid build-up of certain insects and nematodes (Brush, 1983).

Perhaps the most common way in which farmers manipulate the temporal permanence of agroecosystems is through the traditional pattern of slash and burn or shifting cultivation. A parcel of forest is cut and the area burned to release nutrients and eliminate non-crop plants. A mixture of short-term crops, sometimes followed by perennials, is grown until soil fertility becomes inadequate and competition from successional plant species is severe. When that happens, the farmer prepares a new field and the old one returns to long-term fallow. In Nigeria, kabba farmers affirm that an unacceptable weed and insect pest level is the surest sign that a plot of land must be abandoned (Atteh, 1984).

It has been speculated that bush fallows in the swidden system have potential value in controlling insects. Initial pest colonization is slow since the crops are typically planted among vegetation types (forest) that do not share the same pest complex. The great diversity of crops grown simultaneously in swidden cultivation helps to prevent pest build-up on the comparatively isolated plants of each species. Shade from forest fragments still standing in new fields, coupled with a partial canopy of fruit, nut, fire wood, fiber, medicinal and/or lumber tree species, reduces shade-intolerant weed populations and provides alternate hosts for beneficial (or sometimes detrimental) insects. Clearing comparatively small plots in a matrix of secondary forest vegetation permits easy migration of natural control agents from the surrounding jungle (Matteson et al., 1984).

(2) Manipulation of crops in space

Traditional farmers often manipulate plot size, plot site location, density of crops and crop diversity to achieve several production purposes, although most are aware of the links between such practices and pest control.

Overplanting. One of the most common methods of dealing with pests is planting at a higher density than one expects to harvest. This strategy is most effective in dealing with pests that attack the plant during the early stages of growth. When infested plants are detected, they are carefully removed long before actual death so as to avoid contaminating healthy plants.

Farm plot location. In Nigeria many farmers, linked by kinship ties, age grouping or friendship, locate their farm plots lying contiguous to each other but leaving room for the expansion of each farm in a particular direction. In accounting for this practice, farmers reported that all pests in the area will discover and concentrate on an isolated farm. Plots are therefore grouped together to spread pest risk among many farmers (Atteh, 1984). Conversely, in tropical America Brush (1983) reports that farmers deliberately use small isolated plots to avoid pests.

Selective weeding. Studies conducted in traditional agroecosystems show that peasants deliberately leave weeds in association with crops, by not completely clearing all weeds from their cropping systems. This 'relaxed' weeding is usually seen by agriculturalists as the consequence of a lack of labor and low return for the extra work; however, a closer look at farmer attitudes toward weeds reveals that certain weeds are managed and even encouraged if they serve a useful purpose. In the lowland tropics of Tabasco, Mexico, there is a unique classification of non-crop plants according to use potential on the one hand and effects on soil and crops on the other. According to his system, farmers recognized 21 plants in their cornfields classified as 'mal monte' (bad weeds), and 20 as 'buen monte' (good weeds) that serve, for example, as food, medicines, ceremonial materials, teas and soil improvers (Chacon and Gliessman, 1982).

Similarly, the Tarahumara Indians in the Mexican Sierras depend on edible weed seedlings (*Amaranthus, Chenopodium and Brassica*) from April through July, a critical period before maize, bean, cucurbits, and chiles mature in the planted fields in August through October. Weeds also serve as alternate food supplies in seasons when the maize crops are destroyed by frequent hailstorms. In a sense the Tarahumara practice a double crop system of maize and weeds that allows for two harvests: one of weed seedlings or 'quelities' early in the growing season (Bye, 1981). Some of these practices have important insect pest control implications since many weed species play important roles in the biology of herbivorous insects and their natural enemies in agroecosystems. Certain weeds, for example, provide alternate food and/or shelter for natural enemies of insect pests during the crop season but, more importantly, during the off-season when prey/hosts are unavailable (Altieri and Liebman, 1986).

In addition to the type of weed diversity allowed in the field, the timing of field weeding may also impact insect pest dynamics within the agroecosystem. For example, in Nigeria, while the timing of weeding corresponds with growth periods of crops, it is also timed in such a way as to interfere with the egg-laying and breeding times of most of the major pests, thus preventing maturation and consequently reducing their population (Atteh, 1984). Timely removal of weeds that are alternative hosts of specific crop pests is of key importance.

Manipulation of crop diversity. Although most farmers use intercropping mainly because of labor and land shortages or other agronomic purposes, the practice has obvious pest control effects (Altieri and Letourneau, 1982). Many farmers know this and use polycultures as a play-safe strategy to prevent build-up of specific pests to unacceptable levels, or to survive in cases of massive pest damage. For example, in Nigeria, farmers are aware of the severe damage done to an isolated cassava crop by the variegated grasshopper after all other crops have been harvested. To reduce this damage, farmers deliberately replant maize and random clusters of sorghum on the cassava plot until harvest time (Atteh, 1984).

(3) Manipulation of other agroecosystem components

In addition to manipulating crop spatial and temporal diversity, farmers also manipulate other cropping system components such as soil, microclimate, crop genetics and chemical environment to control pests.

Use of resistant varieties. Through both conscious and unconscious selection, farmers have developed crop varieties that are resistant to pests. This is probably the most widely used and effective of all the traditional methods of pest control. Litsinger et al. (1980) found that 73% of the peasant farmers in the Philippines were aware of varietal resistance even if they had not consciously tried to manipulate it. There is evidence in traditional varieties for all the modes of resistance that modern plant breeders select for, including pubescence, toughness, early ripening, plant defense chemistry and vigor.

In Ecuador, Evans (1988) found that infestations of Lepidoptera larvae in ripening corn ears were significantly higher in new varieties than in traditional ones, a factor that influenced the adoption of new varieties by small farmers.

Water management. Manipulation of water level in rice fields is a widely used practice for pest control (King, 1927). Water management is also practiced in many other annual crops for the same purpose. For example, in Malaysia, control of cutworms and army worms is effected by cutting off the tip of infested leaves in a number of annual crops, and raising the water level, which carries the larvae into the field ridges, where birds congregate to eat them.

Plowing and cultivation techniques. Farmers frequently report that they deliberately manage the soil (sometimes using more and sometimes less cultivation) to destroy or avoid pest problems. In Peru, for example, peasants use 'high tilling' of potatoes to protect the tubers from insect pests and diseases (Brush, 1983).

In shifting cultivation, after clearing a piece of land farmers set it on fire after a week or two. Farmers reported that this is done, among other reasons, to reduce weed and pest populations during the first year of cropping (Atteh, 1984).

Use of repellents and/or attractants. Farmers have been experimenting with various natural materials found in their immediate environment (especially in plants) for many centuries, and a remarkable number have some pesticidal properties. Use of plants or plant parts either placed in the field or applied as herbal concoctions for pest inhibition is widespread. Litsinger et al. (1980) interviewed small farmers in the Philippines about materials used in the fields to attract or repel insects. In Aloburo, Ecuador, small farmers place castor leaves in recently planted corn fields to reduce populations of a nocturnal tenebrionid beetle. These beetles prefer castor leaves over corn, but when associated with castor leaves for 12 h or more, beetles exhibit paralysis. In the field, the paralysis prevents beetles from hiding in the soil, which increases their mortality by direct exposure to the sun (Evans, 1988). In southern Chile, peasants placed branches of *Cestrum parqui* in potato fields to repel *Epicauta pilme* beetles (Altieri, 1990). Many times a plant is carefully grown near the household and its sole function is apparently to provide the raw material for preparing a pesticidal concoction. In Tanzania, farmers cultivate *Tephrosin* spp. on the borders of their maize fields. The leaves are crushed and the liquid is applied to control maize pests. In Tlaxcala, Mexico, farmers 'sponsor' volunteer *Lupinus* plants within their corn fields, because those plants act as trap crops for *Macrodactylus* sp. (Trujillo-Arriaga and Altieri, 1990).

Diversity improved pest management systems for traditional farmers

The development and extension of conventional agricultural technology for small farmers throughout the developing world, including better crop protection methods, has met with mixed success. Most technologies have not been made available to small farmers on favorable terms and often have not been suited to the agroecological and socio–economic conditions of small farmers (Altieri, 1984). Understanding traditional farming systems, including the role of biodiversity and the use of effective traditional pest management methods, is the starting point in the design and improvement of traditional pest management systems (Altieri, 1985).

A basic step in the generation of technological modifications appropriate to small farmers is to conduct agro–socio–economic studies which determine the conditions influencing traditional farming systems. The analysis of small farm systems must incorporate farmer knowledge, needs, and production criteria, especially with regard to risk. Risk avoidance is expressed traditionally through, for example, farm diversification and flexibility in planting times. In Mexico, for example, farmers often plant maize at high density and use the thinnings for their animals. This practice can also dilute the attack of specific insect pests. In eastern Guatemala, some farmers let weeds grow in their vegetable fields to increase the food supply for cattle, even though this practice may reduce crop yields. On the other hand the presence of weeds can condition the build-up of natural enemies which in turn reduces insect pests. An understanding of the effects of these interactions of current cropping practices is vital for the development and/or pre-screening of new technologies appropriate to farmers' circumstances (Alticri, 1984).

The goal is to maintain agricultural productivity and ensure crop protection with minimal environmental impact, adequate economic returns, and while providing for the needs of the small farmers. Although there are no recipes on how to achieve sustainability, the idea is to devise an agricultural strategy based on the use of biodiversity and ethnoscience which can bring moderate to high levels of productivity using local resources and skills, while conserving the natural resource base. Given favorable political and ecological circumstances, such systems are sustainable at a much lower cost and for a longer period of time (Altieri, 1991b).

A number of researchers and organizations in developing countries are promoting agroecological techniques in a way which is sensitive to the complexities of local farming methods. This agroecological approach emphasizes properties of food security, biological stability, resource conservation and equity, along with the goal of increased production. Promoted agroecological techniques are culturally compatible since they do not question farmers' rationale, but actually build on traditional farming knowledge, combining it with elements of modern agricultural science.

Linking soil conservation and pest management in Central America

Perhaps one of the major agricultural challenges in Central America is the design of cropping systems adapted to hillside areas in order to maintain yields while reducing erosion (Altieri, 1991b). In Honduras, a couple of non-government organizations (NGOs) have taken on this challenge. One group, in Loma Linda, developed a simple no-till system by which, in a fallow area, weeds are cut initially with a machete without soil being removed. Using a hole or a small plough, small furrows are opened following the contour every 50–60 cm. Crop seeds and chicken manure and/or compost are placed in the furrow and covered with soil. As the crops grow, weeds are kept mowed to avoid excessive competition, with the weed biomass left within the crop row as a mulch for cover and as an addition of organic matter. The mulch also provides a habitat for ground predators and the weed cover can enhance natural enemies or disrupt herbivore colonization by changing the ground color background. Excellent yields have been obtained with this system without the use of chemical fertilizers, and more importantly, without experiencing significant soil loss or pest attack.

In a similar project in Guinope, Honduras, the NGO World Neighbors introduced soil conservation practices among small farmers to control erosion and restore fertility. Various practices such as the use of living barriers and intercropping leguminous plants, increased the vegetational diversity of the farms, with beneficial effects on soil conservation and on crop protection. In the first year, yields increased dramatically from 400 kg ha^{-1} to 1200–1600 kg ha^{-1}. This tripling in per-hectare grain production has assured the 1200 families participating in the program ample grain supplies for the ensuing year.

In both cases, increased productivity at the farm level has meant that most farmers are now farming less land than previously, allowing more land to grow back to natural forest and to be used for planting pasture, fruit or coffee trees. The net result is enhanced diversity within farms, and that hundreds of hectares formerly used for erosive agriculture are now covered with trees, thus enhancing regional diversity. There is some evidence to suggest that at a more regional level, pest problems may diminish in a heterogeneous landscape where agroecosystem mosaics are interspersed among natural vegetation (Altieri, 1991a).

Conclusions

Although it is generally accepted that diversification of cropping systems often leads to reduced herbivore populations, the degree of such reductions varies dramatically within the whole range of diversities and management intensities that polycultures exhibit under farmers' circumstances. If certain existing crop mixtures contain built-in elements of pest control, such elements should be identified and then retained in the course of modernization. In other cases, suboptimal interactions between plants, herbivores, and natural enemies could be improved (i.e. by adding or eliminating diversity) to enhance natural enemy effectiveness in regulating herbivore densities.

Some management ideas can be derived from the ensemble of traditional crop protection practices used by small farmers. Ethnoscience represents a rich resource for modern workers seeking to create pest management systems that are well adapted to the agroecological and socio–economic circumstances of peasants. Clearly, not all traditional crop protection components are effective or applicable; therefore modifications and adaptations may be necessary, but the foundation of such modifications must be based on peasant rationale and indigenous knowledge. Nevertheless, most farmers use a diversity of techniques, many of which fit local conditions well. The techniques tend to be knowledge intensive rather than commodity intensive, i.e. their effectiveness is dependent on detailed knowledge of the agricultural environment rather than the use of expensive inputs. What is difficult is that each agricultural situation must be assessed separately, since herbivore–enemy interactions will vary significantly depending on insect species, location and size of the field, plant composition, the surrounding vegetation, and cultural management. In most cases one can only hope to elucidate the ecological principles governing herbivore dynamics in complex systems, but the polycultural designs necessary to achieve herbivore regulation will depend on the agroecological conditions and socio–economic restrictions of each area. In this regard, farmers' needs and preferences must be considered fully if adoption of new designs is expected. New designs will be most attractive if, in addition to pest regulation, polycultures offer benefits in terms of overyielding, increased

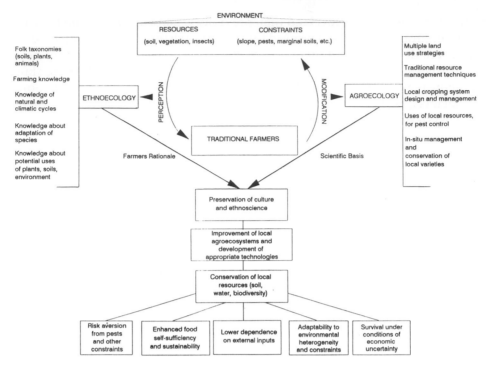

Fig. 1. The role of agroecology and ethnoecology in the retrieval of traditional farming knowledge and the development of sustainable agroecosystems, including appropriate innovations in pest management.

soil fertility, decreased weed competition and diseases and evening-out of labor demands.

Farmers' rationale to maintain certain cropping practices and tolerate a certain amount of yield loss due to pests must be considered if developers desire to convince farmers to adopt improved IPM practices. Otherwise, even if new practices offer reduction in yield loss and cash and labor savings, farmers will not adopt them. For example, in Africa, although burning of the crop residue after harvest is effective to destroy diapausing stem-borers in the dry season, farmers resist the practice because they use maize and sorghum stubble as livestock fodder (Altieri, 1990). A re-evaluation of traditional tropical crop management, which has hitherto assured stable productivity on a long-term basis, will be necessary to develop more sustained and low-input IPM practices and to reverse some of the 'ecological crisis' triggered by modern changes (e.g. large-scale monocultures, promotion of high-yielding varieties and use of pesticides). An increased understanding of the agroecology and ethnobiology of small farming systems can only emerge from integrative studies that determine the myriad factors and conditions influencing farmers' decisions and traditional cropping patterns (Fig. 1). Maintaining the biodiversity of farming systems is essential for the biological functioning of small-scale agriculture. However, biodiversity can only be maintained if cultural diversity is preserved.

References

Altieri, M.A., 1984. Pest management technologies for peasants: a farming systems approach. Crop Prot., 3: 87–94.

Altieri, M.A., 1985. Developing pest management strategies for small farmers based on traditional knowledge. Bull. Inst. Dev. Anthropol., 3: 13–18.

Altieri, M.A., 1990. The ecology and management of insect pests in traditional agroecosystems. In: D.A. Posey and W.L. Overal (Editors), Proc. First Int. Congress of Ethnobiology, 19–22 July 1988, Belem, Brasil, pp. 131–143.

Altieri, M.A., 1991a. Ecology of tropical herbivores in polycultural agroecosystems. In: D.W. Price et al. (Editors), Plant-Animal Interactions: Evolutionary Ecology in Tropical and Temperate Regions. John Wiley, New York, pp. 607–613.

Altieri, M.A., 1991b. Traditional farming in Latin America. Ecologist, 21: 93–96.

Altieri, M.A. and Letourneau, D.K., 1982. Vegetation management and biological control in agroecosystems. Crop Prot., 1: 405–430.

Altieri, M.A. and Letourneau, D.K., 1984. Vegetation diversity and insect pest outbreaks. CRC Crit. Rev. Plant Sci., 2: 131–169.

Altieri, M.A. and Liebman, M.Z., 1986. Insect, weed and plant disease management in multiple cropping systems. In: C.A. Francis (Editor), Multiple Cropping Systems. MacMillan, New York, pp. 183–218.

Andow, D.A., 1991. Vegetation diversity and arthropod population responses. Annu. Rev. Entomol., 36: 561–586.

Atteh, O.D., 1984. Nigerian farmers' perception of pests and pesticides. Insect Sci. Appl., 5: 213–220.

Brown, B.J. and Marten, G.G., 1986. The ecology of traditional pest management in southeast Asia. Working Paper, East–West Center, Honolulu, Hawaii.

Brush, S.B., 1983. Traditional agricultural strategies in the hill lands of tropical America. Culture Agric., 18: 9–16.

Bye, R.A., 1981. Quelites — ethnoecology of edible greens — past, present and future. J. Ethnobiol., 1: 109–114.

Chacon, J.C. and Gliessman, S.R., 1982. Use of the "non-weed" concept in traditional tropical agroecosystems of south-eastern Mexico. Agroecosystems, 8: 1–11.

Evans, D.A., 1988. Insect pest problems and control strategies appropriate to small-scale corn farmers in Ecuador. Ph.D. Dissertation. Univ. of California, Davis, 121 pp.

King, F.H., 1927. Farmers of Forty Centuries. Cape, London, 124 pp.

Litsinger, J.A., Price, E.C. and Herrera, R.T., 1980. Small farmers' pest control practices for rainfed rice, corn and grain legumes in three Philippine provinces. Philipp. Entomol., 4: 65–86.

Matteson, P.C., Altieri, M.A. and Gagne, W.C., 1984. Modification of small farmer practices for better pest management. Annu. Rev. Entomol., 24: 383–402.

Powell, W., 1986. Enhancing parasitoid activity in crops. In: J. Waage and D. Greathead (Editors), Insect Parasitoids. Academic Press, London, 389 pp.

Richards, P., 1985. Indigenous Agricultural Revolution. Westview Press, Boulder, CO.

Toledo, V.M., Carabias, M.J., Mapes, C. and Toledo, C., 1985. Ecologia y Autosuficiencia alimentaria. Siglo Veintiuno Editors, Mexico, Mexico City, 96 pp.

Trujillo-Arriaga, J. and Altieri, M.A., 1990. A comparison of aphidophagous arthropods on maize polycultures and monocultures in central Mexico. Agric. Ecosystems Environ., 31: 337–349.

Agriculture, Ecosystems and Environment, 46 (1993) 273–288
Elsevier Science Publishers B.V., Amsterdam

Environmental and economic effects of reducing pesticide use in agriculture*

David Pimentel**, Lori McLaughlin, Andrew Zepp, Benyamin Lakitan, Tamara Kraus, Peter Kleinman, Fabius Vancini, W. John Roach, Ellen Graap, William S. Keeton, Gabe Selig

College of Agriculture and Life Sciences, Cornell University, Ithaca NY 14853-0999, USA

Abstract

Pesticides cause serious damage to agricultural and natural ecosystems. Thus, there is a need to curtail pesticide use and reduce the environmental impacts of pesticides. This study confirms that it should be possible to reduce pesticide use in the US by 50% without any decrease in crop yields or change in 'cosmetic standards'. The estimated increase in food costs would be only 0.6%. This increased cost, however, does not take into account the environmental and public benefits of reducing pesticide use by 50%.

Introduction

Several studies suggest that it is technologically feasible to reduce pesticide use in the US by 35–50% without reducing crop yields (Office of Technology Assessment (OTA), 1979; National Academy of Sciences, (NAS), 1989). Two recent events in Denmark and Sweden support these assessments. Denmark developed an action plan in 1985 to reduce pesticide use by 50% before 1997 (B.B. Mogensen, personal communication, 1989). Sweden also approved a program in 1988 to reduce pesticide use by 50% within 5 years (National Board of Agriculture, 1988). The Netherlands is developing a program to reduce pesticide use by 50% in 10 years (A. Pronk, personal communication, 1990). These proposals, along with the conclusion by Huffaker (1980) that the US overuses pesticides, prompted us to investigate the feasibility of reducing the annual use of synthetic organic pesticides by approximately one-half.

Farmers in the US use an estimated 320 million kg (700 million lb) of pesticides annually at an approximate cost of $4.1 billion (Table 1). Indeed, investment in pest control by pesticides has been shown to provide significant

---

*Reprinted with permission from BioScience. Pimentel, D. et al., 1991. Environmental and economic effects of reducing pesticide use. BioScience, 41(6): 402–409.
**Corresponding author.

Table 1
US hectarage treated with pesticides (modified from Pimentel and Levitan, 1986)

Land-use category	Total hectares ($\times 10^6$)	All pesticides		Herbicides		Insecticides		Fungicides	
		Treated hectares ($\times 10^6$)	Quantity ($\times 10^6$ kg)	Treated hectares ($\times 10^6$)	Quantity ($\times 10^6$ kg)	Treated hectares ($\times 10^6$)	Quantity ($\times 10^6$ kg)	Treated hectares ($\times 10^6$)	Quantity ($\times 10^6$ kg)
Agricultural lands	472	114	320	86	220	22	62	4	38
Government and industrial lands	150	28	55	30	44	–	11	–	–
Forest lands	290	2	4	2	3	<1	1	–	–
Household lands	4	4	55	3	26	3	25	1	4
Total	916	148	434	121	293	26	99	5	42

Total for hectarage treated with herbicides, insecticides, and fungicides exceeds total treated hectares because the same land area can be treated with several classes of chemicals and several times.

economic benefits. Dollar returns for the direct benefits to farmers have been estimated to range from $3 to $5 for every $1 invested in the use of pesticides (Headley, 1968; Pimentel et al., 1978). However, these figures do not reflect the indirect costs of pesticide chemical use such as human pesticide poisonings, reduction of fish and wildlife populations, livestock losses, destruction of susceptible crops and natural vegetation, honey-bee losses, destruction of natural enemies, evolved pesticide resistance, and creation of secondary pest problems (Pimentel et al., 1980). Moreover, these economic benefits are calculated based on current agricultural practices, some of which actually increase pest problems. Clearly, the direct and indirect benefits and costs of using pesticides in agriculture are highly complex.

The objective of this investigation is to estimate the potential agricultural and environmental benefits and costs of reducing pesticide use by approximately 50% in the US. To estimate the costs and benefits, this study (i) examines current pesticide use patterns in about 40 major US crops, (ii) evaluates current crop losses to pests, (iii) estimates the agricultural benefits and costs of reducing pesticide use by substituting currently available biological, cultural, and environmental pest control technologies for some current pesticide control practices, and (iv) assesses the public health and environmental costs associated with reduced pesticide use.

Extent of pesticide use

Of the estimated 434 million kg of pesticides used annually in the US, 67% are herbicides, 23% insecticides, and 10% fungicides (Table 1). The 320 million kg of pesticides used in agriculture are applied at an average rate of approximately 3 kg ha^{-1} to approximately 114 million ha, i.e. 62% of the 185 million ha that are planted (Pimentel and Levitan, 1986). Thus, a significant portion (38%) of crops receives no pesticides.

The application of pesticides for pest control is not evenly distributed among crops. For example, 93% of all row crops such as corn, cotton, and soybeans are treated with some type of pesticide (Pimentel and Levitan, 1986). In contrast, less than 10% of forage crops are treated. Herbicides are currently being used on approximately 90 million ha in the US, i.e. more than half of the nation's crop land, but nearly three-quarters of these herbicides are applied to just two major crops, corn and soybeans. Field corn alone accounts for 53% of agricultural herbicide use.

The situation is similar for insecticide use. Approximately 62 million kg of insecticides are applied to 5% of the total agricultural land (Table 1). Approximately 25% of all insecticides used are on cotton and corn. Fungicides are used primarily on fruit and vegetable crops (Pimentel and Levitan, 1986). Insecticide use also varies considerably among geographic regions. The warmer regions of the US often suffer more severe pest problems. For example, while only 13% of the alfalfa hectarage in the US is treated with insecticides, 89% of the alfalfa area in the Southern Plains states is treated to control insect pests (Eichers et al., 1978). In the mountain region, where large quantities of potatoes are grown, 65% of the potato crop land receives insecticide treatment, while in the southeast, where only early potatoes are grown, 100% of the potato crop land receives treatment with insecticides (US Department of Agriculture (USDA), 1975). Cotton insect pests such as the boll weevil are also a more serious problem in the southeast than in other regions (USDA, 1983). In the southeast and delta states, 84% of the cotton cropland receives treatment, whereas in the Southern Plains region less that half of the crop (40%) is treated. Moreover some crops (e.g. apples and cotton) can be treated as many as 20 times per season whereas other crops may be treated only once (e.g. corn and wheat).

Crop losses to pests and changes in agricultural technologies

Since 1945, the use of synthetic pesticides in the US has grown 33-fold (Fig. 1). The amounts of herbicides, insecticides, and fungicides used have changed with time mainly due to changes in agricultural practices and adoption of cosmetic standards (Pimentel et al., 1977). At the same time, the toxicity to pests and biological effectiveness of some of the pesticides used have increased at least ten-fold (Pimentel et al., 1991). For example, in 1945, DDT was applied at a rate of about 2 kg ha^{-1}. Today, effective insect control is achieved with pyrethroids and aldicarb applied at only 0.1 kg ha^{-1} and 0.05 kg ha^{-1}, respectively.

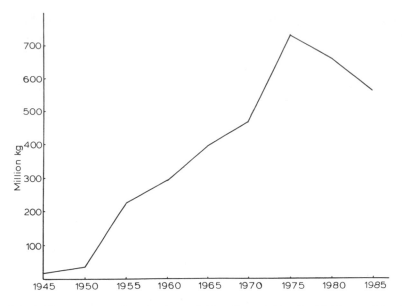

Fig. 1. The amounts of synthetic pesticides (insecticides, herbicides, and fungicides) produced in the US (Pimentel et al., 1991). Approximately 90% is sold in the US. The decline in total amount produced for use is in large part due to the ten- to 100-fold increased toxicity and effectiveness of the newer pesticides.

Currently, an estimated 37% of all crop production is lost annually to pests (13% to insects, 12% to plant pathogens, and 12% to weeds) in spite of the use of pesticides and non-chemical controls (Pimentel, 1986). Although pesticide use has increased during the past four decades, crop losses have not shown a concurrent decline. According to survey data collected from 1942 to the present, losses from weeds have fluctuated but declined slightly from 13.8% to 12% (Pimentel et al., 1991) (Table 2). This is due to improved chemical, mechanical, and cultural weed control practices. During that same period, US losses from plant pathogens, including nematodes, have increased slightly, from 10.5% to about 12% (Table 2). This results in part from reduced sanitation, higher cosmetic standards, and decreases in rotations.

The share of crop yields lost to insects has nearly doubled during the last 40 years (Table 2) despite more than a ten-fold increase in both the amount and toxicity of synthetic insecticide used (Pimentel et al., 1991). The increases in crop losses due to insects have been offset by increased crop yields obtained through the use of higher-yielding crop varieties and the greater use of fertilizers and other inputs (USDA, 1989).

This increase in crop losses despite increased insecticide use is due to several major changes that have taken place in agricultural practices. These include the planting of some crop varieties that are more susceptible to insect pests, the destruction of natural enemies of certain pests (which creates the

Table 2
Comparison of annual pest losses (dollars[a]) in the USA for the periods 1904, 1910–1935, 1942–1951, 1951–1960, 1974, 1989 (after Pimentel et al., 1991)

Period	Percentage of pest losses in crops				Crop value ($\times 10^9$)
	Insects	Diseases	Weeds	Total	
1989	13.0	12.0	12.0	37	150
1974	13.0	12.0	8.0	33.0	77
1951–1960	12.9	12.2	8.5	33.6	30
1942–1951	7.1	10.5	13.8	31.4	27
1910–1935	10.5	NA[b]	NA	NA	6
1904	9.8	NA	NA	NA	4

[a]Not adjusted.
[b]Not available.

need for additional pesticide treatments), the increase in numbers of pests resistant to pesticides, the reduction in crop rotations, the increase in crops grown in monoculture and reduced crop diversity, the lowering of FDA tolerances for insects and insect parts in foods, and the enforcement of more stringent cosmetic standards by fruit and vegetable processors and retailers, the increased use of aircraft application technology, the reduction in crop sanitation (including less attention to the destruction of infected fruit and crop residues), the reduction in tillage, with more crop residues left on the soil surface, the culturing of crops in climatic regions in which they are more susceptible to insect attack, and the use of herbicides that have been found to alter the physiology of crop plants, making them more vulnerable to insect attack (Pimentel et al., 1991). These factors will be explored further in the discussion of alternatives to pesticide use.

Estimated agricultural benefits/costs with a reduction in pesticide use

A reduction in US pesticide use would require improving the efficiency of pesticide application technology supported by alternatives for chemical pest control. Such changes in some cases increase and in other cases decrease control costs. The costs and benefits of alternative controls are examined below.

Crop losses to pests

Losses from pests for 40 major crops grown with pesticide treatments were estimated by examining data on current crop losses, by reviewing loss data based on experimental field tests, and by consulting pest control specialists. Combining these data, however, was often difficult. For example, published data based on experimental field tests usually emphasize the benefits of pesticide use, thus loss data associated with pesticide treatments usually emphasize benefits over costs (Pimentel et al., 1978).

Moreover, field tests often exaggerate total crop losses since the assessments of insect, disease, and weed losses are carried out separately and then combined. For example, on untreated apples, insects were reported to cause a 50–100% crop loss, disease 50–60%, and weeds 6% (Ahrens and Cramer, 1985; Pimentel et al., 1991). This approach yields an estimated total loss of about 140% from all pests combined! A more accurate estimate of losses in the absence of pesticides ranges from 80% to 90% based on current cosmetic standards (Ahrens and Cramer, 1985). Exactly how much overlap exists among insect, disease, and weed loss figures for apples and for other crops is not known.

Our analysis has other important limitations. The figures for current crop losses to pests, despite pesticide use, are based primarily on USDA data and other estimates obtained from specialists. We emphasize that these data are estimates. For certain crops, little or no experimental data are available concerning different yields with and without pesticide use and various alternative control measures. In addition, in some cases, recent data were not available, so for these crops, our estimates were usually extrapolated from data on closely related crops.

Although we recognize the limitations of the data used in this analysis, we believe that there is a need to assemble all available information, in order to provide a first approximation of the potential for reducing pesticide use by one-half without any major sacrifices in yields. We hope that better data will be available in the future, so that a more complete analysis of pesticide costs and benefits can be made.

Reduction of the hazards associated with pesticide use is in itself a complicated issue, particularly because environmental and health-related trade-offs are often associated with changes in technology. Because of the complexity of these trade-offs, they could not be included in the analysis. One example, however, demonstrates the conflict of interest between reducing pesticide use and promoting soil conservation through the adoption of no-till practices or conservation tillage. Although no-till and conservation-till reduce soil erosion significantly (Van Doren et al., 1977), they also significantly increase the need for herbicides, insecticides, and fungicides (Taylor et al., 1984).

However, although reducing pesticide use may conflict with adoption of some no-till systems, highly cost-effective soil conservation alternatives to no-till do in fact exist. These include ridge-till, crop rotations, strip-cropping, contour planting, terracing, wind-breaks, mulches, cover crops and green-mulches (Moldenhauer and Hudson, 1988). Ridge-planting, which includes planting the crop on permanent ridges, 20 cm high along the contour, is growing rapidly in popularity as an effective replacement for no-till for most row crops. It is a form of conservation-till that has many advantages over no-till

(Forcella and Lindstrom, 1988). For example, ridge-till can be adopted without the use of herbicides, and it controls soil erosion more effectively than no-till (Russnogle and Smith, 1988).

Techniques to reduce pesticide use

The increase in crop losses due to pests associated with recent changes in agricultural practices suggests that some alternative strategies might be utilized to reduce pesticide use. Two important practices that apply to all agricultural crops, include widespread use of pest monitoring and the use of improved pesticide application equipment. Currently, many pesticide treatments are applied unnecessarily, and at improper times, due to a lack of programs which indicate when treatments are necessary. Furthermore, pesticides are lost unnecessarily during application (e.g. it is estimated that only 25–50% of the pesticide applied by aircraft actually reaches the target area (Akesson and Yates, 1984; Mazariegos, 1985)). By increasing pest monitoring and improving pesticide application equipment, more efficient pest control can be achieved. These effects are illustrated below by the detailed consideration of several crops.

Insecticides

Corn and cotton account for approximately 25% of the total insecticide use in agriculture in the US. Thus, reducing insecticide use in these two crops by substituting non-chemical alternatives would contribute significantly to a large reduction in overall insecticide use.

Corn. During the early 1940s, little or no insecticide was applied to corn, and losses to insect pests were only 3.5% (USDA, 1954). Since then, insecticide use on corn has grown more than 1000-fold while losses due to insects have increased to 12% (Ridgway, 1980). This increase in insecticide use and the 3.4-times increase in corn losses to insects are due primarily to reduction in crop rotations (Pimentel et al., 1991). Today, approximately 40% of US corn is grown continuously with 11 million kg of insecticide applied to this crop annually (Pimentel et al., 1991). By reinstituting crop rotations, large reductions in pesticide use could be achieved. Rotating corn with soybeans or a similar high-value crop generally increases yields and net profits (Helmers et al., 1986), although rotating corn with wheat or other low-value crops usually reduces net profits per hectare. From a more comprehensive perspective, however, the rotation of corn with other crops has several other advantages including reducing weed and plant disease losses as well as decreasing soil erosion and rapid water run-off problems (Helmers et al., 1986).

By combining crop rotations with the planting of corn varieties resistant to the corn borer and chinch bug, it would be possible to avoid 80% of the total insecticide used on corn whilst concurrently reducing insect losses (Schalk and Ratcliffe, 1977). Such a change is estimated to increase the cost of corn production by $10 per hectare compared with systems in which corn is grown continuously (Pimentel et al., 1991).

A new technique that controls rootworm by combining an attractant with insecticides has been reported to reduce insecticide requirements by 99% (Paul, 1989).

Cotton. The potential for reducing pesticide use in US agriculture is well illustrated by the changes in insecticide use that have occurred in Texas cotton production. Since 1966, insecticide use on Texas cotton has been reduced by nearly 90% (OTA, 1979). The technologies adopted to reduce insecticide use were: monitoring pest and natural enemy populations to determine when to use pesticides, biological pest control, host-plant resistance, stalk destruction (sanitation), adoption of a uniform planting date, water management, fertilizer management, use of rotations, clean seed, and changed tillage practices (OTA, 1979; King et al., 1986).

Currently, a total of 29 million kg of insecticide is applied to cotton, and we estimate that this amount could be reduced by approximately 40% through the use of readily available technologies (Pimentel et al., 1991). By using a monitoring program effectively, one could reduce insecticide use by approximately 20%. Through the use of pest-resistant cotton varieties and the alteration of planting dates, in most growing regions, insecticide use could be reduced by another 3% (Frans, 1985; Frisbie, 1985). An additional 10% reduction in insecticide use could be achieved through the replacement of the price support program with a free market and no price support program for cotton production (NAS, 1989). This change would allow cotton to be grown in regions with fewer insect pests, thereby reducing the need for insecticide use. However, the current price support program would have to change if society is to take advantage of these benefits. At present, however, this is a politically unattractive proposition.

Giving greater care to the type of application equipment employed, especially in reducing the use of ultra low volume application equipment on aircraft (which wastes 75% of pesticide applied), would increase the amount of insecticide reaching the target area. The amount of insecticide reaching the target area could be increased to 75% if ground-application equipment were used instead of aircraft application equipment (Mazariegos, 1985; Pimentel and Levitan, 1986). In addition, covering the spray boom with a plastic shroud can further decrease drift 85% (Ford, 1986), thereby allowing for an additional reduction in pesticide use.

Insecticide use on cotton might be reduced by another 6% through the implementation of other pest control techniques. These include cultivation of short-season cotton, fertilizer and water management, improved sanitation, crop rotations, the use of clean seed, and altered tillage practices (OTA, 1979; Grimes, 1985). Depending on the particular environment, insecticide use on cotton might be reduced by much more than 6%. For example, Shaunak et al. (1982) reported that insecticide use in the Lower Rio Grande Valley of Texas could be reduced 97% by planting short-season cotton under dryland conditions. This practice also resulted in a two-fold increase in net profits over conventional methods.

Thus, by implementing combinations of pest monitoring, improved application technology, short-season cotton, fertilizer and water management, improved sanitation, use of crop rotations, and tillage practices in cotton, insecticide use might be reduced by 40%. These alternative control measures should pay for themselves through direct reduced insecticide and application costs.

Herbicides

Corn and soybeans will be used to illustrate the potential for decreasing herbicide use considerably, because these crops account for about 70% of the total herbicide applied in agriculture (Pimentel and Levitan, 1986).

Corn. More than half (53%) of the herbicides used on crops are applied to corn (Pimentel and Levitan, 1986). More than 3 kg of herbicide are applied per hectare of corn, and more than 90% of the corn planted is treated. By avoiding total elimination of weeds herbicide use can often be reduced by up to 75% (Schweizer, 1989). At present, 91% of the total corn land is also cultivated to help control weeds (Duffy, 1992).

The average costs and returns per hectare from no-till and conventional-till crops have actually been found to be quite similar (Duffy and Hanthorn, 1984). For example, added labor, fuel and machinery costs for conventional-till practices for corn were approximately $24 ha^{-1} higher than those for no-till. However, the costs for the extra fertilizers, pesticides, and seeds in the no-till system were $22 ha^{-1} higher than in the conventional-till system (Duffy and Hanthorn, 1984).

It would be possible to reduce herbicide use on corn by up to 60% if the use of mechanical cultivation and rotations were increased (Forcella and Lindstrom, 1988). Corn and soybean rotations have been found to give substantially higher returns than either crop grown separately and continuously (Helmers et al., 1986). However, we assumed in our calculations that weed control costs might increase by approximately 30% because not all alternative practices and rotations are profitable.

Soybeans. The second largest amount of herbicides applied to any US crop are applied to soybeans, with approximately 96% of soybeans receiving herbicide treatment as well as some tillage and mechanical cultivation for weed control (Duffy, 1982). Several techniques have been developed that increase the efficiency of herbicide applications. The rope-wick applicator has been used in soybeans to reduce herbicide use about 90%, and the applicator was found to increase soybean yields 51% over those from conventional treatments (Dale, 1980). Moreover, a new model of recirculating sprayer saves 70–90% of the spray applied that is not trapped by the weeds (Matthews, 1985). Spot pesticide treatments are a third method of decreasing unnecessary pesticide applications.

Several alternative techniques are available to reduce the need for herbicide use on soybeans. These include ridge-till, tillage, mechanical cultivation, row spacing, planting date, use of tolerant varieties, crop rotations, spot pesticide treatments and reduced pesticide dosages (Tew et al., 1982; Helmers et al., 1986; Forcella and Lindstrom, 1988; Russnogle and Smith, 1988). Employing several of these alternative techniques in combination might reduce overall herbicide use in soybeans by approximately 60%. Despite the results of Tew et al. (1982) that indicate no added control costs for the alternatives, we estimated conservatively that the cost of weed control would increase by $10 ha^{-1} (Pimentel et al., 1991).

Fungicides

When considering the possible reduction of fungicide use, apples and potatoes were selected as demonstration crops. These two crops account for 26% of all the fungicides used in US agriculture (Pimentel et al., 1991).

Apples. Most fungicides are applied to apples and other fruit crops (Pimentel and Levitan, 1986). Integrated pest management data for apples in New York State suggest that fungicide use on apples could be reduced by approximately 10% by improved monitoring and forecasting of disease based on weather data (Kovach and Tette, 1988).

In addition, a recent design in spray nozzle and application equipment demonstrated that the amount of fungicide applied for apple scab control could be reduced by 50% (Van der Scheer, 1984). Thus, by employing better weather forecasting and improved fungicide application technology combined with scouting, fungicide use on apples could be reduced by an estimated 20% (Pimentel et al., 1991).

Potatoes. Approximately 96% of potatoes in the US are treated with fungicides (Pimentel et al., 1991). Without fungicide treatments, losses from potato diseases ranged between 5 and 25%, while the losses with fungicide treatments were reported to be approximately 20% (Teng and Bissonnette, 1985). Shields et al. (1984) reported that the planting of short-season potatoes in Wisconsin reduced the number of fungicide applications by one-third. Correct storage, handling, and planting of seed tubers and proper management of soil moisture and fertility can minimize losses to most diseases (Rich, 1991).

Forecasting and monitoring might also be employed to reduce fungicide use between 15% and 25% (Royle and Shaw, 1988). Monitoring should concentrate on disease incidence and forecasting weather conditions, so that fungicides can be applied before an infection outbreak. Employing a combination of these control measures, it might be possible to reduce overall fungicide use on potatoes by approximately one-third at an estimated cost of $5 ha^{-1} (Pimentel et al., 1991).

Overall pesticide reduction assessment

Substituting non-chemical alternatives for some of the pesticides used on 40 major crops, it is possible that total agricultural pesticide use could be reduced by approximately 50% (Pimentel et al., 1991). The added costs for implementing the alternative pest control measures are estimated to be approximately $1 billion (Table 3). These alternatives would increase total pest control costs by approximately 25% and would increase total food production costs at the farm by 0.6%.

If alternative technologies that result in reduced crop yields were adopted the assessment of benefit and cost relationships would be quite different. For example, each 1% decrease in crop yield in agriculture results in a corresponding 4.5% increase in the farm price of goods. It is also important to note that crop overproduction is the prime reason that the US spends $26 billion annually on price supports (Office of Management and Budget, 1989). Thus, if changes in pest control did in fact reduce yields, it might increase both farm income and decrease government subsidy expenses.

Environmental impacts

Society pays a high price for the use of pesticides. Pesticide-control technology costs approximately $4.1 billion annually in the US, but this does not

Table 3
Current and potential reduced use of pesticides in US crops (million kg). Total pesticide costs (in dollars) plus total added alternative control costs (Pimentel et al., 1991)

	Current	Potential reduced	Total costs ($\times 10^6$)	Added alternative control costs ($\times 10^6$)
Insecticides	62.1	27.2	817.8	156.5
Herbicides	219.6	98.9	3115.8	845.3
Fungicides	37.6	26.6	207.4	16.5
Total	319.3	152.7	4141.0	1018.3

Table 4
Total estimated environmental and social costs from pesticides in the US

Public health impacts	$787 million year^{-1}
Domestic animals deaths and contamination	$30 million year^{-1}
Loss of natural enemies	$520 million year^{-1}
Cost of pesticide resistance	$1400 million year^{-1}
Honey-bee and pollination losses	$320 million year^{-1}
Crop losses	$942 million year^{-1}
Fishery losses	$24 million year^{-1}
Bird losses	$2100 million year^{-1}
Groundwater contamination	$1800 million year^{-1}
Government regulations to prevent damage	$200 million year^{-1}
Total	$8123 million year^{-1}

include the environmental and public health costs. A recent study estimates that the environmental and public health costs of using pesticides cost the nation about $8 billion each year (Table 4).

Conclusion

Pesticides cause serious public health problems and considerable damage to agricultural and natural ecosystems. A conservative estimate suggests that the environmental and social costs of pesticide use in the US are at least $2.2 billion (Pimentel and Levitan, 1986) (Table 1) plus $1.2 billion for monitoring well and groundwater resources) annually, and the actual cost is probably double this amount. In addition to these costs, the nation spends $4.1 billion annually to treat crops with 320 million kg of pesticides.

This study confirms that it should be possible to reduce pesticide use by 50% at a cost of approximately $1 billion. Such a conclusion supports the projections of the OTA (1979) and the NAS (1989) as well as the policies adopted by the Danish and Swedish governments to legislate that pesticide use should be reduced by 50%.

The 50% US pesticide use reduction in our current assessment would help to satisfy the concerns of the majority of the public, who worry about pesticide levels in their food as well as about damage to the environment (Sachs et al., 1987). If pesticide use were reduced by one-half without any decline in crop yield, the total price increase in purchased food is calculated to be only 0.6% and is due to the increased costs of alternative controls. If assured that pesticide residues in food and the environment were greatly reduced, it seems likely that the public would be willing to pay this slight increase in food costs.

In addition, it is clear that the public would accept some reduction in cosmetic standards of food items if it would result in a concurrent reduction in pesticide contamination of food (Healy, 1989). This is confirmed by the growing popularity of organic food stores and supermarkets that guarantee pesticide-free foods (Hammit, 1986; Poe, 1988). Furthermore, the presence of parts of soft-bodied insects in highly processed foods, such as catsup and applesauce, carries no risk to public health and may even be of some nutritional value (Pimentel et al., 1977). The adoption of higher cosmetic standards has resulted in greater quantities of pesticides being applied to food crops. This rapidly growing use of pesticides for cosmetic purposes is detrimental to both public health and the environment, and it is also contrary to public demand (Pimentel et al., 1977).

Although some of the data used in this preliminary investigation have obvious limitations, the estimates presented suggest that it should be possible to reduce pesticide use in the US by up to one-half. It is hoped that more complete data on this issue will be assembled and more detailed analyses made concerning the potential for reducing pesticide use. In particular, more data are needed on identifying those agricultural technologies that have contributed to the increase in pesticide use during the past 40 years whilst simultaneously increasing crop losses to pests.

Implementing a national program to reduce pesticide use in agriculture will require the combined education of farmers and the public and some new regulations. In addition, it will require that the federal government revise its current policies, such as its commodity and price support program, that discourage farmers from employing crop rotations and other sound agricultural practices (NAS, 1989). Several current government policies actually increase the incidence of pest problems and pesticide use (NAS, 1989). At the same time, a greater investment is needed in research on alternative pest control practices. Many opportunities exist to reduce pesticides through the implementation of new environmental, cultural, and biological pest controls (Pimentel, 1991). We strongly support the NAS research recommendations for a search for alternative pest controls (NAS, 1989).

The public seems to be concerned about pesticides contaminating their food and environment, so they must decide to accept the small economic costs necessary to reduce pesticide use. It is hoped that the public, state and federal governments will investigate the ecology, economics, and ethics of pesticide reduction in agriculture. The present analysis suggests that it is essential that a careful assessment be made to evaluate the benefits and risks of pesticides and non-chemical alternatives for society.

Acknowledgments

We thank the following people for reading an earlier draft of this article, for their many helpful suggestions, and, in some cases, for providing additional information: B. Barclay and M. El-Ashry (World Resources Institute), R. Frans (University of Arkansas), J. Hatchett (Kansas State University), M.

Harris (Texas A&M University), H. Hokkanen (Agricultural Research Centre, Jokioinen, Finland), D. Horn (Ohio State University), C. Huffaker (University of California, Berkeley), H. Janzen (Agriculture Canada), P. Johnson (University of New Hampshire), F. McEwen (University of Guelph), B. Mogensen (National Environmental Research Institute, Denmark), I. Oka (Bogor Research Institute for Food Crops, Indonesia), C. Osteen (USDA, Washington, DC), M. Pathak (International Rice Research Institute, Philippines), J. Pierce (Environmental Action, Washington, DC), E. Radcliffe (University of Minnesota), J. Schalk (USDA Veg. Lab, Charleston, NC), D. Wen (Chinese Academy of Sciences), K. Stoner (Connecticut Agricultural Experiment Station), G. Surgeoner (University of Guelph), G. Teetes (Texas A&M University), and at Cornell University, T. Dennehy, E. Glass, R. Roush, J. Tette, and C. Wien. We gratefully acknowledge the partial support of this study by a William and Flora Hewlett Foundation Grant to the Centre for Environmental Research.

References

Ahrens, C. and Cramer, H.H., 1985. Improvement of agricultural production by pesticides. In: F.P.W. Winteringham (Editor), Environment and Chemicals in Agriculture. Elsevier, New York, pp. 151–162.

Akesson, N.B. and Yates, W.E., 1984. Physical parameters affecting aircraft spray application. In: W.Y. Garner and J. Harvey (Editors), Chemical and Biological Controls in Forestry. Am. Chem. Soc. Ser. 238, Washington, DC, pp. 95–115.

Dale, J.E., 1980. Rope wick applicator — tool with a future. Weeds Today, 11(2): 3–4.

Duffy, M., 1982. Pesticide use and practices, 1982. Econ. Res. Serv., Agr. Inf. Bull. No. 462, US Dep. Agric., Washington, DC, 16 pp.

Duffy, M. and Hanthorn, M., 1984. Returns to corn and soybean tillage practices. Econ. Res. Serv., Agric. Econ. Rep. No. 508, US Dep. Agric., Washington, DC, 14 pp.

Eichers, T.R., Andrilenas, P.A. and Anderson, T.W., 1978. Farmers' use of pesticides in 1976. Econ. Stat. Coop. Serv., Agric. Econ. Rep. No. 418, US Dep. Agric., Washington, DC, 58 pp.

Forcella, F. and Lindstrom, M.J., 1988. Movement and germination of weed seeds in ridge-till crop production systems. Weed Sci., 36: 56–59.

Ford, R.J., 1986. Field trials of a method for reducing drift from agricultural sprayers. Can. Agric. Eng., 28(2): 81–83.

Frans, R., 1985. A summary of research achievements in cotton. In: R.E. Frisbie and P.L. Adkisson (Editors), Integrated Pest Management on Major Agricultural Systems. Tex. Agric. Exp. Stn., College Station, TX, pp. 53–61.

Frisbie, R., 1985. Regional implementation of cotton IPM. In: R.E. Frisbie and P.L. Adkisson (Editors), Integrated Pest Management on Major Agricultural Systems. Tex. Agric. Exp. Stn., College Station, TX, pp. 638–651.

Grimes, D.W., 1985. Cultural techniques for management of pests in cotton. In: R.E. Frisbie and P.L. Adkisson (Editors), Integrated Pest Management on Major Agricultural Systems. Tex. Agric. Exp. Stn., College Station, TX, pp. 365–382.

Hammit, J.K., 1986. Estimating Consumer Willingness to Pay to Reduce Food-borne Risk. Rand Corporation, Santa Monica, CA, 77 pp.

Headley, J.C., 1968. Estimating the production of agricultural pesticides. Am. J. Agric. Econ., 50: 13–23.

Healy, M., 1989. Buyers prefer organic food. USA Today, March 20.

Helmers, G.A., Langemeir, M.R. and Atwood, J., 1986. An economic analysis of alternative cropping systems for east-central Nebraska. Am. J. Alternative Agric., 4: 153–158.

Huffaker, C.B., 1980. New Technology of Pest Control. Wiley, New York, 500 pp.

King, E.G., Phillips, J.R. and Head, R.B., 1986. Thirty-ninth annual conference report on cotton insect research and control. In: Proceedings of the Beltwide Cotton Production Research Conference. National Cotton Council, Memphis, TN, pp. 126–135.

Kovach, J. and Tette, J.P., 1988. A survey of the use of IPM by New York apple producers. Agric. Ecosystems Environ., 20: 101–108.

Matthews, G.A., 1985. Application from the ground. In: P.T. Haskell (Editor), Pesticide Application: Principles and Practice. Clarendon Press, Oxford, pp. 93–117.

Mazariegos, F., 1985. The use of pesticides in the cultivation of cotton in Central America. UNEP Ind. Environ., July/August/September, pp. 5–8.

Moldenhauer, W.C. and Hudson, N.W. (Editors), 1988. Conservation Farming on Steep Lands. Soil and Water Conservation Society, Ankeny, IA, 296 pp.

National Academy of Sciences, 1989. Alternative Agriculture. National Academy Press, Washington, DC, 448 pp.

National Board of Agriculture, 1988. Action programme to reduce the risks to health and the environment in the use of pesticides in the agriculture. General Crop Production Division, National Board of Agriculture, Stockholm, Sweden.

Office of Technology Assessment, 1979. Pest Management Strategies. Vol. II. Working papers. Office of Technology Assessment, Washington, DC, 169 pp.

Paul, J., 1989. Getting tricky with rootworms. Agrichem. Age, 33(3): 6–7.

Pimentel, D., 1986. Agroecology and economics. In: M. Kogan (Editor), Ecological Theory and Integrated Pest Management Practice. John Wiley, New York, pp. 229–319.

Pimentel, D. (Editor), 1991. Handbook of Pest Management in Agriculture. Second Edition CRC Press, Boca Raton, FL.

Pimentel, D. and Levitan, L., 1986. Pesticides: amounts applied and amounts reaching pests. BioScience, 36: 86–91.

Pimentel, D., Terhune, E.C., Dritschilo, W., Gallahan, D., Kinner, N., Nafus, D., Peterson, R., Zareh, N., Misiti, J. and Haber-Schaim, O., 1977. Pesticides, insects in foods, and cosmetic standards. BioScience, 27: 178–185.

Pimentel, D., Krummel, J., Gallahan, D., Hough, J., Merrill, A., Schreiner, I., Vittum, P., Koziol, F., Back, E., Yen, D. and Fiance, S., 1978. Benefits and costs of pesticide use in United States food production. BioScience, 28: 772, 778–784.

Pimentel, D., Andow, D., Dyson-Hudson, R., Gallahan, D., Jacobson, S., Irish, M., Kroop, S., Moss, A., Schreiner, I., Shepard, M., Thompson, T. and Vinzant, B., 1980. Environmental and social costs of pesticides: a preliminary assessment. Oikos, 34: 127–140.

Pimentel, D., McLaughlin, L., Zepp, A., Lakitan, B., Kraus, T., Kleinman, P., Vancini, F., Roach, W.J., Graap, E., Keeton, W.S. and Selig, G., 1991. Environmental and economic impacts of reducing U.S. agricultural pesticide use. In: D. Pimentel (Editor), Handbook of Pest Management in Agriculture, Second Edition. Vol. I. CRC Press, Boca Raton, FL, pp. 679–718.

Poe, C.A., 1988. Where cleanliness means profits. Time. September 5, p. 51.

Rich, A.E., 1991. Potato diseases. In: D. Pimentel (Editor), Handbook of Pest Management in Agriculture. Second Edition. Vol. III. CRC Press, Boca Raton, FL, pp. 623–676.

Ridgway, R., 1980. Assessing agricultural crop losses caused by insects. In: Crop Loss Assessment: Proceedings of the E.C. Stakman Commemorative Symposium, 1980, Univ. of Minnesota, St. Paul, pp. 229–233.

Royle, D.J. and Shaw, M.W., 1988. The costs and benefits of disease forecasting in farming practice. In: B.C. Clifford and E. Lester (Editors), Control of Plant Diseases: Costs and Benefits. Blackwell Scientific, Palo Alto, CA, pp. 231–246.

Russnogle, J. and Smith, D., 1988. More dead weeds for your dollar. Farm J., 112(2): 9–11.

Sachs, C., Blair, D. and Richter, C., 1987. Consumer pesticide concerns: a 1965 and 1984 comparison. J. Consum. Aff., 21: 96–107.

Schalk, J.M. and Ratcliffe, R.H., 1977. Evaluation of the United States Department of Agriculture Program on alternative methods of insect control: Host plant resistance to insects. FAO Plant Prot. Bull., 25: 9–14.

Schweizer, E.E., 1989. Weed free fields not key to highest profits. Agric. Res. May: 14–15.

Shaunak, R.K., Lacewell, R.D. and Norman, J., 1982. Economic implications of alternative cotton production strategies in the lower Rio Grande Valley of Texas, 1923–1978. Tex. Agric. Exp. Stn. B-1420, College Station, TX, 25 pp.

Shields, E.J., Hygnstrom, J.R., Curwen, D., Stevenson, W.R., Wyman, J.A. and Binning, L.K., 1984. Pest management for potatoes in Wisconsin — a pilot program. Am. Potato J., 61: 508–516.

Taylor, F., Raghaven, G.S.V., Negi, S.C., McKyes, E., Vigier, B. and Watson, A.K., 1984. Corn grown in a Ste. Rosalie clay under zero and traditional tillage. Can. Agric. Eng., 26(2): 91–95.

Teng, P.S. and Bissonnette, H.L., 1985. Potato yield losses due to early blight in Minnesota fields, 1981 and 1982. Am. Potato J., 62: 619–628.

Tew, B.V., Wetzstein, M.E., Epperson, J.E. and Robertson, J.D., 1982. Economics of selected integrated pest management production systems in Georgia. Res. Rep. 395, Univ. Ga. Coll. Agric. Exp. Stn., Athens, Ga, 12 pp.

United States Department of Agriculture, 1954. Losses in Agriculture. US Dep. Agric., Agric. Res. Serv. 20-1, Washington, DC, 190 pp.

United States Department of Agriculture, 1975. Farmers' use of pesticides in 1971 ... extent of crop use. US Dep. Agric., Econ. Res. Serv., Agric. Econ. Rep. No. 268, Washington, DC, 25 pp.

United States Department of Agriculture, 1983. Agric. Handb. 589. US Dep. Agric., Agric. Res. Serv., Chap. 20, Washington, DC, p. 534.

United States Department of Agriculture, 1989. Agricultural Statistics 1989. US Government Printing Office, Washington, DC (1988 data used), 547 pp.

United States Office of Management and Budget, 1989. Budget of the United States Government. Office of Management and Budget. US Government Printing Office, Washington, DC.

Van der Scheer, H.A.Th., 1984. Testing of crop protection chemicals in fruit growing. In: Annual Report, Research Station for Fruit Growing. Wilhelminadorp, The Netherlands, pp. 70–77.

Van Doren, D.M., Triplett, Jr., G.B. and Henry, J.E., 1977. Influence of long-term tillage and crop rotation combinations on crop yields and selected soil parameters for an Aeric Ochraqualf soil [Maize]. Ohio Agric. Res. Dev. Cent., Res. Bull. 1091, Wooster, OH, 30 pp.

Agriculture, Ecosystems and Environment, 46 (1993) 289–303
Elsevier Science Publishers B.V., Amsterdam

Effects, constraints and the future of weed biocontrol

Peter Harris

Research Station, Agriculture Canada, P.O. Box 3000 Main, Lethbridge, Alta. T1J 4B1, Canada

Abstract

Many pest control approaches are presently promoted as biocontrol, however, in this paper the definition is restricted to 'the utilization of parasites, predators and pathogens to regulate pest populations'. This covers two approaches against weeds: inundative and classical biocontrol. Inundative control applies an organism (usually a fungal disease) in the manner of a herbicide, and like a herbicide, it is usually marketed by industry. Classical biocontrol establishes an organism (usually an insect) from another region, to give continuing control of the pest and is largely done by governments in the public interest. Both types of biocontrol are regulated by legislation originally designed for other purposes, which is, at best, awkward. Both biocontrol approaches, in the right circumstances, are effective means of weed control with few harmful environmental impacts, but their use is limited by ecological, economic and regulatory constraints. These are illustrated by Canadian examples of a fungal disease for inundative control of round-leaved mallow and the white Amur fish for aquatic weeds. The classical biocontrol examples discussed are St. John's-wort, nodding thistle, leafy spurge and knapweed, as well as some new applications. The steps in classical weed biocontrol are outlined as well as the public involvement needed to develop a biocontrol solution for a weed.

Introduction

The term biological control or biocontrol was used by Smith (1919) to describe the use of natural enemies (predators and parasitoids) for the control of pest insects. It is now used to promote all kinds of non-chemical pest control approaches, such as sowing crested wheat-grass to displace range weeds, weed control by grazing management, crop rotation, breeding pest-resistant crops, genetic engineering of crops, and even for organically produced pesticides (Neish, 1988; Garcia et al., 1988, 1990; Gabriel and Cook 1990). To avoid this confusion, the term biorational control is being used collectively for the non-chemical approaches including biocontrol. The use of biocontrol for organic pesticides seems to be an attempt to avoid the poor image of chemical pesticides, even though it is the nature of the chemical and not its means of production that determines its toxicity since some naturally produced fungal toxins are extremely poisonous. Thus, the Environmental

Protection Agency (1986) requires that chemicals produced by microorganisms are subject to the same regulations as those produced by other means.

The author prefers the definition of biocontrol used by Harley (1985): "the utilization of parasites, predators and pathogens to regulate the population of pests". This implies that a biocontrol agent must attack the pest; it can be natural or genetically modified, but domestic animals are excluded here. Weed biocontrol by this definition divides into two approaches, inundative and classical, which are covered by different governmental legislation. Inundative biocontrol involves the application of an organism to the weed, where and when there is a problem, in the same manner as a herbicide. Classical biocontrol involves the establishment of organisms from another geographic region to give control on a continuing basis.

Wapshere et al. (1989) identified two other types of weed biocontrol: conservation and broad spectrum. Conservation of natural enemies is largely theoretical in weed biocontrol. Broad spectrum weed biocontrol involves the use of cattle and other domestic animals, which the author terms grazing management.

This paper describes the components of inundative and classical biological control and gives Canadian examples. The constraints of both approaches is discussed in terms of legislation, economics, impact on the weed, and some new applications.

Legislation

The two approaches to biocontrol are covered by separate Acts although neither uses the term biocontrol. By the Rule of Law principle, enabling legislation is needed for all government action. This protects government by reducing court challenges, and protects the public by limiting government to prescibed procedures. Under North American practice, Acts are written in broad terms, so it is the regulations and guidelines which indicate how they are applied. Most Acts were written some time ago, so they reflect historic needs and thinking. The result, in a rapidly developing field such as biocontrol, is that they often do not meet current needs, so progress is slowed, cost increased, and innovation inhibited without increasing the protection of the public.

Inundative biocontrol can be regulated on a continuing basis at the marketplace. In the US it comes under the Federal Insecticide, Fungicide and Rodenticide Act (FIFRA) (Agricultural Research Service, 1986) as do chemical pesticides. In Canada it is administered by guidelines of the Food Production and Inspection Branch (FPIB) (1990) under the Pest Control Products Act (CPCA) (1985). The Act was designed for chemical pesticides and until new guidelines had been prepared, it was not appropriate for living organisms. The Regina Research Station of Agriculture, Canada studied the indigenous fungus *Colletotrichum gloeosporioides* (Penz.) Sacc. f. sp. *malvae* as a biocontrol agent (Biomal) for round-leaved mallow, *Malva pusilla* Sm. (Mortensen, 1988), in part to encourage the drafting of new Canadian guidelines. Field and laboratory studies were started on this fungus in 1982 (Mortensen, 1988), but registration for commercial use was not obtained until 1992. Similarly in the US, it took 13 years from discovery to registration of *Colletotrichum gloeosporioides* f. sp. *aeschynomene*, for the control of northern joint vetch (Templeton, 1982), which is fast, compared with the time needed for registration of a new herbicide. Thus, the development of inundative biocontrols for weed species is technically possible. The problem is that even though the registration process is less onerous than for herbicides, the development cost is sufficiently high and the market for the control of most single weed species is sufficiently small, that it requires 10-15 years for cost recovery rather than the 4 years desired by most companies. This is reflected by the paucity of registrations: only three inundative biocontrols in North America, and the Canadian registration of Biomal was obtained 10 years after the last registration in the US. If there is going to be rapid development of inundative biocontrol, the rate of return must be increased by simplifying the registration process, providing public subsidies for development, or by restricting development to biocontrol agents for control of the few individual weed species with a large potential market.

The regulatory system can be improved without increasing risks by taking such steps as eliminating the ecozone (major climatic vegetation regions such as the Canadian prairies) restrictions. Natural organisms can be used (but not sold) with less testing within the ecozone in which they occur. If the reason for their absence reflects an inability to overwinter, the prohibition on their use does little for public protection, but it is a major constraint for the many biocontrol organisms with a small use. Changing the regulations is difficult since five federal agencies presently have a veto and jurisdictional differences often seem to be a factor. Also, there are still no Canadian regulations to cover the release of genetically modified biocontrol organisms, but it is a major advance that genetic changes achieved by common breeding techniques such as cross-breeding and exposure to ultraviolet light are now considered to be natural, so they can be used under the present legislation.

The only legislative control possible for classical biocontrol is before release of an organism into a new region. In the US, the enabling legislation is the Federal Plant Pest Act of 1957 (FPPA) and in Canada the Plant Protection Act of 1990 (CPPA). Both Acts are designed to prevent the introduction and spread of plant pests, rather than to regulate biocontrol organisms. Thus, the CPPA classifies organisms that attack noxious weeds as pests, which leads to the nonsensical situation that biocontrol agents are 'beneficial pests'. Similarly, the FPPA in the US prohibits the importation and movement of pests of plants regardless of whether they attack beneficial or harmful plants. Indeed, Ramsay (1973) admitted that the FPPA and the mind set of the regulators was to regard all plant feeding organisms as undesirable. However, in both the US and Canada, weed biocontrol agents can be released with special permission. For weed biocontrol agents this requires a report (about 50 pages) on the target weed and the control agent, which is reviewed both in Ottawa and Washington. The report follows the Environmental Assessment Outline required under the US National Environmental Policy Act of 1969 (NEPA) (Agriculture Research Service, 1986) which was designed to regulate local industrial impact rather than classical biocontrol where the hope is that the agent will have an impact over a large part of the continent. If classical biocontrol is to be a major alternative to chemical pesticides, it needs to be covered by its own regulations, like the Biological Control Act (1984) in Australia which requires that release approval be given for all biocontrol organisms that are likely to be of more benefit than harm. This requires a risk–benefit analysis in contrast to the North American Acts, which prohibit the release of agents likely to cause any harm. The author also thinks that proposals to target a weed for biocontrol should be published and public comment invited. This would help achieve three things: (1) put biocontrol on a sounder moral and probably legal basis since it would ensure that public concerns are considered; (2) allow modification of projects to meet these concerns; (3) provide an opportunity to educate the public and resolve misunderstandings.

Inundative weed biocontrol

Inundative weed biocontrol is an alternative to chemical control. The $3–4 million cost of obtaining registration of a biocontrol organism is considerably less than for a herbicide, and its commercial production can usually be done on a small scale suitable for use against a single weed species, which is needed for species escaping other means of control. Safety is usually high as the host range of the pathogens is extremely narrow and most have no animal toxicity. However, some pathogens are toxic, so, as for chemicals, this must be investigated for registration. The ecological attraction of inundative weed biocontrol is that the agents generally have fewer non-target effects than do chemical herbicides, and companies are prepared to develop and market inundative weed biocontrol agents if they can make a profit.

To be commercially viable an inundative biocontrol agent must (1) produce an abundant and durable inoculum in artificial culture, (2) be genetically stable and specific to the target weed, (3) kill or suppress a high proportion of the target weed over a reasonably wide environmental range (Daniel et al., 1973), (4) have limited ability to spread or survive in nature, so there is a need for continuing sales, and (5) be patentable so that development costs can be recovered. Recently patent protection in the US for natural organisms has been changed to cover the product formulation and the production process instead of biocontrol agent use. Apart from philosophical objections to patenting the use of natural organisms, it restricted the development of inundative biocontrol agents. For example, previously the use patent of the fungus forma specialis of *Colletotrichum gloeosporioides* for northern joint vetch included round-leaved mallow even though it did not attack it, and this prevented the patenting of *Colletotrichum gloeosporioides* f. sp. *malvae*, which is effective against mallow.

The fungus *Colletotrichum gloeosporioides* f. sp. *malvae* meets the commercial requirements for the control of round-leaved mallow. Round-leaved mallow is an increasing in-crop weed problem for which there is no satisfactory chemical or cultural control. The fungus occurs widely in Canada, but its impact in nature is limited because it depends on rain splash for spore spread. Artificial spread by spraying the spores in water is simple and the organism is both safe and effective. Thus, there is now a biocontrol for a weed that was largely escaping chemical and cultural controls.

Unfortunately, most host-specific and otherwise amenable fungi lack the necessary virulence to be effective for inundative biocontrol. For example, the forma specialis of *Colletotrichum gloeosporioides* that attacks the aquatic weed, watermilfoil, (Smith et al., 1989) causes little mortality, so it has no commercial value unless its virulence can be increased. This may not be possible by normal breeding methods and genetically engineered organisms, regardless of safety, cannot be marketed until there are appropriate registration regulations. Some fungi, such as *Fusarium graminearum* Schw. of wheat and corn, produce powerful human and animal toxins (Trenholm et al., 1988) so they are clearly undesirable in food. Thus, toxicity testing is essential before any fungus is registered, which makes it uneconomic to develop fungi such as *Colletotrichum gloeosporioides* f. sp. *hyperici* for St John's-wort control (Hildebrand and Jensen, 1991) as the market is too small. However, present Canadian regulations permit its use in an ecozone in which it occurs naturally, as long as it is not sold. It is presently being applied by government agencies in the Canadian Maritimes.

Plant toxins produced by pathogens are a possible source of a new generation of herbicides. However, the cost of a herbicide registration is about 20 times that of an inundative biocontrol agent, which economically eliminates development of phytotoxins that are specific to most single weed species. Bacteria and viruses are generally of less interest than fungi as inundative biocontrol agents because they need a wound or an insect vector in order to infect the plant, whereas fungi enter through stomata, or with enzymes, directly through the cuticle. However, some rhizosphere bacteria (those that live on exudates around the root) impair seed germination, suppress plant growth, and delay plant development of the host plant. For example, the winter annual grass, downy brome, can be selectively suppressed by fall application of an indigenous bacterium which declines to its natural low level during the following summer (Kennedy et al., 1991). Rhizosphere bacteria will be registered by industry for weed control if they are found to be economic and ecologically safe.

Most insects and higher organisms do not lend themselves to inundative control programs because mass rearing them is too expensive. However, organisms with dense field populations that can be collected cheaply and distributed are exceptions. This, includes the Russian knapweed nematode, *Subanguina picridis* (Kirj.) Brz. (Watson and Harris, 1984), which can be extracted and sprayed in water, or scattered as stem pieces. So far there has been little interest in its commercialization, possibly because the weed is not prevalent enough.

The grasscarp, *Ctenopharyndon idella* Val., is a fish used for the inundative biocontrol of aquatic weeds. It controls aquatic weeds at about one-quarter the cost of mechanical control and half the cost of using herbicides (Zon et al., 1978). In laboratory tests, the fish prefers native and non-weedy aquatic plants to most introduced weedy species (Fedorenko and Fraser, 1978), but in nature it tends to eat plants in the relative proportions in which the different species occur (Van Zon, 1977). This is ecologically desirable since plant diversity is maintained (Van Zon et al., 1978). It is an excellent table fish and has no adverse effects, or has positive effects, on carnivorous game fish (Van Zon et al., 1978). Presently only triploid fish, which are sterile, are used in North America.

Canada is one of the few developed counties in the world not using the grasscarp. A reason may be that federal control of inland fisheries has been ceded to the provinces, so the national use of the fish would involve obtaining approval from nine regulatory bodies working under nine Acts. It is exciting and encouraging that Alberta is now trying the fish, but testing on a national basis would reduce costs and the concerns of adjacent provinces. The author sees a danger in North America that provinces and states will enact their own legislation which will reduce the area and hence the economic viability of biocontrol. This would be particularly detrimental to classical biocontrol.

Classical biocontrol

Classical weed biocontrol selectively reduces the target weed without harm to associated plant species, which consequently increase. In uncultivated situations, this means increased biological diversity (Harris, 1986), which is generally ecologically desirable. Classical biocontrol also meets the sustainable agricultural objectives of minimal non-target effects (Harris, 1990b) and low-input solutions. For example, the establishment of biocontrol agents on skeleton weed in Australia gave control with an amortized return of 112 times its original cost (Marsden et al., 1980) and eliminated an expensive chemical control program. The input is low because the control agent finds the weed on its own, but this means that there is nothing to market commercially. Hence, the private sector cannot be expected to develop classical biocontrol as they would inundative biocontrol. Consequently, developmental investment must come from the public sector or from user organizations.

Classical weed biocontrol can be divided into nine steps (Wapshere et al., 1989), which are summarized as: (1) justification for biocontrol of the weed; (2) target weed approval; (3) exploration for candidate biocontrol agents; (4) biocontrol agent screening; (5) biocontrol agent approval; (6) importation; (7) release and increase; (8) evaluation; (9) distribution. The first steps alternate between the collection of data for reports and obtaining approval for the next step from Agriculture Canada and US Department of Agriculture (USDA) review groups. The steps are not necessarily discrete, for example, the exploration for natural enemies (Step 3) is often done while collecting the first candidate agents for screening. This occurs because funds are short and the client is impatient to release a biocontrol agent as soon as possible, so the 2 years needed for a thorough field survey is intolerable (Waage, 1990). Thus, host specificity tests are started on an organism selected from the literature, but whether or not this makes good economic sense is debatable. Ideally, there is a feed-back from the evaluation (Step 8) of released agents to the selection of new species for screening, but it may be 10 years before the control agent has a major impact. Thus, evaluation (Step 8) continues for this period at the same time as other species are being selected for screening (Step 4).

A complete program requires about 20 scientist-years (Harris, 1979), presently costing about $6 million. Roughly a third of the control agents released fail to become established and a third become established at too low a density to inflict serious damage on the weed. Thus, there is a need to improve success rates, but it is unacceptable to relax safety screening standards because of concerns for desirable plants. However, in over 600 examples of weed biocontrol throughout the world, control agents have remained within their predicted host range, so the trend to make screening tests more extensive, and hence expensive, is not justified by the record.

The impact of classical weed biocontrol

The potential of classical biocontrol for the control of weeds is illustrated by several Canadian projects.

St. John's-wort (Hypericum perforatum L.)

The introduced range weed, St. John's-wort, has been reduced to about 1% of its former density in much of southern British Columbia, Ontario and the Maritimes by the defoliating beetles *Chrysolina quadrigemina* (Suffr.) and

Chrysolina hyperici Forst. (Harris and Maw, 1984). St. John's-wort remains, even in the preferred beetle habitats, but it is no longer the dominant plant that suppresses other herbaceous species. In sites where the beetles have done poorly for climatic reasons, stands of St. John's-wort are still dense (Williams, 1985). On these sites, releases of several other insect species have been made, including a root-feeding beetle, *Agrilus hyperici* (Creutz.). Establishment of *Agrilus hyperici* was tried unsuccessfully several times since 1955 with stocks from a European colony established in California. Over time, the species has gradually been moved north in the US and was established in British Columbia in 1987 with stock from Lewiston, ID. Need for selection in a new region is common unless the agents can be imported directly from a matching climate. For example, the *Chrysolina* beetles in British Columbia required 5–13 years after release before they had a major impact (Harris et al., 1969). The problem was that they stayed on the foliage as the temperature dropped in the fall, and were killed. After natural selection, they now seek shelter in the plant litter and emerge to oviposit when the temperature rises again (Peschken, 1972). The time required for such selection is one reason why (even with unlimited scientific resources) projects commonly take 20 years.

Complaints about this biocontrol program from herbalists seem to arise from a misconception that weed biocontrol results in eradication of the target. The word 'control' does not mean eradication and biocontrol agents merely reduce the abundance of their host, so plenty of St. John's-wort, which is toxic to both cattle and humans, remains for its dubious use as a herb.

This project illustrates that (1) weed biocontrol is slow; (2) several agents may be needed for different habitats; (3) it is necessary to match the climate of the collection and release area or allow time for adaptation; (4) there is room for public misunderstanding; (5) defoliators can be successful biocontrol agents.

Nodding thistle (Carduus nutans L.)

The biocontrol of the introduced, nodding thistle by the seed-head weevil *Rhinocyllus conicus* Froel. has restricted this weed to recently disturbed uncultivated sites (Harris, 1984). Before biocontrol in 1968, the thistle, once established, remained dominant, but now grass returns in about 3 years unless the site is redisturbed. The thistle is still common along railroads, on vacant lots, gravel pits, and overgrazed, drought-stressed pastures, where competition from other plants is low, but is not a problem on a well-managed range. The disturbed sites are beneficial as they provide a sanctuary for the weevil from which it can spread onto the range if needed.

This project illustrates (1) that successful weed biocontrol often needs good competition from other vegetation (Harris, 1986), and (2) that seed destroyers can be successful agents.

*Leafy spurge (*Euphorbia esula *L.)*

Leafy spurge is a toxic European plant that forms up to a 100% cover on infested uncultivated land in the southern Canadian prairies. Direct annual losses in North Dakota are $17 million with indirect losses of $75 million (Thompson et al., 1990) and the smaller area infested in Canada suffers proportionally.

The spurge hawk moth (*Hyles euphorbiae* L.). a defoliator, was tried initially because it was easy to study. Unfortunately, leafy spurge is tolerant of defoliation, the moth larvae are vulnerable to predation and their development requires higher summer temperatures than occur reliably on the Canadian prairies. However, root feeders are a success. The beetle *Aphthona nigriscutis* Foudr. has reduced leafy spurge to a 5% cover at release sites on open, dry coarse soils (Harris, 1990a). *Aphthona cyparissiae* (Koch) has a similar effect in slightly moister soils and its normal habitat is open swales on sandy loam soils. Another species, *Aphthona flava* Guill. thrives in light shade on coarse soils near rivers, but most of the Canadian prairies appears to lack sufficient degree-days for it. *Aphthona lacertosa* Rosh., released in 1991, has survived on loam soils and *Aphthona czwalinai* (Weise) has done well on a clay–loam soil in North Dakota (R. Carlson, personal communication, 1992). The main need now is for species adapted to shady and spring flood sites.

This example illustrates (1) that root feeders may be more effective than defoliators: the loss of leaves is less important to leafy-spurge than the loss of roots as the species is adapted to regrowing from large reserves stored in the roots, and (2) that biocontrol agents are often site specific, so that several control agents are needed to control the weed over its range.

*Knapweed (*Centaurea maculosa *Lam. and* Centaurea diffusa *Lam.)*

Diffuse and spotted knapweed are introduced plants of the dry grasslands in the northwest US and the interior of British Columbia where they displace up to 85% of the grass and other herbaceous plants. This competitive weed is disastrous for ranching as well as to the native fauna and flora. Presently over 100 000 ha of grasslands are infested in Canada and 10 million ha are threatened (Harris and Cranston, 1979), but the main problem is in Montana with 12 million ha infested and 17 million threatened, which represents 40% of the state (Chicoine et al., 1988). With funding from British Columbia and Montana, 12 biocontrol agents, which cover all feeding guilds in the flower heads and roots (different feeding sites such as flower buds, soft achenes, ripe achenes, roots) have been approved for release.

In Europe the seed-heads of these knapweeds are commonly attacked by seven or more insect species at a site. Biocontrol agents which occur together may be exploitative or interference competitors (Keddy, 1989). The former compete only through their ability to consume the weed, which is desirable in biocontrol. The latter kill or directly suppress the competing species and so may be undesirable. Competition theory (Akcakaya and Ginzburg, 1989) suggests that exploitative competitors specialize in what they do best with little overlap between them. If there are several exploitative competitors, as in knapweed seed-heads, several will probably be needed to optimize seed reduction. Paradoxically, this means that the cost of biocontrol is likely to be higher if the weed, in its native region, is exploited by many rather than the single species that dominates the exploitation of nodding thistle seed-heads in Europe.

Two seed-head flies, *Urophora affinis* Frau. and *Urophora quadrifasciata* (Meig.), reduced spotted knapweed seed from 40 000 m^{-2} to 1500 m^{-2}, which is clearly progress, although not enough to achieve control. However, in the dry zone at White Lake, B.C., the two seed-head flies and a root beetle, *Sphenoptera jugoslavica* Obenb., have gradually reduced diffuse knapweed cover on grazed pasture from 52% in 1986 to 13.7% in 1992 (A. Sturko, personal communication, 1992). Both spotted and diffuse knapweed control should improve as more of the agents become abundant. It is also apparent that the grazing regime is an important factor in control as fence lines separating pastures have different knapweed densities.

This example illustrates that (1) it may be necessary to use several exploitative competitors for weed contról, and (2) that there is a need to integrate biocontrol with pasture management.

New directions for classical weed biocontrol

As of 1990, there have been classical weed biocontrol projects in Canada on 22 weed species with 54 agents. All of the projects have been against abundant introduced weeds on uncultivated land. Two new directions for classical biocontrol involve its use against weeds in cultivated crops, and against native weeds.

Australia has had spectacular success, with an amortized value of $25 million a year, against skeleton weed, *Chondrilla juncea* L., an introduced weed of wheat, with a rust disease (Marsden et al., 1980). In the USSR, the establishment of a defoliating beetle, *Zygogramma suturalis* L., from Canada, on ragweed, *Ambrosia artemisiifolia* L., initially increased by two- to three-fold the yields of infested crops (Kovalev and Vechernin, 1986). The final results were not as good since, after 10 years, Reznik et al. (1990) reported that *Z. suturalis* did not control the weed in cultivated crops because it was a poor disperser and that the weed populations were not annually stable enough to permit the build-up of the beetle populations. Presumably, the initial *Ambrosia* control in crops resulted from an overflow from the massive beetle populations on the uncultivated areas. After the beetle controlled these, the overflow disappeared. Nevertheless, the example suggests that dense stable weed populations in a crop could be controlled by a biocontrol agent that accepts the agronomic practice. Also, dense unstable populations should be amenable to control by a good disperser.

The classical biocontrol of the native North American snakeweeds, *Gutierrezia* spp., is being investigated by the USDA. These toxic rangeland weeds of Texas and adjacent regions, which seem to have increased since European settlement, have close relatives in South America attacked by genus-specific insects (C.J. Deloach, personal communication, 1990). There are many examples in which native plants, such as chestnut and elm, have been selectively reduced by accidentally introduced organisms, so technically the biocontrol of a native weed is the same as that of an introduced species. The main difference is a philosphical feeeling that it is unethical to depress the population of native plants and a fear that the native plant supports a coevolved community, which would collapse with its control. This view was advanced by Pemberton (1985), but refuted by Johnson (1985) who suggested that many native weed species have increased as a result of human activity, so biocontrol would merely be restoring a historically more normal density. Also, most species (plants and animals) are dependent on a habitat, such as forest, and survive as long as a forest exists even if there are changes in relative abundance of tree species. Clearly, the biocontrol of each native plant species needs to be examined closely, as do introduced plant species; but it is wrong, in the author's opinion, to a priori exclude the use of classical biocontrol to reduce the density of native weeds.

Responsibilities of classical biocontrol user groups

Formerly Agriculture Canada, on it own initiative, developed classical biocontrol solutions for problem weed species. The situation has gradually changed, as it probably has in most other countries, so that the support from a client is now the driving force. Some producer or other client groups are more effective than others at getting government to initiate and sustain a program for the necessary 20 years, so there is a need to discuss the process.

A producer group can get a government to start biocontrol programs with a well-organized lobby; however, unless this is supported by a sound economic and ecological justification for the program, it is unlikely to survive the 20 years needed to complete the program. For example, a North American program against toadflax was started in the 1960s in response to widespread concern, but without economic data. Canada obtained approval for release of one defoliating moth and none of the agents studied by the US reached the approval stage. In the late 1980s, the Canadian program was revived with economic data, as a result of concern from the B.C. Cattlemen's Association and the province of Alberta. Their interest, and associated funding, resulted in approval for release of four biological control agents by 1992, including two root-feeders which are likely to be more effective than the defoliator. The economic and ecological data quantifies both the problem from the weed and its benefits, such as honey flow or use as a herbal medicine as well as the risks from biocontrol to related desirable plants. Thus, the justification report is a benefit–risk analysis of using biocontrol. A good report also determines the biocontrol agent feeding-guild with the greatest impact on the weed. Where the public view the weed as pretty or desirable, it may be essential to publish the study. Thus, publication of the purple loosestrife, *Lythrum salicaria* L., study (Thompson et al. 1987) was responsible for gaining public support by explaining that besides being pretty, the weed causes serious problems.

The 20 years needed for the biocontrol of a weed extends over the life of five or more governments and it is almost guaranteed that without continued interest from a client, the program will be cancelled to redirect resources to new priorities. An interested client requires progress reports and meets with the project scientists to ensure that there is progress through the nine steps described previously. If progress is hampered by jurisdictional difficulties between the levels of government and government agencies, the intervention of the client can be helpful.

Agriculture Canada funds much of the overhead facility and permanent staff costs of the European studies for Canadian weed biocontrol projects, but the overseas project costs are largely provided by the client or by government on the clients' behalf. As governments wrestle to balance budgets, there is a need for new funding sources such as user check-off fees or, as in Montana, a tax on pesticides to fund the development of alternatives to chemical control. The Montana tax plus funding from the province of British Columbia has supported the screening of most of the knapweed biocontrol agents released since the mid-1980s. Screening is essential to demonstrate that candidate biocontrol agents can be released without threat to desirable plants. The host specificity of an insect is determined by a hierarchy of steps with initial ones such as the selection of habitat and oviposition site being more firmly held than the suitability of a plant species for larval development (Harris and McEvoy, 1992). It is this sequence of requirements that makes biocontrol agents safe, but it also contributes to the cost of testing. It costs about two scientist-years, which is currently about $600 000 if the work is done in North America and $400 000 if done in the insect's native region in Europe, where it is not necessary to work in quarantine. About a third of this money must come from the client in project funds and is most easily obtained by forming a consortium of several interested parties.

The final involvement of the client is to help distribute the biocontrol agents. The first releases are used to determine whether there is enough impact to warrant distribution. The Canadian practice is for promising species to be distributed by the provinces to establish regional colonies on the weed. The regional colonies are then used for distribution to the public through field days to explain the needs and expected impact of the agent. This process is aided if the client producer-group publishes articles on the progress of the biocontrol project and on the individual agents and helps arrange the field days.

Conclusions

Both inundative and classical biocontrol of weeds offer exciting possibilities for solving certain weed problems in an economic and environmentally acceptable manner. However, this will require enabling legislation that allows maximum freedom for the development of biocontrol as well as protection of the public. Inundative biocontrol will not become a solution for most single weed species unless the registration costs can be reduced. We need to increase our success rate with classical biocontrol agents and the participation of client groups is essential if the required effort in it is to be maintained.

References

Agricultural Research Service, 1986. Final regulations for implementing National Environmental Policy Act (NEPA). Fed. Reg., 51(186): 34190–34192.

Akcakaya, R.H. and Ginzburg, L.R., 1989. Niche overlap and the evolution of competitive interactions. In: A. Fontdevila (Editor), Evolutionary Biology of Transient Unstable Populations. Springer Verlag, Berlin, Germany, pp. 32–42.

Biological Control Act 1984, No. 139 of 1984, Commonw. Gov. Print. Cat. 15668/84 5, Commonwealth Government Printer, Canberra, A.C.T., Australia, 25 pp.

Chicoine, T., Fay, P. and Nielsen, J., 1988. Predicting spotted knapweed migration in Montana. Mont. Agresearch, 5: 25–28.

Daniel, J.T., Templeton, C.E., Smith, Jr., R.J. and Fox, W.T., 1973. Biological control of northern jointvetch in rice with an endemic fungal disease. Weed Sci., 21(4): 303–307.

Environmental Protection Agency, 1986. Statement of Policy: Microbial products subject to the Federal Insecticide, Fungicide, Rodenticide and Toxic Substances Control Act. Fed. Reg., 51: 23313.

Fedorenko, A.Y. and Fraser, F.J., 1978. A review of the biology of grass-carp (Ctenopharyndon idella, Val.) and its evaluation as a potential weed control agent in British Columbia. Fish. Mar. Serv. Tech. Rep. No 786, Environmental Protection Agency, Washington, DC, 15 pp.

Food Production and Inspection Branch, 1990. Guidelines for registration of naturally occurring microbial pest control agents. Agric. Can., R-90-03, 53 pp.

Gabriel, C.J. and Cook, R.J., 1990. Biological control — the need for a new scientific framework. Bioscience, 40: 204–206.

Garcia, R., Caltagirone, L.E. and Gutierrez, A.P., 1988. Comments on a redefinition of biological control. Bioscience, 38: 692–694.

Garcia, R., Caltagirone, L.E. and Gutierrez, 1990. Biological control — the need for a new scientific framework. Bioscience, 40: 207.

Harley, K.L.S., 1985. What is biological control. In: P. Ferrar and D.H. Stechmann (Editors), Biological Control in the South Pacific. Min. Agric. Fish. For., Tonga, pp. 32–36.

Harris, P., 1979. The cost of biological control of weeds by insects in Canada. Weed Sci., 27: 242–250.

Harris, P., 1984. *Carduus nutans* L. nodding thistle and *C. acanthoides* L., plumeless thistle (Compositae). In: J.S. Kelleher and M.A. Hulme (Editors), Biological Control Programmes Against Insects and Weeds in Canada 1969–1980. Commonw. Agric. Bur., Farnham Royal, pp. 115–126.

Harris, P., 1986. Biological control of weeds. Fortschr. Zool., 32: 123–138.

Harris, P., 1990a. The Canadian biocontrol of weeds program. In: B.F. Roche and C.T. Roche (Editors), Range Weeds Revisited. Washington State Univ., Pullman, pp. 61–68.

Harris, P., 1990b. Environmental impact of introduced biological control agents. In: M. MacKauer, L.E. Ehler and J. Roland (Editors), Critical Issues in Biological Control. Intercept, Andover, UK, pp. 289–300.

Harris, P. and Cranston, R.S., 1979. An economic evaluation of control methods for diffuse and spotted knapweed in western Canada. Can. J. Plant. Sci., 59: 375–382.

Harris, P. and M. Maw, 1984. *Hypericum perforatum* L. St. John's-wort (Hypericaceae). In: J.S. Kelleher and M.A. Hume (Editors), Biological Control Programmes Against Insects and Weeds in Canada 1969–1980. Commonw. Agric. Bur., Farnham Royal, pp. 171–177.

Harris, P. and McEvoy, P., 1992. The predictability of insect host plant utilization from feeding tests and suggested improvements for screening weed biocontrol agents. In: E.S. Delfosse and R.R. Scott (Editors), Proc. 8th Int. Symp. Biol. Contr. Weeds, DSIR/CSIRO, Melbourne.

Harris, P., Peschken, D. and Milroy, J., 1969. The status of biological control of the weed *Hypericum perforatum* in British Columbia. Can. Entomol., 101: 1–15.

Hildebrand, P.D. and Jensen, K.I.N., 1991. Potential for the biological control of St. John's-wort (*Hypericum perforatum*) with an endemic strain of *Colletotrichum gloeosporioides*. Can. J. Plant Pathol., 13: 60–70.

Johnson, H.B., 1985. Consequences of species introductions and removal on ecosystem function — Implications for applied ecology. In: E.S. Delfosse (Editor), Proc. VI Int. Symp. Biol. Contr. Weeds, 19–25 August 1984, Vancouver, Canada. Agriculture Canada, Ottawa, Ont. pp. 27–56.

Keddy, P.A., 1989. Competition. Chapman and Hall, London, UK, 202 pp.

Kennedy, A.C., Elliot, L.F., Young, F.L. and Douglas, C.L., (1991). *Rhizobacteria* suppressive to the weed downy brome. Soil Sci. Soc. Am. J., 55:

Kovalev, O.V. and Vechernin, V.V., 1986. Description of a new wave process in populations with reference to introduction and spread of the leaf beetle *Zygogramma suturalis* F. (Coleoptera, Chrysomelidae). Entomol. Rev., 65: 93–112.

Marsden, J.S., Martin, G.E., Parham, D.J., Ridsdill Smith, T.J. and Johnston, B.G., 1980. Skeleton weed control. In: Returns on Australian Agricultural Research. CSIRO, Canberra, pp. 84–93.

Mortensen, K., 1988. The potential of an endemic fungus, *Colletotrichum gloeosporioides*, for biological control of round-leaved mallow (*Malva pusilla*) and velvetleaf (*Abutilon theophrasti*). Weed Sci., 36: 473–478.

Neish, G., 1988. Standing Committee on Environment and Forestry. Minutes of proceedings. Canada, Parliament, Vol. 28, p. 24.

Pemberton, R.W., 1985., Native weeds as candidates for biological control research. In: E.S. Delfosse (Editor), Proc. VI Int. Symp. Biol. Contr. Weeds, 19–25 August 1984, Vancouver, Canada. Agriculture Canada, Ottawa, Ont. pp. 869–877.

Peschken, D.P., 1972. *Chrysolina quadrigemina* (Coleoptera: Chrysomelidae) introduced from California to British Columbia against the weed *Hypericum perforatum*: comparison of behaviour, physiology, and colour in association with post-colonization adaptation. Can. Entomol., 104: 1689–1698.

Pest Control Products Act, 1985. R.S., c. P-10, s.l. Min. Supply and Services, Ottawa, Canada, 7 pp.

Plant Protection Act, 1990. Bill C-67. Canada Gov. Publ. Centre, Ottawa, 25 pp.

Ramsay, M.J., 1973. Beneficial insect or a plant pest? The regulatory agency's dilemma. Proc. 2nd Int. Symp. Biol. Contr. Weeds, 4–7 October 1971, Misc. Publ. 6. Commonw. Inst. Biol. Contr., Farnham Royal, UK, pp. 40–46.

Reznik, S.Ya, Belokobyl'sky, S.A. and Lonanova, A.L., 1990. Effect of agroecosystem stability on Ambrosia leaf beetle *Zygogramma suturalis* (Coleoptera, Chrysomelidae) population density. Zool. Zh., 69: 54–59.

Smith, C.S., Slade, S.J., Andrews, J.H. and Harris, R.F., 1989. Pathogenicity of the fungus, *Colletotrichum gloeosporioides* (Penz.) Sacc., to Eurasian watermilfoil (*Myriophyllum spicatum*. L.). Aquat. Bot., 33: 1–12.

Smith, H.S., 1919. On some phase of insect control by the biological method. J. Econ. Entomol., 12: 288–292.

Templeton, G.E., 1982. Biological herbicides: discovery, development, deployment. Weed Sci., 30: 430–433.

Thompson, F., Leitch, J.A. and Leistritz, F.L., 1990. Economic impact of leafy spurge in North Dakota. N. D. Farm. Res., 47: 9–11.

Thompson, D.Q., Stuckley, R.L. and Thompson, E.B., 1987. Spread, impact and control of purple loosestrife (*Lythrum salicaria*) in North American Wetlands. Vol. 2. US Fish Wildl. Res., Washington, DC, 55 pp.

Trenholm, H.L., Prelusky, D.B., Young, J.C. and Miller, J.D., 1988. Reducing mycotoxins in animal feeds. Agric. Can. Publ. 1827/E, Agriculture Canada, Ottawa, Ont., 22 pp.

Van Zon, J.C.J., 1977. Grass carp (*Ctenopharyngodon idella* Val.) in Europe. Aquat. Bot., 3: 143–155.

Van Zon, J.C.J., van der Zweerde, and Hodges, B.J., 1978. The grass-carp, its effects and side-effects. In: T.E. Freeman (Editor), Proc. IV Int. Symp. Biol. Contr. Weeds, 30 August–2 September 1976 Univ. Florida, Gainsville, FL, pp. 251–256.

Waage, J.K., 1990. Ecological theory and the selection of biological control agents. In: M. Mackauer, L.E. Ehler and J. Roland (Editors) Critical Issues in Biological Control. Intercept. Andover, pp. 135–157.

Wapshere, A.J., Delfosse, E.S. and Cullen, J.M., 1989. Recent developments in biological control of weeds. Crop Prot., 8: 227–250.

Watson, A.K. and Harris, P., 1984. *Acroptilon repens* (L.) DC., Russian knapweed (Compositae). In: J.S. Kelleher and M.A. Hulme (Editors), Biological Control Programmes Against Insects and Weeds in Canada 1969–1980. Commonw. Agric. Bur., Farnham Royal, pp. 105–110.

Williams, K.S., 1985. Climatic influences on weeds and their herbivores: biological control of St. John's wort in British Columbia. In: E.S. Delfosse (Editor), Proc. VI Int. Symp. Biol. Contr. Weeds, 19–25 August 1984, Vancouver, Agriculture Canada, Ottawa, Ont., pp. 127–132.

Agriculture, Ecosystems and Environment, 46 (1993) 305–324
Elsevier Science Publishers B.V., Amsterdam

Developing an environmentally sound plant protection for cassava in Africa

J.S. Yaninek*, F. Schulthess

Biological Control Program, International Institute of Tropical Agriculture, B.P. 08-0932, Cotonou, Benin

Abstract

Cassava is a food crop of increasing importance for the rapidly growing rural population in Africa. Easy to grow even under harsh agronomic conditions, cassava is the primary source of carbohydrates for more than 200 million people, including the poorest on the continent, and provides food security to most subsistence farmers. Several pests including phytophagous arthropods, plant pathogens and weeds constrain cassava production on the continent. The exotic species introduced accidentally from the Neotropics constitute the largest group of pests. The long cropping cycle exposes cassava to relatively few additional pests in Africa, and enhances sustainable pest management interventions such as biological control, host plant resistance and cultural practices. Historically, cassava plant protection focused on resistance breeding, but since the invasion of several devastating exotic pests in the 1970s, a wider range of pest management solutions are now being pursued. Plant protection interventions that are developed and tested by teams of multi-disciplinary scientists with input from extension agents and client farmers are urgently needed. Appropriate technology development requires an understanding of key pest–host–agroecosystem interactions in the context of the agronomic practices of the farmer and the socio-economic importance of the crop. A regional project to develop, test and implement ecologically sustainable cassava plant protection in West Africa is presented as a general model for developing appropriate pest management in Africa.

Introduction

Cassava, *Manihot esculenta* Crantz (Euphorbiaceae), a woody perennial shrub from the Neotropics, is grown principally for its edible storage roots. This crop occurs worldwide, generally between latitudes 30°N and 30°S from sea level up to 2300 m near the equator, and is common in lowland tropical regions receiving between 750 and 3000 mm rainfall (Cock, 1985). Although cassava grows best in loamy sandy soil of reasonable fertility, it can be produced in soils too depleted to support most other food crops; consequently it is often planted in marginal areas. Cassava is cultivated by planting vegetative cuttings during the wet season and is harvested anywhere from 8 to more than 36 months after planting. Consequently, cassava has been widely adopted

---

*Corresponding author.

as an important and dependable source of calories in the developing world. Once rooted, cassava can withstand prolonged periods of drought and pest attacks by reducing biomass production and remobilizing photosynthate reserves in the stems and roots (Cock, 1979). The presence of potentially dangerous levels of hydrocyanic glucosides in many varieties requires special processing for human consumption, but provides protection against some arthropod and vertebrate pests (Jones, 1959).

Cassava production in Africa

Cassava was introduced into Africa by the Portuguese during the last half of the sixteenth century (Jones, 1959). Except in some coastal locations, cassava was of little importance as a food crop on the continent prior to this century. It then spread rapidly and increased tremendously in importance as its value as a food reserve during famine and locust outbreaks became evident (Jones, 1959). It is now found from sea level up to 1800 m and between latitudes 15°N and 15°S where rainfall exceeds 750 mm per annum. Cassava continues to expand into dry savannah throughout the continent.

With 40% of the world's total production, more cassava is now produced in Africa than on any other continent. Cassava is a key component of the traditional cropping system in most of the lowland humid and subhumid tropics of the continent. About 90% of this production is used for human consumption (Dorosh, 1988). Cassava is a food staple and the most important source of carbohydrate for more than 200 million of the poorest people on the continent (Herren and Bennett, 1984). Its leaves are consumed in many regions as an important source of protein and vitamins (Silvestre and Arraudeau, 1983). It is grown generally by small-scale farmers on plots of less than 0.5 ha where production rarely exceeds 7 t ha^{-1} (Dorosh, 1988).

The rapidly growing rural and urban populations in Africa are accelerating demand for cassava. The result is that the productive land available for cultivation is declining as the traditional fallow periods are being shortened (Nweke and Polson, 1991). Such poor agronomic practices ultimately degrade the natural resource base of the cassava agroecosystem. Restoring and maintaining this resource base in a sustainable equilibrium with production demand requires a major effort and immediate attention.

Cassava pests in Africa

"Manioc in Africa is occasionally damaged by various sorts of insects, including ants, beetles, caterpillars, and grasshoppers, sometimes by red spiders, and by a number of larger vermin, of which goats and wild pigs are probably the most troublesome among animals that vary in size from rats to hippopotami. ..... But on the whole, none of these pests causes serious trouble."

So reported Jones (1959) in an early comprehensive study of cassava production in Africa. In addition to the exotic nature of the host plant on the continent, Bernays et al. (1977) attributed the absence of significant pests, partly to the presence of hydrocyanic glucosides in the plant. However, the pest situation changed as cassava cultivation intensified and exotic pests were introduced. Although it is now widely accepted that cassava in Africa is attacked by a number of serious pests, only a few in-depth studies of the ecological and economic importance of any of these species have been carried out.

The major cassava pests in Africa include relatively few phytophagous arthropods, plant pathogens and weeds (Table 1) compared with the pest complex found in the Neotropics (Bellotti and Schoonhoven, 1978). Together, these species reduce cassava production by an estimated 50% (Herren and Bennett, 1984). The most severe pests are the exotic species accidentally introduced into areas where the local germplasm is susceptible to attack, effective natural enemies are absent, and where the tradition of practices to cope with the introduced pests has not had time to evolve. In addition, pest problems are caused when intensification of cassava production erodes the environmental stability inherent in balanced agroecosystems.

Table 1
Major pests of cassava in Africa and their areas of origin

Pests	Origin
Arthropods	
Cassava mealybug — *Phenacoccus manihoti* Mat.-Ferr.	Neotropics
Cassava green mite — *Mononychellus tanajoa* (Bondar)	Neotropics
Variegated grasshopper — *Zonocerus variegatus* (L.)	Africa
Whiteflies — *Bemisia tabaci* (Genn.)	?Mid. East
Larger grain borer — *Prostephanus truncatus* (Horn)	Neotropics
Plant pathogens	
Cassava mosaic virus	?Africa
Bacterial blight — *Xanthomonas campestris* pv. *manihotis* (Arthaud-Berthet) Starr	?Neotropics
Anthracnose — *Colletotrichum gloeosporiodes* f. sp *manihotis* Henn.	?
Root rots (e.g. *Sclerotium, Fusarium* and *Phytophthora* spp.)	?
Weeds	
Imperata cylindrica (Anderss.) C.E. Hubbard	Africa
Chromolaena odorata (L.) R.M. King & Robinson	Neotropics

Arthropod pests

Many of the important arthropod pests of cassava in Africa are exotic neotropical species. The cassava mealybug, *Phenacoccus manihoti* Mat.-Ferr., was the most serious arthropod pest of cassava on the continent. In many regions, cassava production ceased altogether when the mealybug invaded as both field crops and local sources of planting material were destroyed. Although the mealybug is now controlled satisfactorily in most infested agroecosystems by the introduced parasitoid *Epidinocarsis lopezi* (DeSantis) (Herren and Neuenschwander, 1991), the parasitoid appears unable to control mealybug hosts from cassava grown on extremely poor, sandy soils (less than 5% of the total cassava area in West Africa (Neuenschwander et al., 1990). These mealybug populations subsequently explode during the dry season and cause conspicuous plant damage.

The cassava green mite, *Mononychellus tanajoa* (Bondar), and the larger grain borer, *Prostephanus truncatus* (Horn), are also neotropical pests introduced into Africa in the early 1970s and 1980s, respectively (Nyiira, 1972; Hodges et al., 1983; Krall, 1984). The cassava green mite spread across the cassava belt in a relatively short 10 years and, following the spectacular control of the cassava mealybug, is now the most serious cassava pest in many regions of the continent (Yaninek, 1988; Yaninek and Herren, 1988). The mite causes yield losses ranging from 30 to 80% depending on variety, cultural practices and local agroecological conditions, threatening food security in many regions of the continent (Yaninek et al., 1989). The more recently introduced larger grain borer is a serious pest of stored maize (Hodges et al., 1983; Krall, 1984), but also attacks harvested and processed cassava. Losses of up to 74% after 4 months of infestation have been reported (Hodges et al., 1985).

The whitefly, *Bemisia tabaci* (Genn.), has been a pest of cassava and many other crops in Africa for a long time, but it is believed to be an exotic species of mid-Eastern origin (Greathead, 1989). This homopteran causes only minor damage as a plant feeder, but can be devastating as the principal vector of the African cassava mosaic virus (Leuschner, 1979).

The only major indigenous arthropod pest is the polyphagous variegated grasshopper, *Zonocerus variegatus* (L.), found across the humid and sub-humid tropics in Africa. This species emerges as a synchronized aggregation of nymphs from soil-borne eggs at the end of the wet season and early in the dry season (Chapman et al., 1986). It feeds on a wide variety of green vegetation including cassava, often the only significant crop found during this period. Grasshopper populations reach maturity by the beginning of the wet season when adult females concentrate their eggs in humid and shaded soils of undisturbed habitats. Large populations of the grasshopper can defoliate a cassava field causing conspicuous and often significant damage (Page et al., 1980), and can kill cassava if the bark is chewed off the stems.

Plant pathogens

A number of foliar and root diseases commonly affect cassava production in Africa. Prior to the 1920s, cassava was apparently free of significant foliar pathogens, although root rots were a problem locally (Jones, 1959). The most important pathogen in Africa is the African cassava mosaic virus, a pathogen found throughout the cassava belt. This virus is so common in some parts of Africa that the farmers do not consider leaves exhibiting virus symptoms to be diseased. Early estimates placed African cassava mosaic virus losses across the continent at 11% (Padwick, 1956), while losses in individual cultivars were reported to range from 24 to 69% (Terry and Hahn, 1980). This susceptibility in many cultivars is a constraint to the introduction of exotic cassava germplasm into Africa. In the more humid regions, cassava bacterial blight and anthracnose damage can exceed that of mosaic virus. The rapid spread of cassava bacterial blight in parts of West and Central Africa during the 1960s and 1970s suggests that the pathogen was introduced (Terry, 1979). Although cassava bacterial blight can cause total yield loss when the plants are young, most obvious attacks result in significant yield reductions, but rarely destroy the crop (Terry and Perreaux, 1984). Comparable data on anthracnose are not available. The preharvest root rots (e.g. *Sclerotium*, *Fusarium*, and *Phytophthora* spp.) in Africa have long been known to be a problem locally, but most observations are anecdotal and precise loss data are unavailable.

Weeds

Numerous weed species can cause severe cassava production losses, some estimated as high as 80% if left unchecked (Akobundu, 1980). Cassava is most susceptible to weeds during the first 6 weeks after planting and periodically thereafter during the rainy season. Although traditional practices have evolved in most regions to keep weed problems under control, the labor required, about 45% of the total production costs, can be a limiting factor (Dorosh, 1988). Little is known about the effect of individual weed species and their importance as a refuge for natural enemies found in the cassava agroecosystem. The two most important weeds in cassava are *Imperata cylindrica* and *Chromolaena odorata*, while other troublesome weeds include *Panicum* spp., *Andropogon* spp., *Hyperrhania* spp., *Pennisetum* spp., *Mimosa* spp. and *Commelina* spp.

Plant and pest phenology

Cassava is grown principally for the starch stored in specialized storage roots. Since cassava is a plant of indeterminate growth, both the aerial portion of the plant and the storage roots increase in biomass simultaneously. There is no distinct critical period during which fruit and/or seeds must develop prior to harvest, therefore cassava is more tolerant to pest attacks then are many other crops (Cock, 1979; Schulthess, 1987; Yaninek et al., 1990). However, there is considerable genetic variation in the production and allocation of dry matter within plants between cassava cultivars (F. Schulthess, personal communication, 1990). Thus, the impact of the major cassava pests on cassava production depends largely on the time of attack relative to the age of the host plant, the genetic susceptibility of the cultivar, length of drought stress and soil conditions. A schematic outline of cassava phenology and the expected timing and abundance of major cassava pests is presented in Fig. 1.

Cassava growth begins with shoots and green stems using reserves from the planted stem cutting until fine roots access nutrients in the soil. The leaf area develops rapidly and provides more photosynthate than is needed to meet the metabolic demands of the plant. Consequently, the excess supply is shunted to storage stems and roots beginning between 6 and 8 weeks after planting, depending on environmental conditions and variety. There is usually a positive correlation between leaf area and the dry matter allocated to the storage

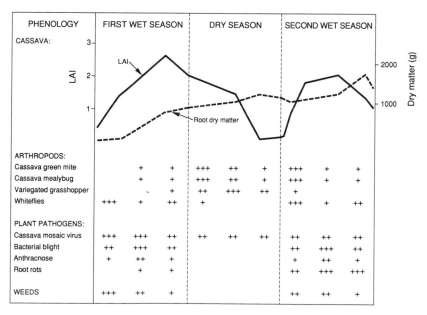

Fig. 1. Phenology of cassava leaf area and root dry matter in association with common arthropod, plant pathogen and weed pests.

roots during the initial growth period. This relationship can be curvilinear, indicating that the plant produces more leaf area than is needed for maximum storage root growth (Cock et al., 1979), perhaps as insurance against pests that affect the canopy. Most cassava reaches a maximum leaf area by the end of the wet season, that declines as smaller and fewer leaves are produced. Meanwhile, the dry matter in the roots continues to increase.

At this stage of growth, cassava production can be severely constrained by weeds and plant pathogens. The rains bring weeds which can rapidly over-grow and shade young cassava plants, and compete directly for nutrients in the soil. This reduces growth rates, and sometimes kill the crop if left unat-tended. Plant pathogens are usually the most important pest problem during the wet seasons. Poorly managed cassava planting material is usually infected with the non-systemic African cassava mosaic virus, or is quickly inoculated by the ubiquitous homopteran vector *B. tabaci*. Cassava bacterial blight is a severe problem in the humid ecologies, and can kill the crop given sufficient inoculum and adequate virulence.

During the dry season, as the rains subside and the air humidity drops, the smaller, slower growing leaves contribute to a precipitous drop in leaf area. When the metabolic demand exceeds the supply of photosynthates provided by the leaf area, dry matter stored in the stems and roots is remobilized. This metabolic stress usually breaks the plants' apical dominance and leads to the production of lateral shoots. Leaf area declines more rapidly with the com-bined effect of drought and large arthropod populations. Most dry season pests are foliage-feeding arthropods that contribute directly to reducing the leaf area which indirectly accelerates the decline in stem and root dry matter. Large populations of the cassava mealybug can kill cassava by depleting the reserves in the stems and roots. Cassava green mites and the variegated grasshopper can defoliate the crop, causing significant reductions in yield, but rarely cause lethal plant damage. The losses in stems and roots during this period are usu-ally proportional to the amount of dry matter present at the start of the dry season (F. Schulthess, personal communication, 1990).

Cassava harvest often begins near the end of the dry season and early in the subsequent wet season when root starch content is low, but little else is avail-able. Plants that are harvested later compensate for drought losses if given sufficient time for recovery. The smaller, slower-growing leaves produced by the plant during this period are augmented by the production of many lateral shoots. The new leaves are smaller than in the previous wet season and are produced mainly on lateral shoots initiated during the dry season at the ex-pense of the stems and roots. Severe pest attacks during the dry season places so much demand on the plant's reserves that stem and root dry matter contin-ues to be remobilized, 6–8 weeks into the wet season, even though pest pop-ulations are usually very low by this time. An epizootic of the variegated gras-shopper on cassava during this period would be catastrophic. Fortunately, most grasshoppers are killed by an entomopathogen before this can occur (Chapman et al., 1986). The major pests during the regrowth period are the plant pathogens. Their impact on cassava is reduced compared with the initial wet season when the plants were more vulnerable. The root quality of plants attacked during the dry season declines significantly during this period. This is often expressed as an increased incidence of root rot proportional to the severity of pest attack, and by undesirable changes in root texture, color, taste and processing characteristics.

Cassava pest management in Africa

Until recently, cassava was neglected as an object of agricultural research in Africa. Compared with cash crops such as oil palm, cocoa and cotton, and food crops such as maize, cowpeas and rice, cassava has been perceived as a low-value food crop with little prestige as a subject for research. In particular, pest management and plant protection considerations were largely ignored. This can be attributed in part to the crop's exotic origin, its relatively minor importance as a food crop until earlier this century, and to the absence of serious arthropod pests at that time. When research on cassava finally did start, breeding for resistance to pathogens dominated the early plant protection research. The demand for food security as populations increased and the threat of famine prompted colonial governments to develop widely adapted cassava cultivars. The most important pest problem then was the presence of African cassava mosaic virus and the dearth of resistant germplasm. In the 1930s, the British working in colonial East Africa began a program to develop resistant cultivars to African cassava mosaic virus using wild cassava, *Manihot glaziovii*, as a source of resistance (Bock and Guthrie, 1979). Although immunity was never found, good resistance was obtained.

Following the wave of independence which swept across the continent in the late 1950s and early 1960s, the first of several new pests of cassava were observed. The Siam weed, *C. odorata*, spread rapidly across the cassava belt during this period, apparently augmenting populations of the seasonally abundant variegated grasshopper (Greathead, 1989). However, the polyphagous feeding behavior of the grasshopper and the novelty of this new weed generated relatively little research activity, particularly where cassava was concerned. An outbreak of cassava bacterial blight in parts of West and Central Africa during the same period prompted breeders to search for resistant germplasm. The germplasm with resistance to African cassava mosaic virus introduced from East to West Africa appeared to be resistant to cassava bacterial blight as well (Jennings, 1979). It was during this period that research on cassava in Africa expanded beyond disease resistance and yield improvement to include agronomy and early farming systems research.

In the 1970s, research on cassava became multi-disciplinary and the farming systems approach was widely adopted to measure the production constraints confronted by peasant farmers. This research paradigm was also used to test technologies being developed by research stations in farmers' fields. Many of these technologies were developed without particular regard to farmers and national pest managers. The approach was top down, most technologies were too difficult to transfer, and many pest constraints were overlooked. Meanwhile, two more exotic pests, the cassava green mite and the cassava mealybug, reached the continent with devastating consequences.

Because of the problems encountered earlier in transferring cassava technologies, research to develop, test and adopt 'appropriate' on-farm technologies flourished during the 1980s. Increasing concerns over the widespread devastation caused by several exotic pests and the need for an urgent, sustainable solution provided the attention needed to attract support for pest management research. Slowly, but inevitably, entomologists, plant pathologists and weed scientists were given the opportunity to address cassava pests as subjects of pest management research, and not simply as objects of resistance breeding. Meanwhile, another exotic pest of maize and harvested and processed cassava, the larger grain borer, *P. truncatus*, found it way into Africa (Hodges et al., 1983).

The spectacular success of the Biological Control Program of the International Institute of Tropical Agriculture (IITA) in controlling the cassava mealybug drew attention to the benefits that a well-conceived and carefully implemented plant protection research project can provide (Herren, 1989). Norgaard (1988) estimated the overall cost–benefit ratio of the mealybug biological control effort to be 1:149, while Neuenschwander et al. (1989) showed empirically that the introduced parasitoid improved root yields by an average of 2.5 t h^{-1} in the savannah region of West Africa. Now in the 1990s, the cassava mealybug project has become a paradigm for developing environmentally sound and economically feasible plant protection for basic food crops. The momentum gained in research, training and implementation was soon capitalized upon as other pests of cassava were targeted for control. These activities gave rise to an overall management strategy for food crop production based on sustainable plant protection and led to the formation in 1988 of an entire program (formerly the Biological Control Program, now the Plant Health Division) at IITA dedicated to these principles.

A sustainable plant protection research philosophy

To manage arthropod pests in the context of ecologically sound plant protection requires a thorough understanding of the pest's biotic potential in the affected farming systems. Traditional systems of cassava cultivation are ideal starting points for developing ecologically sound plant protection practices. Selective aspects of a pest's biology and key interactions in the surrounding agroecosystem are identified before corrective measures can be developed and tested. Such an approach favors a pest management philosophy which minimizes the need for artificial control tactics in favor of ecologically oriented methods. This often requires an inter-disciplinary, as well as a multi-institutional, team effort before limiting factors can be properly identified and corrective measure taken.

Plant protection based on ecological principles aims to prevent the need, and consequently, the use of pesticides, while Integrated Pest Management (IPM) integrates biological and cultural control practices with the judicious use of pesticides. IPM may be appropriate where pesticides are a part of the production system, but in the case of cassava production in Africa, pesticides cannot be afforded. This approach conserves the efficacy of natural enemies by avoiding the lethal contact and residual toxicity of most pesticides, and preserves the environmental integrity of water resources and the food-chain within the targeted agroecosystem.

Cassava plant protection technologies

The cassava plant protection technologies, available for testing and adaptation, can be grouped into three categories of interventions: biological control, host plant resistance and cultural practices (Table 2). Chemical controls are impractical, uneconomical, non-sustainable, and hazardous to the environment and to the farmers, and therefore, are not considered. Biological control consists of three basic strategies. These include classical biological control where ecologically adapted natural enemies are introduced from out-

Table 2
Sustainable intervention technologies available for testing, adaptation and implementation

Pests	Interventions technologies		
	Biological control	Host plant resistance	Cultural control
Arthropods			
Cassava mealybug	●		●
Cassava green mite	●	●	●
Variegated grasshopper	●		●
Whiteflies			●
Larger grain borer	●		●
Plant pathogens		●	●
Weeds			●

side the target area, conservation of natural enemies present in the system through cultural practices which enhance their activity, and augmentation where local natural enemies are multiplied and released to increase their impact.

Cassava pests have been studied for many years in view of developing host plant resistance. Plant breeding for resistance was the earliest plant protection technology pursued, and it continues to be widely practiced by both national and international research institutes. Although much of the recent effort has been put into developing varieties which are high yielding and fast producing, selection for pest resistance is receiving new attention (Hahn et al., 1989). The resistance found against African cassava mosaic virus appears to be stable across a wide range of agroecosystems, and to a lesser degree for cassava bacterial blight. The search for cassava green mite resistance is more recent. Presently, there are efforts to evaluate leaf pubescence and drought tolerance as deterrents to mite damage.

The role of cultural practices in enhancing cassava production is well known, but poorly documented where pests are concerned. Pest constraints to cassava production are frequently related to other production practices, although their effect on pest populations in Africa are only now being investigated. Appropriate cultural practices can already be identified based on systems research, even though much work remains to be done. Good cassava production starts with quality planting material free of avoidable plant pathogen and pest contaminants. Other potential production constraints such as weeds, mulching, time of planting, spacing, intercrops and time of harvest are usually moderated by appropriately timed and properly implemented agronomic practices. Fallow management can reduce undesirable weeds, while maintaining desirable refugia for beneficial natural enemies.

Cassava green mite

In South America where the cassava green mite is indigenous, locally selected cultivars and well-adapted natural enemies keep the pest in check in many areas. Local natural enemies alone prevent yield losses of about 30% (Braun et al., 1989) (G.J. de Moraes, personal communication, 1990). Natural enemies of the mite family Phytoseiidae have been identified as the most promising predators of cassava green mites (Bellotti et al., 1987). Consequently, an extensive foreign exploration effort was launched, and among the more than 50 phytoseiid species found associated with the mite in the Neotropics, a dozen promising species have been shipped to Africa for experimental releases (Yaninek et al., 1989). Two of these species have been identified as viable field candidates based on establishment and preliminary impact studies. These include *Amblyseius idaeus* (Denmark and Muma) and *Ty-*

phlodromalus limonicus sensu lato (Garman and McGregor) from Brazil (Yaninek et al., 1992; J.S. Yaninek, unpublished data, 1991).

Another group of natural enemies being exploited for spider mite control are pathogens. In the case of the cassava green mite, a pathogenic fungus has been found attacking the mite in the drier areas of northeast Brazil. The fungus, *Neozygites* sp., is an Entomophthorales with a high degree of host specificity. Preliminary results from epidemiology, infectivity and specificity studies show the fungus as a promising classical biological control agent in ecologies receiving between 800 and 1200 mm rainfall a year (Moraes et al., 1990).

The incorporation of sources of cassava green mite resistance into agronomically acceptable cultivars offers potential for the control of the cassava green mite. Several national and international programs are developing promising cultivars with moderate levels of resistance which will be available soon for on-farm trials in Africa. Appropriate cultural practices for cassava green mite can be identified based on systems research, even though much work remains to be done. Field studies recently corroborated by computer simulations indicate that cassava planted early in the wet season suffers low mite-induced yield reductions compared with cassava planted later in the season, and that a positive relationship exists between soil fertility (organic matter, nitrogen, water), plant vigor, and mite density. The relationship between mite density and yield is non-linear, with the greatest mite-induced yield losses occurring in plants of intermediate vigor. This identifies mite-infested cassava grown on soils of intermediate fertility as the most likely target for cultural practices that improve soil fertility and water-holding capacity.

Variegated grasshopper

Most of the damage caused by this grasshopper is compensated by regrowth if the harvest can be delayed a few months after attack (Page et al., 1980; F. Schulthess, unpublished data, 1989). However, this does not help farmers who rely on foliage as a vegetable or who harvest roots shortly after a grasshopper attack. In these cases, microbial agents could be used to control the grasshopper. As the humidity increases toward the end of the dry season, the pathogenic fungus *Entomophaga grylli* attacks the grasshopper and usually devastates the population soon after the start of the wet season. Augmentation of *Entomophaga grylli* or the application of another well-adapted, virulent fungus at the time of attack should prove efficacious. Promising virulent strains of *Entomophaga grylli* and another fungus, *Metarhizium anisopliae*, are presently available for on-farm testing. Cultural practices that reduce breeding sites and the availability of preferred host plants such as *C. odorata* should also reduce source populations.

Cassava mealybug

The cassava mealybug, *Phenacoccus manihoti*, the target of one of the most successful classical biological control campaigns, is now controlled satisfactorily in most agroecosystems by the introduced parasitoid *Epidinocarsis lopezi* (Herren and Neuenschwander, 1991). However, in areas where cassava is grown on poor, sandy soils, the parasitoid appears unable to develop properly because the mealybug hosts are inferior. In mulched fields, adjacent to such outbreak populations, no mealybug damage was observed (Neuenschwander et al., 1990). Therefore, the efficacy of *Epidinocarsis lopezi* in these situations may be improved by promoting cultural practices that enhance the fertility and water-holding capacity of the soil (e.g. mulching, weed management, selected intercrops and proper fallow periods). This should improve the quality of the mealybug hosts and the efficiency of the parasitoid. In addition, releases of exotic predators into these pockets of infestation are planned. Two coccinellids from South America, *Hyperaspis notata* (Mulsant) and *Diomus* sp. have already been established locally, in several countries, and more species are being released.

Larger grain borer

Recently, an introduced histerid beetle predator *Teretriosoma nigrescens* Lewis from Costa Rica has been released against this pest in Togo, West Africa (R.D. Markham, personal communication, 1991). The IITA is also involved in a project to identify, import and release promising natural enemies of the larger grain borer from Central America into Africa. Promising intervention technologies that are available will be tested in this project.

Plant pathogens

Host plant resistance and cultural practices can alleviate the impact of some plant pathogens. Presently, a number of cassava varieties resistant to several plant pathogens have been developed by national and international institutes and are available for on-farm testing in Africa. Using resistant sources originally developed in East Africa, the IITA has further developed and distributed to African national programs a number of cassava lines that are highly resistant to African cassava mosaic virus and cross-resistant to cassava bacterial blight (Jennings, 1979; Hahn et al., 1980). Some national programs are now incorporating these sources of resistance into locally acceptable cultivars which are available for on-farm testing, although some local sources of resistance appear to be available (Rossel and Thottappilly, 1985).

Sanitation of diseased planting material could have an immediate and significant positive effect on cassava production in a continent where most serious diseases are endemic. The low transmission efficiency of the African cassava mosaic virus vector, *B. tabaci*, could be exploited by using thermotherapy and meristem culture to clean up planting material, and roguing of infected plants to eliminate sources of infection. Cropping systems influence the epidemiology of plant pathogens. There are indications that intercropping cassava appears to reduce cassava bacterial blight (Terry and Perreaux, 1984). Measures to restrict the movement of cutting material between regions and countries in the continent would help reduce the spread of infected cassava, and allow for the local control of some plant pathogens. However, this will be very difficult to implement in a continent where planting material is often in short supply and where government enforcement is ineffective.

Weeds

The principle method of weed control in cassava is by hand-weeding early in the growing season, and periodically thereafter, depending on the species, rainfall, and cassava variety. But because of the amount of labor required, alternate methods of control would be attractive if appropriate. Commercial herbicides are potential risks to the environment and usually beyond the economic reach of most cassava farmers. Biological control has a great theoretical potential, but remains largely unexplored for tropical weeds. Cultural practices hold the greatest promise. Profusely branching cassava cultivars, adjusting planting densities, and carefully selected and properly timed intercrops significantly reduce the amount and frequency of weeding required. In a similar manner, some noxious weeds in cassava have been successfully managed with selected low-growing herbaceous legumes (Akobundu, 1987).

Implementation

There are many ecological and socio–economic constraints to agricultural production in Africa (Harrison, 1987) that also affect the implementation of new plant protection technologies. Weather is variable in an unpredictable manner. Soils are low in clay and organic matter, and highly acidic in humid regions. Biological constraints include the toll that harsh environments and poor health conditions take on the human population. Access to sufficient water limits the productive regions, and from a socio–economic point of view, farm size is small, sometimes fragmented, with uncertain land tenure.

Small-scale farmers are reluctant to accept changes that increase their exposure to risks; they usually take actions to avoid risk such as overstocking livestock or having many children. These farmers are always short of cash for investment into their farms and few have access to credit. Labor has become an acute problem in many rural areas because of the exodus of young people seeking opportunity in the big cities. The status of women (the majority of farmers in Africa) in many communities also impedes agricultural development. Women are often subjected to more uncertain land tenure, less access to credit and restrictions in their ability to make decisions concerning the crops they grow.

The client farmer is both a target and a resource when implementing plant protection technologies. Most farmers cannot afford to pay for plant protection technologies, but are often eager to participate in on-farm trials if involved in the process of developing and adapting the technology. However, farmer participation is no panacea given the problems of implementation. Their involvement must be balanced with the biological and socio–economic constraints being addressed (Bentley and Andrews, 1991).

The implementation of most plant protection technologies require the active participation of farmers and agroecosystem managers such as extension agents and research scientists. Proper implementation of these technologies, often 'the cutting edge of modern agriculture' (Goodell, 1984), is among the most demanding activities of IPM. In developing countries, this requires specially trained extension agents able to use unconventional methods to reach poorly educated farmers (Kushwaka, 1982). However, most national programs lack sufficient capacity to train, test and adapt plant protection technologies, a cadre of skilled agroecosystems managers including farmers, extension workers and national program researchers needed to pass the knowledge and technologies along to farmers.

Cassava plant protection in West Africa: a model program

A regional project is being developed to test and adapt sustainable plant protection technologies for the most important cassava pests found in selected ecologies in four West African countries — Ghana, Benin, Nigeria and Cameroon. The project is expected to establish crop protection practices that preserve environmental quality and human health. The essential logistic support needed to test, adapt and implement crop protection technologies, develop a nucleus of trained and experienced plant protection researchers and practitioners, and generate timely information resources required by the crop protection networks will be established by the project collaborators within each target country. Participation by extension workers and farmers will be encouraged throughout the project to assure the appropriateness of suggested technologies and to increase the probability of technology adoption.

The project will be a collaborative effort between the IITA, national plant protection staff, extension workers and farmers in targeted countries in West Africa, with a parallel component in South America involving the Centro Internacional de Agricultura Tropical and Brazil (CIAT-IITA, 1991). A special effort will be made to enhance the status and representation of women in plant protection activities. The activities will be divided into three interrelated and partially concurrent phases covering a period of 4 years. The first phase will refine the existing knowledge base on major pests of cassava through diagnostic surveys. This information will be linked to databases generated by a Rockerfeller-funded Collaborative Study of Cassava in Africa (Nweke, 1988). In the second phase, farmers will participate in the development and testing of a range of crop protection technologies. At the same time, formal training of farmers, extension workers and researchers in the principles and practices of sustainable crop production and protection will be provided. In the third phase, progress in achieving training objectives and technology implementation will be evaluated.

Information resources will be developed to facilitate processing, summarization, interpretation and communication of the large amount of multi-disciplinary data anticipated during the project. A systems approach will be used to identify critical interactions and to evaluate the impact of tested technologies. Accurate taxonomic information is essential to pest management, thus the project will generate taxonomic redescriptions and keys for important species of cassava pests. Together, these information resources will become part of the plant protection legacy of this project that remains with participating national and international institutes. The paradigm developed for sustainable plant protection in this project should also serve as a model for plant protection technologies and pest management strategies for other cropping systems.

Conclusion

Cassava plant protection and pest management research is slowly receiving the attention it deserves in Africa. After a prolonged period, where the only significant effort to increase cassava production came from plant breeders, a small but increasing national and international effort is being put into developing ecologically appropriate cassava plant protection technologies. The initial effort is promising, but there remain many gaps in the agroecological baseline data, required trained personnel and necessary logistic support.

The prospect of developing a sustainable cassava plant protection paradigm based on the success of previous cassava pest management thrusts is encouraging. However, IPM as increasingly practiced in the West, where the 'P' in IPM emphasizes 'pesticides' more than 'pests', has little promise in Africa. Fortunately, there is virtually no market in subsistence agriculture in Africa for non-sustainable inputs such as pesticides that are not state or donor subsidized. Unfortunately, much of the missing information needed to develop proper IPM tactics is also required for ecologically sound plant protection.

The large number of exotic pests found on cassava in Africa and the neotropical origin of the crop give reason to be hopeful that effective and sustainable pest management technologies can be developed. The success in developing cultivars resistant to African cassava mosaic virus and cassava bacterial blight suggests that similar resistance can be found for other pests either locally or in introduced germplasm. The spectacular control of the cassava mealybug throughout the continent and recent positive experiences with the cassava green mite demonstrate the potential of classical biological control in this agroecosystem. A large national capacity in biological control has been assembled throughout the cassava belt over the past 10 years. This capacity provides a basis for new work in classical biological control and for developing and adapting other intervention technologies.

The proposed regional cassava plant protection project in West Africa could be a model for developing and implementing sustainable plant protection anywhere in Africa. The proposed approach brings with it the advantages of multi- and interdisciplinary research. The emphasis on extensive and intensive pest diagnosis, followed by on-farm evaluation of the pest technologies provides the experimental rigor often lacking in most field trials. The systems approach advocated in this project provides a basis for making informed decisions concerning implementation and subsequent evaluation. Finally, the database to be developed by this project will provide the basis for evaluating the impact of the tested technologies and become a tangible part of the legacy left with the national programs long after the project is completed.

While the future of ecological pest management in Africa may be uncertain, the future of sustained crop production in the continent will certainly depend on the adoption of environmentally sound plant protection principles.

References

Akobundu, I.O., 1980. Weed science research at the International Institute of Tropical Agriculture and research needs in Africa. Weed Sci., 28: 439–444.

Akobundu, I.O., 1987. Weed Science in the Tropics: Principles and Practices. John Wiley, New York, 522 pp.

Bellotti, A.C. and Schoonhoven, A., 1978. Mite and insect pests of cassava. Annu. Rev. Entomol., 23: 39–67.

Bellotti, A.C., Mesa, N., Serrano, M., Guerrero, J.M. and Herrera, C.J., 1987. Taxonomic inventory and survey activity for natural enemies of cassava green mites in the Americas. Insect Sci. Appl., 8: 845–849.

Bentley, J.W. and Andrews, K.L., 1991. Pests, peasants, and publications: anthropological and entomological views of an integrated pest management program for small-scale Honduran Farmers. Hum. Organ., 50: 113–124.

Bernays, E.A., Chapman, R.F., Leather, E.M., McCaffrey, A.R. and Modder, W.D., 1977. The relationship of *Zonocerus variegatus* (L.) (Acridoidea: Pyrgomorphidae) with cassava (*Manihot esculenta*). Bull. Entomol. Res., 67: 391–404.

Bock, K.R. and Guthrie, E.J., 1977. African mosaic disease in Kenya. In: T. Brekelbaum, A. Bellotti and J.C. Lozano (Editors), Proc. Cassava Protection Workshop, 7–12 November

1977, International Agriculture in the Tropics (Series CE-14), CIAT, Cali, Colombia, pp. 41–44.

Braun, A.R., Bellotti, A.C., Guerrero, J.M. and Wilson, L.T., 1989. Effect of predator exclusion on cassava infested with tetranychid mites (Acari: Tetranychidae). Environ. Entomol., 18: 711–714.

Chapman, R.F., Page, W.W. and McCaffery, A.R., 1986. Bionomics of the variegated grasshopper, *Zonocerus variegatus*, in west and central Africa. Annu. Rev. Entomol., 31: 479–505.

CIAT-IITA, 1991. Ecologically sustainable cassava plant protection in South America and Africa, an environmentally sound approach. UNDP Project Document, CIAT, Cali, Colombia and IITA, Ibadan, Nigeria, 92 pp.

Cock, J.H., 1979. A physiological basis of yield loss in cassava due to pests. In: T. Brekelbaum, A. Bellotti and J.C. Lozano (Editors), Proc. Cassava Protection Workshop, 7–12 November 1977, International Agriculture in the Tropics (Series CE-14), CIAT, Cali, Colombia, pp. 9–16.

Cock, J.H., 1985. Cassava, New Potential for a Neglected Crop. Westview Press, Boulder, CO, USA, 191 pp.

Cock, J.H., Franklin, D., Sandoval, G. and Juri, P., 1979. The ideal cassava plant for maximum yield. Crop Sci., 19: 271–279.

De Moraes, G.J., de Alencar, J.A., Wenzel Neto, F. and Mergulhao, S.M.R., 1990. Explorations for natural enemies of the cassava green mite in Brazil. In: R.H. Howeler (Editor), Proceedings of the Eighth Symp. of the International Society of Tropical Root Crops, Bangkok, Thailand, October 30–November 5, 1988, pp. 351–353.

Dorosh, P., 1988. The economics of root and tuber crops in Africa. RCMP Research Monograph No. 1, IITA, Ibadan, Nigeria, 68 pp.

Goodell, G., 1984. Challenges to international pest management research and extension in the third world: do we really want IPM to work? Entomol. Soc. Am., Fall: 18–26.

Greathead, D.J., 1989. Present possibilities for biological control of insect pests and weeds in tropical Africa. In: J.S. Yaninek and H.R. Herren (Editors), The Search for Sustainable Solutions to Crop Protection in Africa. IITA Publication Series, IITA, Ibadan, Nigeria, pp. 173–194.

Hahn, S.K., Terry, E.R. and Leuschner, K., 1980. Breeding cassava for resistance to cassava mosaic disease. Euphytica, 29: 673–683.

Hahn, S.K., Isoba, J.C.G. and Ikotun, T., 1989. Resistance breeding in root and tuber crops at the International Institute of Tropical Agriculture (IITA), Ibadan, Nigeria. Crop Prot., 8: 147–168.

Harrison, P., 1987. The Greening of Africa: Breaking Through in the Battle for Land and Food. An International Institute for Environment and Development–Earthscan Study. Paladin Grafton Books, London, 380 pp.

Herren, H.R., 1989. The biological control program of IITA: from concept to reality. In: J.S. Yaninek and H.R. Herren (Editors), The Search for Sustainable Solutions to Crop Protection in Africa. IITA Publication Series, IITA, Ibadan, Nigeria, pp. 18–30.

Herren, H.R. and Bennett, F.D., 1984. Cassava pests, their spread and control. In: D.L. Hawksworth (Editor), Advancing Agricultural Production in Africa, Proceedings of CAB's First Scientific Conference, Arusha, Tanzania, 12–18 February 1984, Commonwealth Agricultural Bureaux, Slough, UK. pp. 110–114.

Herren, H.R. and Neuenschwander, P., 1991. Classical biological control of cassava insects and mites in Africa. Annu. Rev. Entomol., 36: 257–283.

Hodges, R.J., Dunstan, W.R., Magazini, I. and Golob, P., 1983. An outbreak of *Prostephanus truncatus* (Horn) (Coleoptera: Bostrichidae) in East Africa. Prot. Ecol., 5: 183–194.

Hodges, R.J., Meik, J. and Denton, H., 1985. Infestation of dried cassava (*Manihot esculenta*

Crantz) by *Prostephanus truncatus* (Horn) (Coleoptera: Bostrichidae). J. Stored Prod. Res., 21: 73–77.

Jennings, D.L., 1979. Inheritance of linked resistances to African cassava mosaic and bacterial blight diseases. In: T. Brekelbaum, A. Bellotti and J.C. Lozano (Editors), Proc. Cassava Protection Workshop, 7–12 November 1977, International Agriculture in the Tropics (Series CE-14), CIAT, Cali, Colombia, pp. 45–50.

Jones, W.O., 1959. Manioc in Africa. Stanford University Press, Stanford, California, 315 pp.

Krall, S., 1984. A new threat to farm-level maize storage in West Africa, *Prostephanus truncatus* (Horn) (Coleoptera: Bostrichidae). Trop. Stored Prod. Inf., 50: 26–31.

Kushwaka, K.S., 1982. Towards supervised pest management. Indian J. Entomol., 44: 393–402.

Neuenschwander, P., Hammond, W.N.O., Gutierrez, A.P., Cudjoe, A.R. and Baumgaertner, J.U., 1989. Impact assessment of the biological control of the cassava mealybug, *Phenacoccus manihoti* Matile-Ferrero (Hemiptera: Pseudococcidae) by the introduced parasitoid *Epidinocarsis lopezi* (DeSantis) (Hymenoptera: Encyrtidae). Bull. Entomol. Res., 79: 579–594.

Neuenschwander, P., Hammond, W.N.O., Ajuonu, O., Gado, A., Echundu, N., Bokonon-Ganta, A.H., Allomasso, R. and Okon, I., 1990. Biological control of the cassava mealybug, *Phenacoccus manihoti* (Hom., Pseudococcidae) by *Epidinocarsis lopezi* (Hym., Encyrtidae) in West Africa, as influenced by climate and soil. Agric. Ecosystems Environ., 32: 39–55.

Norgaard, R.B., 1988. The biological control of cassava mealybug in Africa. Am. J. Agric. Econ., 70: 366–371.

Nweke, F.I., 1988. Collaborative study of cassava in Africa (COSCA): project description. COSCA Working Paper No. 1, IITA, Ibadan, Nigeria, 68 pp.

Nweke, F.I. and Polson, R., 1991. The dynamics of cassava production in Africa: preliminary analysis of indicators. A report from Phase I of the Collaborative Study of Cassava in Africa (COSCA) for discussion at the Third Meeting of the Steering Committee of COSCA in London, 13–15 August 1990, IITA, Ibadan, Nigeria, 100 pp.

Nyiira, Z.M., 1972. Report of investigation of cassava mite, *Mononychus tanajoa* Bondar. Unpublished report, Kawanda Research Station, Kampala, Uganda, 14 pp.

Padwick, G.W., 1956. Losses caused by plant diseases in the colonies. Phytopathol. Pap., 1: 1–60.

Page, W.W., Harris, J.R.W. and Youdeowei, A., 1980. Defoliation and consequent crop loss in cassava caused by the grasshopper, *Zonocerus variegatus* (L.) (Orthoptera: Pyrogomorphidae) in southern Nigeria. Bull. Entomol. Res., 70: 151–163.

Rossel, H.W. and Thottappilly, G., 1985. Virus diseases of important food crops in tropical Africa. International Institute of Tropical Agriculture, Ibadan, Nigeria, 61 pp.

Schulthess, F., 1987. The interactions between cassava mealybug (*Phenacoccus manihoti* Mat.-Ferr.) populations and cassava (*Manihot esculenta* Crantz) as influenced by weather. Ph.D. Dissertation, Swiss Federal Institute of Technology, Zurich, 136 pp.

Silvestre, P. and Arraudeau, M., 1983. Le Manioc. Maisonneuve & Larose, Paris (Techniques agricoles et productions tropicales, XXXII), 262 pp.

Terry, E.R., 1979. Cassava bacterial blight. In: T. Brekelbaum, A. Bellotti and J.C. Lozano (Editors), Proc. Cassava Protection Workshop, 7–12 November 1977, International Agriculture in the Tropics (Series CE-14), CIAT, Cali, Colombia, pp. 75–84.

Terry, E.R. and Hahn, S.K., 1980. The effect of cassava mosaic disease on growth and yield of a local and improved variety of cassava. Trop. Pest Manage., 26: 34–47.

Terry, E.R. and Perreaux, D., 1984. Cassava diseases, their spread and control. In: D.L. Hawksworth (Editor), Advancing Agricultural Production in Africa, Proceedings of CAB's First Scientific Conference, Arusha, Tanzania, 12–18 February 1984, Commonwealth Agricultural Bureaux, Slough, UK, pp. 105–110.

Yaninek, J.S., 1988. Continental dispersal of the cassava green mite, an exotic pest in Africa, and implications for biological control. Exp. Appl. Acar., 4: 211–224.

Yaninek, J.S. and Herren, H.R., 1988. Introduction and spread of the cassava green mite, *Mononychellus tanajoa* (Bondar) (Acari: Tetranychidae), an exotic pest in Africa and the search for appropriate control methods: a review. Bull. Entomol. Res., 78: 1–13.

Yaninek, J.S., de Moraes, G.J. and Markham, R.H., 1989. Handbook on the Cassava Green Mite *Mononychellus tanajoa* in Africa: A Guide to their Biology and Procedures for Implementing Classical Biological Control. IITA Publication Series, IITA, Ibadan, Nigeria, 140 pp.

Yaninek, J.S., Gutierrez, A.P. and Herren, H.R., 1990. Dynamics of *Mononychellus tanajoa* (Acari: Tetranychidae) in Africa: impact on dry matter production and allocation in cassava, *Manihot esculenta*. Environ. Entomol., 19: 1767–1772.

Yaninek, J.S., Mégevand, B., de Moraes, G.J., Bakker, F., Braun, A. and Herren, H.R., 1991. Establishment of the neotropical predator *Amblyseius idaeus* (Acari: Phytoseiidae) in Benin, West Africa. BioControl Sci. Technol., 1: 323–330.

Agriculture, Ecosystems and Environment, 46 (1993) 325–326
Elsevier Science Publishers B.V., Amsterdam

325

Author Index

Announcement from the Publisher

ELSEVIER SCIENCE PUBLISHERS

prefers the submission of electronic manuscripts

Electronic manuscripts have the advantage that there is no need for the rekeying of text, thereby avoiding the possibility of introducing errors and resulting in reliable and fast delivery of proofs.

 The preferred storage medium is a $5\frac{1}{4}$ or $3\frac{1}{2}$ inch disk in MS-DOS format, although other systems are welcome, e.g. Macintosh.

 Your disk and (**exactly matching**) printed version (printout, hardcopy) should be submitted together. In case of revision, the same procedure should be followed such that, on acceptance of the article, the file on disk and the printout are **identical.**

 Please follow the general instructions on style/arrangement and, in particular, the reference style of this journal as given in the Guide for Authors.

 Please label the disk with your name, the software & hardware used and the name of the file to be processed.

 Further information can be found under 'Instructions to Authors - Electronic manuscripts'.

Contact the Publisher for further information.

ELSEVIER EDITORIAL SERVICES
Mayfield House, 256 Banbury Road
Oxford, OX2 7DH, UK
Fax: (+44-865) 516120 or 56472

EES/92